The Paraphysical Principles of Natural Philosophy:

ParaPrincipia and a New Comprehensive Unified Physics

James E. Beichler, PhD

CONTENTS

FOREWORD

The word 'paraphysics' has never been precisely defined. To establish paraphysics as a true science, the word is first defined, and its scope and limits identified. The natural phenomena that are studied in paraphysics, psi phenomena, are distinguished by their common physical properties. The historical roots of paraphysics are also discussed. Paraphysics can be defined by a specific body of natural phenomena, and it has a historical basis. Therefore, paraphysics is a distinguishable science. It only needs a theoretical foundation.

Rather than using a quantum approach, a new theory of physical reality can be based upon a field theoretical point of view. This approach dispels philosophical questions regarding the continuity/discrete debate and the wave/particle paradox. Starting from a basic Einstein-Kaluza geometrical structure and assuming a real fifth dimension, a comprehensive and complete theory emerges. The four forces of nature are unified, as are quantum and relativity.

Life, mind, consciousness, and psi emerge as natural consequences of physics. The scientific concept of consciousness, ambiguous at best, has become an increasingly important factor in modern physics. No one has ever defined consciousness in an acceptable manner let alone develop a workable theory of consciousness while no viable physical theories of life and mind are even being considered even though they are prerequisites of consciousness. In the five-dimensional model, life, mind, consciousness are explained as increasingly complex 'entanglements or patterns of potential density variation within the single unified field. Psi is intimately connected to consciousness, giving the science of paranormal phenomena a theoretical basis in the physics of hyperspace. Psi results from different modes of consciousness interacting non-locally via the fifth dimension.

Several distinct areas of future research are suggested which will lead to falsification of the theory. A new theory of the atomic nucleus is clearly indicated, as is a simple theory of the predominant spiral shape of galaxies. A quantifiable theory of life is also suggested. And finally, this model strongly implies a direct correspondence between emotional states and psi phenomena that should render the existence of psi verifiable.

Author, 1999

This book was originally written in 1999 as my Doctoral Dissertation at the Union Institute and University in Cincinnati, Ohio. I graduated in June of 1999 with a PhD in Paraphysics, the only such legitimate degree in the world. I taught Freshman and Sophomore physics in the years after that at two small schools in West Virginia. I was laid off at the second school, WVUP, in May 2008 as a cost cutting move because I did not have enough students. Their statistical study was purposely flawed and showed that I only had six students per required core course when in fact the correct number was closer to fourteen, which would have been more than enough.

The Monday after I left, my physics Lab and Storage Room were gutted and remodeled as Art classroom. My physics courses were shopped out to local high schools. My Division Chair could not change the President' minds and the next year the Natural Science and Mathematics Division itself was dissolved, again using the 'cost cutting' excuse, which was ridiculous. The real problem was not 'cost' but twofold. Our President's PhD was in Literature, Black Children's Literature, and she knew nothing about science or the need for and importance of Science and Mathematics, let alone Physics. The other problem was that two or three of the teachers in my own Division were prejudiced and jealous of me and they had access to the President.

When I was first hired at the university, by an earlier President, I was told that those teachers had started a rumor that my PhD was not real. My answer to that was that they were welcome to look up my Dissertation (this book) on the internet since it was nationally registered and copyrighted. at That at least stopped the rumor but they got even with me later. In my Divisional peer review a year before I was laid off, I received recommendations of 'excellent' but two of the teachers in my division added a note that the Physics I was teaching was not real physics.

They were jealous of my PhD (they were both master's degree Biology teachers) and the fact that I was attending and presenting my research at international conferences. They also complained that the professional organizations that I belonged to (the SSE and ARPR) were not the proper organizations in their opinion. Of course, they forgot that I was also a member of the APS and gave presentations there but had not published anything with them as I had the other organizations. My school, a Community College branch of WVU was not even a 'publish or perish' university and I was the only person in the Division that was publishing anything.

So, in my division peer reviews the year before I would have been awarded tenure and been protected against being laid off, they added the note that I was not teaching correct Physics (whatever that might be) in my classes. I responded by putting a long letter in my personnel file defending myself. I explained that I taught normal and proper physics, with my students rating me very highly, and my detractors were Biologists who had no idea what constituted proper Physics.

As a PhD level Historian of Science, I was somewhat of an expert on this issue. I also wrote that their complaints were BS (Bad Science). My file with my letter went to the main campus of WVU for review and was sent back to our President for extra review of my teaching. She took offense at my use of BS. However, as I stated to her, it was in my personnel file and the letter was not for public view. It seems that the other teachers could criticize me without retribution, but I was unable to criticize them or correct them. So, I was denied tenure and laid off after seven years of highly rated teaching all because my fellow teachers were jealous of me and saw my research in the physics of consciousness as non-normal physics.

I later sued the University for classes I taught but was never paid for. After two years in the courts, the school decided to settle out of court and pay me off for far less than the lawsuit requested. My lawyer recommended that we take the offer because the school claimed they had no records of my having taught the courses because they had lost my personnel file and requested that I turn over my own evidence that I had taught the courses and not been paid. The lawyer did not want to fight a 'he said, she said' court case with them without my personnel files so we settled out of court.

Needless to say, for whatever reason I was never able to find a teaching job again despite an excellent record of teaching. I was laid off at fifty-nine years of age and was forced into retirement at sixty-two when I was forced to apply for Social Security because I had run out of my meager retirement funds. It's a good story in that it demonstrates the prejudice against any real scientist who conducts fringe or alternative-science research, as if some areas of Physics are off-limits to real and proper science. but I hold nothing against any of these people for their prejudices and biases. They were the losers, not me.

I have spent the last decade and a half since I was laid off completing my research, writing, presenting my research at conferences and publishing some of it. As far as my research is concerned, I have accomplished more than any other living person. But the discrimination against me continues and my ideas have not only not been accepted by the scientific community (or even a small part of it. They haven't even given my theoretical work a fair hearing. And this Dissertation was the beginning of it all. It laid out an outline where I was headed with my future research and what I hoped to accomplish.

I have completely and comprehensively merged the relativity and the quantum theories, unified them, forged then together like oxygen and carbon with iron to make steel. Science now has a single theoretical framework with which to understand everything in the universe that we presently now of and new things which are predicted by the theoretical model. This merging has come from considering them equally correct, equally wrong, and equally misunderstood. By correcting the Riemannian geometry to include (discrete) zero point-elements and not just metric elements approaching but never reaching zero, I have expanded the most generalized geometry used in physics to include the discrete point used in quantum physics, and the rest is history, or maybe someday it will be.

Unfortunately, the scientific and philosophical communities have considered relativity and the quantum as mutually incompatible and therefore impossible to unify as equals or as two parts of the same structure for the past century. Nothing is as far from the truth as this mistaken belief. This bias seems to be absolute, so it is hard to break through the barrier it represents. Scientists who do alternative science reject my work as based on normal physics which they believe cannot be used to explain the paranormal and mainstream scientists who dismiss the paranormal aspects of our shared reality altogether. Anomalous and psychic phenomena may be normal, thus paranormal, but they are still perfectly natural.

The scientific community rejects my work as 'woo-woo' physics and will not even bother to consider the possibility that my work, even my unification research, which is normal, is valid. So, I am equally shunned in science and academia, by nearly everyone. That is now changing as consciousness studies become more acceptable, but it is too slow a process for me to take advantage of during what I have left of my lifetime.

So, I give the world my first attempt to completely explain our true physical reality, the *ParaPrincipia*. This Dissertation outlines the whole process of nature and has been printed here almost as it was written in 1999. It is very nearly exactly as written two and a half decades ago. It was originally meant to be a plan for my future research and proved to be just that. The only changes that I have made with this book are those recommended by the grammar and spelling checker in a more modern computer word processor.

Otherwise, I have added about twenty or thirty short sentences needed for clarification of some issues. As a professional historian of science, I want to keep my Dissertation and copy of the *ParaPrincipia* as they were when I wrote them so people can see how my ideas have developed over time, which is very important for reasons that I will not explain at this time.

In fact, I have been amazed at how well I said things back then and how little the fundamental ideas and concepts of my theoretical work have changed over the years. This Dissertation was more of a list of what I wanted to accomplish with my theory. The pale dim image or reality that I had come to learning and experience back then has turned into a bright picture of everything I could have ever hoped for. Yet I was never sure that I would accomplish anything. So, in the end I must thank everyone, without reserve, that I have known and come into contact or interacted with, for you are all part of this final product. I even thank those people who have screwed up my life so badly, since they have given me the time and desire to get my work finished before I die. I leave the rest to history, the history of the future.

James E. Beichler, 2023

CHAPTER 1
It's all in the word!
The logical necessity of paraphysics

Paraphysics can never be established as a legitimate branch of science until it is properly defined. Its definition must be precise enough to establish the scope of paraphysics as well as distinguish it from its closest relatives, physics, and parapsychology. Since it deals with psi, a rather controversial subject in its own right, the precision of the definition is all the more important.

The scope and methods of paraphysics must be firmly established independent of psi so scientists will accept the legitimacy of paraphysics in the absence of absolute proof that psi exists. This will not be difficult since paraphysics does not depend on psi alone, nor is it dependent on any one theory of psi, although a successful physical theory of psi would greatly enhance the acceptance and further development of the science of paraphysics.

Any scientific understanding of paraphysics not only requires a valid definition of paraphysics, but also a comprehensive development of what constitutes normal physics. Within this context, anyone with so much as a rudimentary knowledge of physics would first define physics by listing the fields that it contains, such as thermodynamics, classical mechanics, electrodynamics, quantum theory and relativity.

While this definition of physics is far from complete, it does at least indicate the central idea behind physics that allows the development of a simplified notion of paraphysics. Paraphysics could thus be deemed the field of physics describing phenomena which are not dealt within the regularly accepted fields of normal physics. While this definition is understandable at a rudimentary level, it is neither sufficient nor adequate for any real understanding of paraphysics.

In this regard, defining paraphysics is a far more serious problem than might be thought at first. It is fraught with academic danger and controversy. On the one hand, normal physics represents a respectable and well-defined academic discipline, so anything paraphysical, beyond the norm for everyday physics, could be classified as paraphysics. This is the way that most people, presently regard paraphysics but science very nearly completely shuns it.

But this view is grossly inaccurate since paraphysics is not a dumping ground for those phenomena, natural or otherwise, which physics cannot explain. For example, Professor Tenti of the University of Toronto has recently placed a short web page on the Internet. He defines paraphysics as the physics of natural phenomena beyond the scope of normal physics, but he illustrates paraphysics with science fiction phenomena that have no reality in the world of nature.

On the other hand, physics means 'nature', so physics must include anything and everything, which is natural, and thus includes all natural phenomena. If paraphysics is to be considered a branch of physics, it must include phenomena that are both natural and paranormal which seems to be a contradiction in terms. This definition introduces a potential logical contradiction and begs the question of definition since the word paranormal has itself been given no meaningful definition.

Paraphysics must be regarded as either a branch of physics or an extension of physics; it must be 'of physics' yet 'beyond' physics. The difficulty defining paraphysics is further complicated by the fact that physics progresses and changes as time passes, consuming new ideas and areas of knowledge. Normal physics is thus undergoing a constant change. As the scope of physics expands into the future, what is beyond physics today may not be beyond the physics of tomorrow. However, the fundamental tenets of physics remain constant against this changing historical background of knowledge, so paraphysics must ultimately be defined with respect to the basic, unchanging, underlying concepts upon which physics is based rather than be regarded as an extension of physics 'beyond' its present knowledge base.

Since some factors remain the same, consistent throughout the history of physics despite the progressive nature of science, founding paraphysics upon fundamental unchanging tenets proves to be the only valid method of distinguishing between physics and paraphysics, or normal and paranormal physics, while remaining within the realm of natural phenomena. These considerations raise the question of whether paraphysics is a separate discipline or a branch of physics. But it must be assumed that paraphysics is a sub field of physics because physics covers all naturally occurring phenomena which includes the paranormal as well as the normal.

This potential contradiction and its solution have a precedent in the mathematical research of Bernhard Riemann. Before he rethought geometry, the universe was thought to be either infinite and unbounded or finite and bounded. But Riemann developed a geometric model in which the universe could be finite yet unbounded, a completely new viewpoint that circumvented the older infinite versus finite argument.

As in normal physics, paraphysics deals with the nature of physical reality although it goes beyond the notion of „matter in motion" as the fundamental quantity that is used to explain natural phenomena in normal physics. Ideas and concepts such as life, thought, mind, and consciousness are beyond the scope of normal physics, yet they are well-established norms in science whose existence would never be questioned. Since physics does not normally include explanations of these concepts, physics cannot be as all-inclusive of nature and natural phenomena as many would believe. Thus, there are already precedents for paraphysics to maneuver within the plenum of natural phenomena outside of the direct realm of physics.

These concepts do not seem to fit the physics model of 'matter in motion', so they may well be paraphysical in nature. Physics may or may not break these down, reduce them to a common physical quantity for the purpose of explanation at some unspecified future date, but even if it does their nature will probably remain unique enough, different from the other quantities defined and measured in physics, that they will be 'beyond' presently accepted norms in physics as well as future extensions of those norms. So, although the scope of physics is wide and comprehensive within nature, it is not complete to all "that is in nature" or all "that is possible." The need for paraphysics thus becomes apparent beyond its simple definition as "the physics of psi" or "the physics of paranormal phenomena."

The concepts of life, thought, mind and consciousness are normally dealt with in other academic disciplines such as the life sciences and psychology, but these sciences are also incomplete in their understanding of these concepts. The paranormal concept of psi, a general term describing the quantity which acts in both extrasensory perception (ESP) and psychokinetic (PK) phenomena, is intimately bound to these quantities and thus lies outside of normal physics.

Psi also lies outside the scope of the life sciences and psychology, so the new academic field of parapsychology was developed six decades ago to deal with the mental aspects of psi. By so introducing a new science to deal with the psychological aspects of psi, it has become necessary to distinguish between the mental and physical aspects of psi and thus differentiate between parapsychology and paraphysics.

As science approaches a complete realization of the most fundamental aspects of reality, in so far as such a realization is even possible, it may well be found that there is no difference between paraphysics and parapsychology, just as there may be no real difference between either mind and matter or consciousness and physical reality in the final analysis. But at this point in the evolution of science there is no direct evidence that either consciousness and physical reality or mind and matter are at their most fundamental level the same thing. There are only suspicions. So, the difference between paraphysics and parapsychology must stand, at least until scientific evidence indicates otherwise.

It is quite possible that drawing a distinction between what is purely paraphysical and what is purely parapsychological very probably may aid in the search for a consistent and accurate definition of mind and consciousness as well as matter and physical reality. There have been many cases in science where concepts have been defined as much by 'what they are not' as by 'what they are.' Deciding what is purely physical about psi may even help science to find the limits of consciousness and thus define consciousness. The definition of paraphysics as an independent field of science must therefore be deemed an important step in the evolution of science. Within our normal worldview, as defined by the fundamental bases of classical physics, relativity, and quantum mechanics, all of nature is defined in the terms of matter, motion, or matter in motion.

So, any possibility that there is a more fundamental reality underlying these concepts implies a paraphysical explanation which goes 'beyond' the common concept of 'matter in motion' to establish paraphysics within the science of physics.

Physics in perspective

Before the scope of physics can be extended to include the 'paranormal,' it must be clearly understood what 'normal' physics encompasses. Part of the problem in developing a precise definition of physics and thus deciding between "normal" and "paranormal" rests in the dual use of the word. The word physics can be directly translated from ancient Greek as "nature," so physics can be considered either the nature of an object or thing (an intrinsic meaning) or how that object works or fits within the framework of all of nature (an extrinsic meaning) and thus interacts with other material objects.

At the most fundamental level, physics represents a logical study of phenomena in the world around us. Within this context, physics was originally an attempt to discover the 'nature' of things in our world, the 'natural' world. Objects were once thought to seek their natural positions at the center of the earth and universe, so they fell downward. All material objects had the intrinsic property of gravity. But later, gravity became an extrinsic relationship between all the different material bodies that constitute the universe. The common thread in all such explanations is their reliance on logic and reason once a first premise has been established. So, above all, physics is a logical study of nature based upon some first principle or premise.

This logical study has always been conducted by finding relationships between the most fundamental physical quantities that have been observed in nature. The existence of material objects allows us to determine the relative positions of such objects, from which we have developed a concept of relative space. On the other hand, changes of position of material objects in the relative space of position have allowed us to develop a concept of relative time. Nothing in the physical world seems more fundamental or basic than this relative space-time framework, yet the material objects upon which this relative framework is dependent must still be otherwise represented by their materiality, their "quantity of matter."

So, physics has evolved as the study of matter and the "motion of matter" only within the context of relative space and time. As a logical mental endeavor, physics has come to follow a pattern of development that we know as the scientific method while mathematics, which is itself a logical system of thought, has become a major tool for exploration and exposition in the study of physics. The first physics that can be recognized from today's perspective was that of Aristotle as proposed in his book *Physics,* written about four hundred and fifty years before the Common Era. His efforts to explain nature were based upon the fundamental quantity of motion as a change in position of a material object. Earlier philosophers had noted the significance and explanatory power of "change" within nature, but Aristotle seems to have been the only one to use change of position and thus motion as the basis of his science.

Thales (550 BCE), Anaximenes (550 BCE), Anaximander (550 BCE), Heraclitus (500 BCE), and Anaxagoras (475 BCE) all saw the change of some substance as the basis for science. Aristotle relied on this previous work only to the extent that he borrowed and applied the concept of change, albeit in a different form than his predecessors. He understood the rate of change of position of a body, but never really developed a concept of speed. Therefore, his physics was seriously flawed by today's standards, but an ingenious creation in his own era.

During the Middle Ages, the concepts of speed, average speed, instantaneous speed and acceleration were developed in the intellectual centers of Paris, Cambridge and Oxford. These concepts were adopted and further refined by Galileo, René Descartes, Christian Huygens, and acceleration were developed in the intellectual centers of Paris, Cambridge and Oxford. These concepts were adopted and further refined by Galileo, René Descartes, Christian Huygens, and Isaac Newton in the seventeenth century with which the new science finally replaced the older Aristotelian view of the world.

In the seventeenth century, Descartes first and then Newton formed the first modern concept of mass, although Descartes' laws of motion were erroneous, leaving the job of building the first successful physics of 'matter in motion' to Newton. Newton's Principia or the *Mathematical Principles of Natural Philosophy* in English translation, published in 1687, established the concept of 'matter in motion' as the basis of physics so thoroughly that it has seldom been challenged and never seriously threatened, even given the non-Newtonian advances made in physics during the twentieth century. Despite these modern changes, physics has retained the same dependence or essence of meaning as in Aristotle's day in that physics still reduces all of nature to simple configurations of 'matter in motion.'

Today's common dictionaries define physics in terms of energy and matter. For example, *Webster's* defines physics as …

> … a science that deals with matter and energy and their interactions in the fields of mechanics, acoustics, optics, heat, electricity, magnetism, radiation, atomic structure and nuclear phenomena. (Webster's, 1707)

The *Oxford Dictionary* defines physics, in a similar manner, as …

> … restricted to the science, or group of sciences, treating of the properties of matter and energy, or of the different forms of energy on matter in general (excluding Chemistry, which deals specifically with the different forms of matter, and Biology, which deals with vital energy). (Oxford, XI, 747)

Both definitions depend heavily on the concept of energy. This dependence dates to the late nineteenth century when a group of prominent physicists, called energeticists, tried to base all of physics on energy instead of force and thus break with their Newtonian heritage. In fact, the Oxford Dictionary refers to Tyndall's 1860 definition, G.F. Barker's 1892 definition, J.B Stallo's 1900 definition, and W. Watson's 1900 textbook definition as the etymology for its definition of physics. It gives no later definition to which it can refer its readers.

This seems strange since the Second Scientific Revolution occurred between the time that these definitions were made and the present. While these definitions are somewhat dated, it can be argued, at least to support the modern dictionaries, that modern quantum mechanics and general relativity also stress the fundamental nature of energy in physics. However, these definitions remain inadequate despite their obvious popular appeal. Energy is just one of two fundamental ways in which 'matter in motion' can be packaged. The current dictionary definitions of physics merely reflect the prevalent thought and fashionable attitude that energy and momentum are somehow more fundamental than 'matter in motion' and thus mimic modern quantum mechanical prejudices.

According to the Heisenberg uncertainty principle, the momentum or energy and time can never be less than Planck's constant. Written mathematically in Heisenberg's original form, we have the following well-known equations.

$$\Delta x \, \Delta p \geq h/2\pi$$

$$\Delta E \, \Delta t \geq h/2\pi$$

These two relationships imply the fundamental nature of space and time as well as momentum and energy. So, quantum philosophers argue that energy and momentum are more fundamental and are thus prior to matter or motion as represented by mass and velocity, respectively. Yet, both kinetic energy ($\frac{1}{2}mv^2$) and momentum (mv) are no more than ways of packaging matter (represented by the mass m) in motion (represented by the velocity v).

Even general relativity (GR), the other mainstay of modern physics, expresses a similar idea in the form of the stress-energy tensor, according to the basic formula which models the structure of space-time. The curvature of space, represented by the curvature tensor $G_{\mu v}$, is equivalent to the stress-energy stress energy tensor, $T_{\mu v}$, which represents matter, energy, or both.

$$G_{\mu v} = kT_{\mu v}$$

The equivalence of matter and energy are also an important consequence of the special theory of relativity where they are related by Einstein's most famous equation.

$$E = mc^2.$$

In the modern scientific view energy and matter are interchangeable under the proper physical conditions and circumstances, although nature does not allow easy access to the vast amount of energy stored in common matter. Matter, in the form of atoms and molecules as we know it, is quite stable. But many commoners seem to believe that everything is just energy, there is nothing else, which is not true. Some scientists also hold this mistaken belief as if energy is the most fundamental element in the universe.

But GR goes beyond quantum mechanics by suggesting a possible reality beyond 'matter in motion' whether it acts as energy or pure matter. The above formula equating matter and curvature offers only a mathematical equivalency that is still wide open to philosophical and physical interpretation. Simply put, GR implies the possibility that physical reality reduces 'matter in motion' to a space-time curvature within the geometry of the universe. During his own lifetime, Einstein's attitude on this question quavered. He changed his opinion on the reality and fundamental nature of curvature at least twice. Early in his career, he thought that matter represented physical reality and matter curved the space-time continuum.

Later, he came to believe that matter and curvature were both as real as each other, with matter curving space-time and variations in the spacetime curvature causing the motion of matter. But toward the end of his career, Einstein concluded that curvature was the true reality, and that curvature merely manifests itself as matter and energy in our four-dimensional spacetime continuum. (Graves, 186-187) In Einstein's final estimation, the curvature of space-time is the ultimate reality.

So, are 'mass and velocity' more primary, are 'momentum and energy' more primary, or are 'space and time' more primary? Quantum mechanics says the answer is momentum and energy while relativity theory still gives primacy to mass and velocity (which represent 'matter in motion') in space-time. Quantum mechanics implies no reality beyond the wave packet or wave function while GR implies a specific fundamental reality beyond 'matter in motion.' In either case, matter and motion are the more fundamental quantities since energy and momentum are no more than convenient ways of packaging the physical reality of 'matter in motion. Therefore, a definition of physics based upon the fundamental nature of 'matter in motion' is more accurate than modern dictionary definitions based upon the interaction of energy and matter.

It may presently be fashionable to define physics upon notions of energy and matter, but it is not accurate. If matter is no more than packaged energy, then any statement regarding an interaction between matter and energy makes no sense. Physics just boils down, even today, to the study of 'matter in motion' no matter how they are bundled together for the purpose of physical analysis. Yet there are those who would argue that this notion is too simplistic. After all, light waves and photons do not have mass and mass seems to be the singular noteworthy characteristic by which we define matter. However, light waves and photons cannot be defined, measured, or even completely understood without reference to matter (the source of the electric fields) or 'matter in motion' (the source of magnetic fields).

In the meantime, there is presently a serious scientific movement to measure the mass of photons even though the very idea is contrary to the theoretical expectations of commonly held modern scientific beliefs. The discovery of some infinitesimally small but measurable mass for photons would completely alter the theoretical basis of many segments of modern science. Still other scientists refuse to define the mass of the photon as zero on principle and merely state, "to within present measurements it is zero."

Even the act of defining matter by mass alone raises serious problems. If matter is equated to mass, and mass is defined by inertia while inertia is given as a property of matter, the whole argument for basing the definition of matter on mass becomes circular. Yet twentieth century physics has offered no better or more fundamental definition of matter except for equating matter in some way to the curvature of space-time. Otherwise, physics still depends upon a purely Newtonian definition of matter. Other properties by which matter could be defined are its extension in space and time as well as the fact that two pieces of matter cannot occupy the same position in space-time.

In the best and most thorough analysis, no matter how we look at 'normal' physics, 'matter in motion' is still the basis of physics. Therefore, 'normal' physics remains at least semi-Newtonian as opposed to the perceived non-Newtonian or non-classical nature of quantum theory. Since physics is still at least partially Newtonian in its dependence upon the concepts of matter, motion and "matter in motion," paraphysics would constitute a physics that is completely non-Newtonian in its abandonment of 'matter in motion' for a more fundamental explanation of natural reality. So paraphysics is far broader based than just a physics of psi or paranormal phenomena in general.

Some scientists and philosophers who have been trained within the quantum mechanical tradition wield the title 'classical physics' like an ax to bludgeon relativity theory. They say that relativity theory is 'classical' and thus imply that quantum theory is more relevant and modern while relativity theory is not wholly 'up to date' with modern physics. These slights are based upon the fact that relativity is deterministic as is Newtonian mechanics as opposed to the non-determinism of quantum mechanics. But 'matter in motion' is prior to both quantum mechanics and determinism. Quantum mechanics reduces all that can be known about a physical event to 'matter in motion' upon collapse of the wave packet and refuses to look beyond that reduction except to its probabilistic interpretation of the wave function.

Quantum mechanics therefore ignores any reality beyond 'matter in motion' as far as our physical world is concerned. Quantum philosophers have traditionally neglected both the philosophical and the practical problems imposed by their interpretations of nature by ignoring the situation of reality before collapse, although this attitude has been changing within the past two decades. So, traditional quantum theory is even more classical than GR because GR reduces 'matter in motion' to a totally different entity that is expressed as a field with the curvature of the space-time continuum affecting the field structure. Although there is no necessity to rid science of matter and adopt curvature as more fundamental as represented in the extended GR model, at least there is an implication that curvature could represent an ultimate reality.

In this regard, quantum mechanics is not completely ambivalent to searching for a physical reality and some theorists have sought to go beyond the "collapse of the wave packet" to discover an ultimate physical reality, but they have been a minority within the scientific community until recently. The notion that there is a physical reality beyond the fundamental quantity of 'matter in motion' is more than a century old, predating both quantum theory and GR by several decades. Karl Pearson summarized and analyzed the different views on the fundamental nature of matter and motion during the latter decades of the nineteenth century. Pearson rejected both Lord Kelvin's aether vortex atoms and Roger Boscovich's atoms of force as merely replacing one form of 'matter in motion' with another form of 'matter in motion.'

However, he accepted William Kingdon Clifford's reduction of 'matter in motion' to variations in space curvature as a true change in the basis of physics and physical reality. (Pearson, 1885, 27-30) Pearson referred to Clifford's theoretical research on what would today be called a unified field theory. Clifford reached his conclusions regarding matter as space curvature nearly a half century before Einstein even developed GR using a similar model of matter. When the line of development from Bernhard

Riemann, who formulated the generalized model of non-Euclidean geometry used by both Clifford and Einstein, and Clifford to the present, is considered, it would seem evident that physics has been progressing toward something beyond its traditional form for more than a century, or at least beyond its own common fundamental tenets and principles.

The adoption of space or space-time curvature to explain electromagnetism (in Clifford's case) or gravitation (in Einstein's model), respectively, implies an explanation of physical reality beyond the most common fundamental elements incorporated in the concept of 'matter in motion.' In the final analysis, paraphysics is a branch of 'physics' but only as long as we use the broader definition of physics whose scope is all of nature and natural phenomena, whether normal or paranormal. The fact that phenomena are either normal or paranormal while they remain natural, as opposed to supernatural, is not a contradiction in terms.

The primary respect by which paraphysics differs from 'normal' physics is its association with those paranormal phenomena, such as psi and psi related events, which defy explanation within the physics paradigm of explaining nature only in terms of 'matter in motion.' But the association of paraphysics with psi is just that, an association. Paraphysics is defined by neither the paranormal nor psi. Paraphysics also includes any and all attempts to explain normal phenomena in more fundamental terms than simple 'matter in motion,' which is the mainstay of 'normal' physics. At least a tacit recognition of paraphysics in this second sense of the word has been granted by J.G. Gallimore who has published a book entitled *Transverse paraphysics: the new science of space, time, and gravity control.*

His book has nothing to do with psi phenomena. Gallimore uses the term paraphysics in the sense of going beyond 'normal' physics to explain known physical quantities and phenomena. According to this definition, the relationship of paraphysics to psi and paranormal phenomena is left intact although the science of paraphysics does not depend on just the existence of psi.

The relationship between paraphysics and psi is a historical fact as well as a fact of nature and dates back more than a century to the very time when the scientific investigation of psi first developed. This was the era in which the term 'paraphysics' was coined. But within this context, paraphysics had the restricted meaning of a physics of psychic or spiritualistic phenomena, unlike its broader meaning today.

The use and abuse of 'paraphysics'

The etymology of the word paraphysics is quite interesting as is the history of paraphysics itself. The word was first used in the late nineteenth century by the German psychiatrist and psychical researcher Baron Albert F. von Schrenck-Notzing. (Beal, 486) It dates from the same historical period as the word parapsychology. Both terms emerged from the scientific research conducted on the psychic phenomena associated with modern spiritualism. However, unlike the term parapsychology, paraphysics was largely ignored until the 1970s when it was finally revived and popularized. The word parapsychology has been in general use within a scientific context since the 1930s when it was first revived and popularized.

Since then, parapsychology has become an accepted branch of science although there are still many who claim to be parapsychologists even though they have neither formal scientific training nor a scientific background in the field. So today there is still a group of people who corrupt the use of the word parapsychology. Parapsychology and Paraphysics are not the same thing. Despite those who misuse parapsychology as well as those who remain skeptical as to the validity of parapsychology as a science, parapsychology has fared far better than paraphysics. The term paraphysics does not have the same respectability as parapsychology nor is use of the term as widespread while the term is far more often abused.

For example, the word parapsychology is well defined and can be found in any standard dictionary or encyclopedia. There are even encyclopedias dedicated solely to parapsychology. The history of the word parapsychology can be found in many general books on the subject, encyclopedias dedicated to parapsychology and/or the paranormal as well as standard dictionaries and encyclopedias. The popularity of the term parapsychology and its general acceptance as the name of a field of science, although that acceptance has been given begrudgingly by some scientists and other scholars, no doubt stems from the seminal work done by Joseph B. Rhine in the 1930s and thereafter. Since Rhine first began his simple experimental research on ESP, many other scientists have followed his lead and conducted research on all the various aspects of psi within the methodological structure of normal science.

On the other hand, paraphysics has had no real champion of the stature of Rhine so the word is used only sporadically by a few within the scientific community. Paraphysics is far more often misused within the pop culture of „occult" phenomena and the 'New Age' movement. The use and misuse (if not outright abuse) of the word paraphysics by 'New Age' religionists and the occultists accounts for some of the reluctance of more traditional scientists to recognize paraphysics as a legitimate science.

A similar problem occurred during the late nineteenth century when the few scientists involved in psychic research were forced to differentiate between the fantastic and semi-religious manifestations of mediums and the purely psychic phenomena in which they were primarily interested. More traditional scientists of the era shunned any connection with psychic research because of its close relationship to the modern spiritualism movement if not due to an outright prejudice against all such work.

Even when they were first developed and became commonly known, the non-Euclidean geometries were often commonly associated with the spiritualism movement. They were thus shunned by many reputable scientists even though the non-Euclidean geometries were purely mathematical and thus philosophically valid constructs. The association of non-Euclidean geometries with the spiritualism movement and the occult did more harm to the possibility of developing a physics based on a non-Euclidean space than any other factor. (Beichler, 1996)

Given these circumstances, it is plain to see that the use of the word paraphysics has been ambiguous at best and the scientific field of paraphysics has never been clearly defined. This lack of definition has allowed paraphysics to become a dumping ground for all types of paranormal, semi-religious and occult phenomena. Even many of the physicists who conduct basic research on psi are regarded as working in parapsychology rather than paraphysics. A new emphasis on the physical aspects of psi evolved by the 1970s and a popularization of the term paraphysics began within a more scientific context. In 1969, the Parapsychological Association was admitted to the American Association for the Advancement of Science after four decades of basic research.

The acceptance of parapsychology as a legitimate science by the scientific community, as marked by this event, together with the rising popularity of parapsychology and the optimism of the period were essential factors in the renewed interest in paraphysics as a separate and distinct scientific endeavor. The new interest in parapsychology also initiated a greater emphasis on the physical characteristics of psi at just the time when more scientists and thus more physicists became acquainted with the fledgling science of parapsychology.

Other factors that influenced this new look at paraphysics include the changing attitudes of physicists toward both quantum mechanics and relativity theory and a more open-minded attitude toward the acceptance of concepts such as mind, thought and consciousness and their direct role in natural phenomena as studied in physics. In Edgar Mitchell's 1974 book *Psychic Exploration*, both James B. Beal and Brendan O'Regan contributed articles on "The Emergence of Paraphysics" under a section heading of "The Emergence of a New Natural Science." Their articles could be viewed as harbingers of the new science of paraphysics or perhaps a 'call to arms' for interested scientists, but the new science they foresaw has not yet fully emerged.

These titles underscored the hopes of those scientists involved in the physical research of psi that a new era was dawning for paraphysics. In his introduction to the book, Mitchell defined paraphysics as "a new field within noetics that is extending the laws and methods of physics in an attempt to explain some paranormal phenomena." (Mitchell, 25-26) He did not open paraphysics to all the various forms of paranormal phenomena, but limited paraphysics to psi and psi-related phenomena. He also addressed the emergence of paraphysics as a science.

> Today the emergence of paraphysics is paralleling and overlapping much of psychic research and parapsychology and will probably eventually embrace them both. (Mitchell, 41)

Despite his announcement of the emergence of the science of paraphysics, the new science remains stillborn today.

Gerald Feinberg, a well-known and respected physicist who works in neither parapsychology nor paraphysics but seems to have an open mind regarding these subjects, penned the foreword to Mitchell's book. In his role as an establishment physicist, passing judgment on the role of physics in psi research, Feinberg clearly defined the preconditions under which the scientific community could accept psi phenomena as legitimate subjects for research in physics.

Feinberg stated that detailed studies of the properties of paranormal phenomena would be more appropriate than continuing the prevalent scientific effort of parapsychology "whose primary purpose is to convince others that the phenomena exist." (Feinberg, 9) Feinberg realized that far too much effort had already been devoted to proving the existence of ESP, PK and other associated phenomena as well as the reality of psi. If psi exists, then it has properties which should be well identified and studied, then the phenomena would stand on their own merits without an independent proof of existence.

Physicists have no interest in proving the reality of psi, as do the parapsychologists. The physicists' concern with reality is different from that of psychologists in general. Physicists normally work with concrete definable quantities rather than abstract ideas and quantities such as mind. If there is any evidence of a physical effect of the human mind on its physical environment and the properties of this interaction can be identified, such as is the case with psi phenomena, then physicists will develop a theory or theories to explain the effect according to the physical properties displayed.

Physical properties cannot exist for a 'thing' that does not itself exist. These physical properties would define the quantity psi in lieu of a physical theory of psi whose development would be the next step in the scientific process. Once a theory or even a group of theories is developed, scientists can conduct experimental tests for the veracity and validity of the concept by way of proving or disproving the theories. Establishment physicists would pay far more attention to paraphysics if a testable theory of psi were offered. This process was indirectly implied in Feinberg's statements. What Feinberg did not state, but implied, was the fact that isolating, identifying and defining the properties of psi is a necessary precondition for deriving a physical theory of psi. He thus implied that an acceptable physical theory of psi would be necessary for physicists to accept psi as real.

A few years later, the mathematician C. Musés went still further and stated that "some scientists still shun [paraphysics] and act as though it does not exist." His concern not only reflected the circumstances surrounding the current status of paraphysics, but also was also somewhat prophetic. He continued his discussion by giving a simple definition of paraphysics, stating that "Other scientists who are seeking to enlarge the scope of scientific thinking use the term paraphysics to denote the field of phenomena covering interactions of nonphysical things (such as consciousness) with physical bodies and objects." (Musés, 280)

In his article "Paraphysics: A New View of Ourselves and the Cosmos," Musés clearly demonstrated his belief that paraphysics is a legitimate science. The article appeared in 1977 in the book *Future Science: Life Energies and the Physics of Paranormal Phenomena.* The mere existence of this book and the fact that it was even published offers some support to the argument that the study of paranormal phenomena represents a specific and important problem for physics.

The co-editor of this book, John W. White, had already edited and published another collection of articles with a similar theme, entitled *Frontiers of Consciousness.* This other book included a small section on paraphysics, which was introduced with a short explanation of the science by the editor. White cited and repeated a definition that had been offered by E. Douglas Dean of the Newark College of Engineering in his 1967 Presidential Address to the Parapsychological Association. He simply and succinctly defined paraphysics as "the study of those paranormal phenomena which can be viewed as extensions and generalizations of physical phenomena." (White, 1974, 183)

It is no coincidence that Dean was himself a physical scientist, a fact that reflected the growing interaction of physics and psi. He continued to state that paraphysics "is especially concerned with the relationships between psychic research and physics, including phenomena such as dowsing, radiesthesia, physical mediumship, and the physical aspects of parapsychological research and of paranormal healing." (White, 1974, 183) While others working in paraphysics might contest the basic phenomena that were included in this group, the list is not that far out of the realm of what is normally considered paraphysical. Dean did not include work done with plants and polygraphs in his list, but White followed the introduction with an article on such research.

The scope of paraphysics, as stated in this definition was significant more for the phenomena that were left out, such as psychokinesis and poltergeist activity, which are normally included in any list of PK phenomena. Also missing were all ESP phenomena, which must now be included within the scope of paraphysics. These are not isolated references to paraphysics. Still others have noted or briefly discussed the science of paraphysics. In yet another instance, Professor Wilbur Franklin repeatedly spoke of paraphysics during open discussions at the 1975 Parapsychological Association conference dedicated to issues concerning education in parapsychology. He also presented a paper on "The Role of Paraphysics in

Physics Education," which was commended by all present. He opened his paper with a challenge to all parties interested in psi phenomena.

> The choice of whether paraphysics has a justified role to play in physics and physics education or, rather, in the newly developing and often criticized pseudoscience is a significant contemporary choice. The question of whether the hierarchies of conservative thought in physics education will restrain the emergence of new findings and educational programs in parapsychology/ paraphysics because of conditioned heritage or whether the strength of the chrysalis of advocators will be sufficient to bring the powerful techniques of physics and physics education to the students of the occult fields is a question with holistic significance. (Franklin, 229)

Franklin also found that three college physics departments offered courses in paraphysics/parapsychology and another non-physics department offered course work in "Physics and Parapsychology." In addition, Franklin offered a course in "teleneural" interactions in his own physics department.

While the total sum of five colleges and universities do not set the trend for the academic establishment in the United States, they do demonstrate a growing toleration and open-mindedness toward paraphysics that was beginning to emerge by 1974. Franklin concluded from his survey and own experiences as follows:

> The potential of the field of paraphysics/ parapsychology for social usefulness, the understanding of information transfer in cellular and neural systems, and the basis of the conservation laws and forces in the physical universe seem to warrant serious consideration of a major thrust into this interesting area by physical scientists and educators at this time. (Franklin, 241)

Despite this optimistic assessment of the role of paraphysics in science and education, the last quarter of a century has brought only minimal advances in paraphysics research and education.

Defining moments

While the use of the term paraphysics is slowly growing within the scientific community, it is still far from a popular term. None of the standard dictionaries of the English language define paraphysics. The word cannot be found in Webster's, *Funk and Wagnall's* or even the full *Oxford Dictionary of the English Language*. Nor is paraphysics listed as a subject in any standard encyclopedias such as the Britannica or the Americana.

The closest thing to a definition for paraphysics that can be found in a standard dictionary is that given for the adjectival form of the word, 'paraphysical,' in the *Oxford Dictionary*. According to the 1989 edition, paraphysical is "subsidiary or collateral to what is physical; of, pertaining to, or designating physical; phenomena for which no adequate scientific explanation exists." (Volume XI, 173) Parascience is also defined in the same dictionary as "the study of phenomena assumed to be beyond the scope of scientific inquiry or for which no scientific explanation exists." (Volume XI, 206)

From these two definitions, it is easy enough to extrapolate an approximate meaning for paraphysics as it might appear in a future edition of the *Oxford Dictionary*. Paraphysics could be defined as "the study of physical phenomena which seem to be beyond the scope of ordinary physics, or for which no apparent physical explanation can be found." The interesting question that this situation implies is "why do the editors of the dictionary find it necessary to define paraphysical and parascience but ignore paraphysics?" Perhaps they have realized that no science of paraphysics has been developed even though there is clear enough evidence of anomalous phenomena that defining the terms paraphysical and parascience is warranted. The same is true for the 1969 and 1993 editions of *Webster's* dictionary. Paraphysical is defined, but paraphysics is not to be found.

In this case, paraphysical refers to anything "resembling physical phenomena but without recognizable physical cause (~phenomena such as levitation, telekinesis, materialization - Fred Sommers)." (*Webster's,* 1639) In this definition, there is no limitation of paraphysical to 'natural' phenomena, which is unfortunate if not erroneous. However, the reference to "physical cause" would seem to limit the type of phenomena covered to real events in our world.

The cited examples of "levitation, telekinesis and materialization" are also somewhat lacking. Once again, ESP events such as clairvoyance, telepathy, and precognition, as well as remote viewing are not mentioned, but should have been included. Paraphysics is not just limited to what are usually classified as PK phenomena. ESP phenomena are not as overtly physical as PK phenomena, but they must have a paraphysical basis even if it only occurs at the interface between mind and matter where the human brain comes to recognize the psi event that has occurred.

In a random selection of books on parapsychology, the word paraphysics is found to be missing in nearly all cases. It is not mentioned in either Irwin's or Broughton's books even though they are both considered textbooks in the study of parapsychology and Irwin even dedicates a section on physical theories of psi. The term does not appear in the *Foundations of Parapsychology,* which is also a fundamental book in parapsychology education, even though one of the authors, Joseph Rush, is a physicist and contributed a whole chapter of the book, "Physical and quasi-physical theories of psi," to analyzing the various physical theories from the standpoint of a physicist. Nor does either Jule Eisenbud (in *Parapsychology and the Unconscious*) or Patrick Grim (in *Philosophy of Science and the Occult*) mention paraphysics in their books. Grim has even included a section of his book on "Quantum Mysticism."

And finally, Jeff Mishlove's second edition of the *Roots of Consciousness* ignores paraphysics even though it includes a section on the physical aspects of psi titled "Consciousness and the New Physics." His only mention of paraphysics comes in reference to a Defense Intelligence Agency report entitled "Paraphysics Research and Development - Warsaw Pact." Someone in the American government seems to be taking paraphysics seriously. While this list is far from exhaustive, it does demonstrate the extent to which paraphysics is ignored or underestimated by parapsychologists even when they are writing about physical theories of psi.

On the other hand, a definition of paraphysics can be found in *The Donning International Encyclopedic Psychic Dictionary* that is edited by Dr. June G. Bletzer. She defines paraphysics as:

> … the study of PHYSICS in relationship to psychic phenomena; 1. Study of: (a) physics of paranormal processes; activity that resembles physical phenomena but is without recognizable physical cause; (b) (B. Herbert); anomalous physical effects not explained by current physical theories; 2. (Laboratory) an approach to PSYCHIC ENERGY above and beyond the usual study of physics; 3. Investigations made on the borderline of both

physics and psychic phenomena. Syn. PARAPSYCHOLOGY. (Bletzer, 454)

Her first and last definitions do resemble the ideas set forward in the *Oxford Dictionary* but having limited paraphysics to "psychic phenomena" and "paranormal processes" grossly restricts her definition.

Nor does she address the fact that paranormal processes, which are not psychic in nature, may not belong in paraphysics. Nor does she mention or make any reference to the fact that physics and paraphysics are limited to "natural" phenomena, whether normal or paranormal. The definition attributed to Benson Herbert does not take into account or consider the fact that physics progresses over time and absorbs newly discovered phenomena.

In the second definition, Bletzer introduces the phrase "psychic energy." The use of such a phrase implies a theory for the explanation of paranormal phenomena that does not exist. Although an energy exchange must be considered for any physical action according to modern "normal' physics, there is no physical evidence upon which a new form of energy such as psychic energy could be postulated. Even then, her definition of paraphysics does not correspond to her definition of paraphysical a few entries above. In his definition for paraphysical, Bletzer introduces new quantities such as "etheric world intelligences" and "nature spirits." Paraphysical is the adjectival form of the noun paraphysics, so there must be an intimate and total correspondence between the two terms, but there is not in Bletzer's definitions.

Bletzer's definitions present a far more significant problem for scientists who wish to recognize paraphysics as a legitimate science since they are found in a dictionary that seems to be dedicated to the 'New Age' religions and philosophies rather than a strict scientific discipline. Scientists and scholars pay far less attention to publications such as Bletzer's, whether that bias is justified or not, because of an implied association with non-scientific beliefs in superstition and the occult.

Science claims to have banished superstition and the occult from its own practices during the First Scientific Revolution and does not wish to bring superstition under its banner once more. Bletzer's definitions also lack clarity and distinction because Bletzer claims that paraphysics and parapsychology are synonymous. These two words may seem synonymous in the minds of members of the popular culture that has grown around occult phenomena and 'New Age' thinking, but there is a very strict difference between paraphysics and parapsychology.

Both the use and misuse of the term paraphysics as well as the abuse of the word beg the question of why the term has not been more precisely defined within its scientific context. Even scientists are prone to misuse the term paraphysics. Professor Tenti of the University of Toronto has recently offered a seminar on "The Paraphysics of Star Trek." He defines paraphysics as "the study of natural phenomena outside the sphere of ordinary physics." (Tenti, 1) Although this definition is oversimplified and incomplete, it does contain some of the elements of a sound definition.

However, Tenti proceeds to portray purely fictional phenomena, such as sound in the vacuum of space, as illustrating paraphysics. By so doing, he contradicts his own definition of paraphysics as well as all other definitions of paraphysics. While it is encouraging that Tenti deals seriously enough with paraphysics to offer a seminar on the subject, it is discouraging that he associates paraphysics with purely fictional phenomena. Such misuse and abuse of the word forms the greatest deterrent to developing a science of paraphysics that can be taken seriously by the scientific world. Parapsychology also suffers from this same malady, but not to the same degree because parapsychology has long been recognized as an academic discipline.

In contrast to parapsychology, paraphysics deals strictly with the physical aspects of psi. Parapsychology is interested in exploring the mental aspects of psi. Most of the standard parapsychology dictionaries and encyclopedias shun all mention of paraphysics and its explanation. *The Encyclopedia of Occultism and Parapsychology* is one of the few such publications that at least mention the paraphysical aspects of psi.

Paraphysics itself is not included in this encyclopedia, but the Paraphysical Laboratory is at least mentioned. This English laboratory dates from the 1960s and is run by the Society for Psychical Research. Its mandate is to study the physical aspects of psychic phenomena and to this end it publishes the *International Journal of Paraphysics*. This journal is the only professional publication dedicated directly to paraphysics although several other journals, both inside and outside of the parapsychology community, include articles dealing with paraphysics.

However, two recent dictionaries have included definitions of paraphysics, a fact that is heartwarming if not an indication that perhaps the study of paraphysics is again moving toward a small renaissance. In Alex Jack's *The New Age Dictionary*, paraphysics is simply defined as "the study of laws that govern psychic phenomena" while paranormal is "faculties and phenomena beyond normally understood causal factors." These definitions can be contrasted to parapsychology, which is the "science of measuring and investigating extrasensory perception and behavior." (Jack, 148) While these definitions appear in a dictionary of 'New Age' terms, they are far more accurate than definitions found elsewhere.

Jack's definitions leave much to be desired, but he does not seem to mix paraphysics with parapsychology and his short definitions are free of 'New Age' jargon and misinterpretations of scientific topics, as found in Bletzer's dictionary. Michael Thalbourne's *a Glossary of Terms Used in Parapsychology* has a different purpose. His definitions relate only to the science of parapsychology, not to 'New Age' interpretations and one would expect his definitions to have a more precise and scientific ring to them.

Yet Thalbourne's concept of paraphysics is not without its own problems. He defines paraphysics as the part of parapsychology which "deals with the physical aspects of psi phenomena" while parapsychology is the "scientific study of paranormal phenomena." (Thalbourne's, 51) Paraphysical is a "synonym for 'psychokinesis'" and paranormal is the term applied to "any phenomenon which in one or more respects exceeds the limits of what is deemed physically possible on current scientific assumptions." (Thalbourne, 50)

From these definitions, it should be quite clear that Thalbourne does not really distinguish between the mental and physical aspects of psi and further thinks that paraphysics is a branch of parapsychology. One might thus wonder if he believes that physics is a branch of psychology. Since he links paranormal phenomena to physical impossibilities, one might also wonder how he regards thought, love, hate, mind and consciousness, which, although not physically impossible, certainly have no physical explanations given "current scientific assumptions."

Despite the obvious differences between parapsychology and paraphysics, there is a great deal of interaction between them. There is a common ground where paraphysics and parapsychology come together and this common ground has never been properly explored, so both scientists and laymen tend to cloud or muddy the distinction between the two. From a strictly logical point of view, if paranormal phenomena such as

ESP and PK are real and psi exists, then they act in the same physical world as that which is explained by physics, so a physical explanation of psi is demanded by its very physical existence and the physical explanation of these paranormal phenomena falls within the branch of physics known as paraphysics. It is this sense of the word that Bletzer addresses. She makes no mention of the other broader context of paraphysics, which would deter from her occultism.

Yet the occult is no stranger than some of the more recent discoveries of 'normal' physics, even when what is usually considered 'normal' is being stretched to the breaking point. Since parapsychologists and physicists have failed to define paraphysics and determine its scope, whether through prejudice or ignorance, it is quite evident that they have surrendered their right to define paraphysics to anyone who is interested in the subject.

The physics of other worlds

Within its broader context, paraphysics can also be considered a logical study of those phenomena occurring in nature that cannot be understood within the context of our 'normal' concepts of space and time since 'matter in motion' is explained against the common reference frame of space and time. This broader extension of the context necessitates a further refinement of the new definition of paraphysics that emphasizes the role of paraphysics as either the physics of other worlds or the physics governing the extension of our world into realms beyond our normal relative space-time reference frame.

If a new and more fundamental concept than 'matter in motion' is discovered or developed, it would strongly imply either a corresponding development of new concepts of space and time or the existence of other universes while any extension of our normal concept of space and/or time would conversely necessitate a new definition of 'matter in motion' and thus physics.

While 'New Age' practitioners and 'occultists' do nothing to advance paraphysics as a science, if anything their attitudes are anti-scientific in general and they do more harm to the science of paraphysics than good, many scientists and scholars remain dubious if not outright skeptical about psi, parapsychology, and paranormal phenomena in general. Yet, the recent development of alternate worlds and universes within normal science is just as far-fetched as psi in many cases.

One well known skeptic can be found in the person of the physicist Victor Stenger. It seems that he has defined paraphysics within the new context of other worlds, but for totally different reasons. "If physics is the study of the nature of matter, then we might term the study of a world beyond matter paraphysics." (Stenger, 8) Stenger is not referring to other worlds as an extension of our physical reality. He believes that our world, the world of matter, is all that exists. To Stenger, there is only physics and metaphysics, so he would lump all paraphysical phenomena with the occult and the various religions into the field of metaphysics, which covers his non-material world. This would include the science of paraphysics.

Stenger's definition of paraphysics is meant to obfuscate the issue rather than clarify it. His short definition represents at the very least a gross an oversimplification if not an outright tautology. It is strictly and logically improper to use a word to define itself. If paraphysics is a form of physics, then you cannot use the word physics to define paraphysics. Another more accurate definition is required unless physics is itself defined in the process of defining paraphysics.

And finally, if the new science which is evolving depends upon a radical break with the concepts of normal physics, although Stenger is not suggesting such a radical departure from the physics with which he is familiar, then we will be dealing with a paraphysics of the normal in the future instead of dealing with a physics of the paranormal. It should be clear from these considerations that the definition of paraphysics as merely the physics of paranormal phenomena is now inadequate.

The definition of paraphysics as a study of the nature of other worlds indicates a new way of thinking in physics that reflects a fundamental change in attitude and worldview. Stenger's definition reflects this change, although such an observation certainly does not mean that he would ever condone or accept the consequences of the change. Paraphysics could never have been a valid science within either the present or past worldviews of science. Both the Newtonian and modern worldviews oppose the very notion of paranormal phenomena such as psi. But the changing attitudes of physicists over the past three decades have opened new vistas and possibilities that include a more generally acceptable attitude toward paraphysics. Psi phenomena represent only one type of paranormal phenomena which will someday be discovered as we learn more of worlds other than our own, such as those considered by scientists.

Science has now reached the point where many physicists believe that a grand unification of all the natural forces is inevitable and one popular contending theory has as many as ten (or even twenty-six) dimensions of spacetime. This or any other hyperspace theory would go so far beyond our normal concepts of matter and motion that such theories arguably fall within the realm of paraphysics rather than physics. And, like quantum mechanics, the fact of a real physical hyperspace opens possibilities for physics that may account for paranormal phenomena.

Science is not now, nor has it ever been, static. It is progressive and progressive is its existence. One would think that a challenge to the basic concepts of physics would be inevitable at some point in time if science were to continue advancing. So, it is not beyond reason to believe that a new branch of science such as paraphysics, as herein described, could, would and should still evolve. Thermodynamics is only a century and a half old while electromagnetism is nearly as old and quantum mechanics and relativity are still younger. Each of these fields of physics began by challenging some of the basic concepts which science holds, or held, dear. Physics itself did not emerge out of the various disciplines that existed under the single banner of natural philosophy until the middle of the nineteenth century.

And when it did emerge, it came from a synthesis of two separate areas of study, heat theory and mechanics, which had previously been independent of one another. Such changes become evident when problems arise which cannot be solved within previously existing theories. Cracks have now appeared in the edifice of modern physics that probably foretell if not directly imply a newly emerging shift in science. There is a growing concern within the scientific community that relativity theory and quantum theory are still at odds after six decades of philosophical debate and scientific research and this concern is feeding the attitude that a new paradigm is emerging.

Despite the unprecedented successes of modern physics in explaining our world and extending the range of phenomena that we witness daily; many scientists are uneasy with the present state of physics. Not only does the fundamental split between relativity and the quantum remain, but there are philosophical problems with quantum theory itself. Quantum theory has also failed to provide an expected grand unification of the natural forces after several decades of efforts.

In his book *A Unified Grand Tour of Theoretical Physics*, Ian D. Lawrie has confirmed the uneasiness felt by physicists.

> While the mathematical developments which constitute quantum mechanics have been outstandingly successful in describing all manner of observed properties of matter, it is fair to say that the conceptual basis of the theory is still somewhat obscure. I myself do not properly understand what it is that quantum theory tells us about the nature of the physical world, and by this, I mean to imply that I do not think anybody else understands it either, though there are respectable scientists who write with confidence on the subject. (Lawrie, 95)

Paul Budnik quoted this same passage before he elaborated still further on the idea.

> Most [physicists] are complacent. Many revels in the thought that modern physics has transcended classical ideas. The recent books by [Roger] Penrose are a good example. This though is far more flattering to the ego than to believe something is fundamentally flawed in one's understanding and to have no idea how to address the problem. The intellectual modesty and honesty of physicists like Lawrie is far more likely to lead to new physics than the arrogance of those who would dismiss the conceptual problems in physics. (Budnik, 1)

Clearly, these physicists believe that there are serious conceptual problems in physics that are largely ignored by a more complacent majority of physicists and philosophers. All such problems have been glossed over if not outright ignored for decades by the successes of quantum mechanics, but they have not disappeared. They come to the front of scientific thought whenever there is a lull in the advance of quantum theory or a new problem for which quantum mechanics does not have an answer.

Most of the problems in physics either stem from, or are in some way related to, Einstein's refusal to accept quantum mechanics as complete. Indeed, quantum mechanics is not complete. No theory that ignores one of the four fundamental forces (or interactions as they are now called) of nature can be considered complete and quantum theory ignores gravitation and space-time curvature on the atomic level.

It is true that the gravitational effect within an atom is extremely small in comparison to the other forces acting within the atom, but the mass of the elementary particles which interact to form an atom is itself curved space according to the general theory of relativity, so even if the gravitational forces are negligibly small, the space curvature of elementary particles within the atom's structure cannot be ignored. Scientists have just never thought it worthwhile to explore the relationship between curvature on the atomic level and the quantum mechanical explanation of atomic structure that ignores that curvature. That is unfortunate. This strategic failure of science introduces a paradox into physics that has itself gone unrecognized.

Many of the basic disparities between the discrete and continuous aspects of reality are well recognized but unspoken in absolutely every attempt to unify relativity and the quantum as well as every attempt to derive a quantum theory of gravity. At this time, there is every reason to believe that fundamental changes in physics are in the offing and will have to be accepted before any such unification can take place. These fundamental changes in physical concepts will more than likely be so radical as to mark a limitation to physics as it is normally understood and thus represent a break between the physical and paraphysical.

Given the recognition that a new paradigm in physics is very likely emerging as well as the fact that the quantum and relativity must be unified before physics can progress, there is no guarantee that the unification of relativity and the quantum will follow any neat pattern that fits into our present conceptions of physics, as adequately demonstrated by the development of ten-dimensional theories of space-time. By necessity, any unification or synthesis of relativity and the quantum will constitute a direct challenge to the fundamental concepts upon which physics is based because relativity and the quantum represent two fundamentally different approaches to nature, continuity and the discrete. Only a fundamental conceptual change could unify these two opposing ideas into a single theory.

Given the existence of such theories as well as others yet to be discovered, it might even seem that the word paraphysics would have to be invented if it did not already exist. All these factors indicate that the waiting period is nearly over and the birth of paraphysics as a legitimate science is finally at hand. All of these factors indicate that the waiting period is nearly over and the birth of paraphysics as a legitimate science is finally at hand.

And given recent trends in physics, especially those dealing with matter and mind as well as the role of consciousness in physical reality, it seems likely that an explanation of psi will emerge from the physics of reality rather than a strict and direct application of physics to solving the problem of so controversial a subject as psi.

CHAPTER 2
Strange facts in search of a theory
The definitive psi

The science of physics is built upon observation, definition, measurement and noting patterns of action which can be explained by hypotheses, testing those hypotheses, building theories and expanding those theories to cover more and more phenomena. Definition is an important and integral part of the process. In fact, definition lies at the very basis of the scientific method.

Without strict and precise definitions, the quantities to be measured in the pursuit of science could not be measured and more precise definitions allow measurements to be further refined. If a quantity cannot be precisely defined, then precise measurements cannot be taken, and patterns of action cannot be described in enough detail to build accurate theories. In a broad sense, the history of any branch of science can be characterized and portrayed as the progression and refinement of the definitions that form the basis of that field, and the parasciences are no different. So, a precise definition of psi, the quantity to be measured in parapsychology and paraphysics, is necessary for the further development of these sciences.

Earlier this century, Rhine established the basis of the new science of parapsychology by defining psi accurately enough to be measured, if only by statistical methods. Before Rhine accomplished this feat, no direct measurements could be made and most of the psychic studies depended on anecdotal evidence, observation alone.

Although the early investigators had carried out experiments, psychical research was not at that time primarily an experimental science. The tendency was to regard the accumulation and validation of large numbers of reports of ostensibly paranormal events as the right way of finding out about psi, while experiments served the subordinate purpose of providing convincing evidence of the occurrence of telepathy or other psi-phenomena. A more completely experiment oriented approach was set into motion in 1927 when the first university laboratory for the experimental study of the subject was started at Duke University in North Carolina under J. B. Rhine. From that time, parapsychology, like other branches of scientific research, used experiment as a means not merely of verifying the fact of psi but of finding out about its nature and properties. (Thouless, 16)

The evolution of psychic studies to an experimental science moved parapsychology to the third step of the scientific method as listed above.

The previous study of psychic phenomena had only progressed to the observational phase although there were attempts to develop hypotheses to explain the observed phenomena. Even today, the studies of psi phenomena and the fledgling attempts to develop a theoretical basis for psi have only moved into the pattern recognition and hypothesis building stages of a science. The field of parapsychology is littered with hypothetical structures to explain the psi process, or rather some features of it, but there is no comprehensive theory of psi. The lack of such a theory has caused scientists to waste valuable time on "proving" the existence of psi. Gerald Feinberg expressed this opinion over two decades ago (Feinberg, 24), yet scientists are still trying to "prove" the existence of psi. But there is no basic precept in science that will not allow any proof of the existence of psi, nor should it.

Nothing can be proven to exist from a philosophical point of view, so why should psi, which seems to be the subtlest of all influences, be susceptible to „proof" when nothing else is so susceptible. And knowing that nothing can be absolutely "proven," why is a "proof" of the existence of psi even being demanded? There is absolutely, without question, something fundamentally wrong with scientific attitudes toward psi phenomena that need to be addressed by the whole scientific community. In view of this lack of both a "proof" as well as a comprehensive theory of psi, scientists have sometimes resorted to redefining psi or inventing an alternative word or phrase to replace the term. There are criticisms of the use of the term psi that reflect the concern that words like psi, ESP and PK are already theory laden. Such words are thought to convey a particular but unwarranted point of view, attitude or theory that may or may not be valid.

Use of the term psi is associated wholly with parapsychology and thus stresses the purely mental aspects of the phenomena in question, while the physical aspects of the process are rendered subsidiary or secondary to the mental manifestations. Edwin May has recently criticized use of the term psi as not precise enough for quantitative measurement, (May 1996, 3) but his alternate use of the term "anomalous" lacks specificity and even the precision of so scientifically vague a term as psi.

The modern trend toward the use of the phrase 'anomalous phenomena' instead of 'psi phenomena' suffers the same problem as the older idea of 'paranormal phenomena. Anomalous phenomena are those phenomena and events that do not fit the preconceived theoretical framework of normal science. On the other hand, what is paranormal goes beyond what is normal within the wider framework of naturally observable (and thus perceivable) phenomena, regardless of any theoretical framework induced upon normalcy by science? In either case, not all paranormal and anomalous phenomena are psychic in nature. Psi, no matter what the faults with the term, distinguishes a particular class of anomalous and paranormal phenomena that are associated with mind.

Psi phenomena or events follow specific patterns as reflected in the fact that they these phenomena have distinguishable properties. These properties separate psi from other paranormal and anomalous phenomena, so psi is a more precise term than either "anomalous" or "paranormal." The problem is that psi is not, nor has it ever been, designated as a measurable quantity.

Some scientists do not like the term because it does not specifically refer to a "thing" that is directly measurable. Instead, psi is a classification of subtle processes that have common characteristics. The processes are characterized in the first approximation by their association with mind, but the term does not imply any specific theory of mind. Nor does it limit theories of psi exclusively to a quantity of mind alone.

Further problems with psi are inherent in the possible dual nature of psi phenomena as manifested in the physical world. Some psi processes exhibit a mind/mind interaction while others exhibit a mind/matter interaction. This inherent duality will probably have a simple unified explanation as is assumed by use of the word psi to cover both types of phenomena, but the phenomena are qualitatively different at the common level of reality where psi manifests.

These two types of psi are termed Extra Sensory Perception (ESP) and Psycho-Kinesis (PK) phenomena, respectively. These classifications of psi predate the concept of psi as a unification of the common elements of all such phenomena. These classifications of psi actually predate the concept of psi as a unification of the common elements of all such phenomena. ESP and PK phenomena were found to have enough common characteristics that it was decided to group them together under the single banner of psi.

> Another term that will sometimes be met with is one originally suggested by Dr. Wiesner and myself. We were inclined to think that there might be no real difference between what were called telepathy, clairvoyance, and precognition, that they might be the same capacity working under different circumstances. We suggested that this capacity might be indicated by the Greek letter Ψ (psi) which would also have the advantage over such terms as ESP that it implied no theory about the psychological nature of the process but could be used merely as a label. This suggestion has been widely taken up and the term 'psi-phenomena' is as well understood (at least in the United States) as the term 'paranormal phenomena'. We also suggested that this Greek symbolism might be extended to cover the difference between ESP and PK, the former being called $\Psi\gamma$ (psi-gamma) and the latter $\Psi\kappa$ (psi-kappa). This suggestion has not, however, been generally accepted. (Thouless, 1-2)

By Robert Thouless" own testimony, he and Wiesner adopted the term psi because it was not "theory laden" and thus "implied no theory about the psychological nature of the process." The term psi filled a perceived need, so the scientific community adopted it.

The use of other terms to replace psi is premature at best. To use another term now would be just as improper as using the term psi if it were ever demonstrated that a more accurate term was available. Granted, psi is bound to a psychological viewpoint of the phenomena in spite of Thouless' intent, but substituting another term would only cloud the issue and detract from the obvious connection between mind and psi. It has not been demonstrated to anyone's satisfaction or notion of sufficiency that psi is a physical quantity rather than psychological, even though the two could someday prove be one and the same at a more fundamental level of reality. So, a more physically oriented term than psi is unwarranted. Psi is whatever scientists choose to make it, within the restrictions set by nature, but not within the restrictions set by human philosophical notions.

The only thing that is truly evident is that psi must be determined by its properties, in lieu of a precise definition of psi. Without reference to its properties, psi is just an empty category of phenomena that may or may not be related, even though the existent evidence strongly implies that a single idea rests behind all psi phenomena. Both Feinberg and Henry Margenau have stated that the properties of psi must be emphasized rather than the statistical "proofs" of its existence that have dominated research because science will never accept the reality of psi based upon experimental "proofs" of its existence, especially when these proofs are statistical in nature.

> My point is simply this: in order to study these obscure things, you must practice selectivity and concentrate your attention upon instances where positive results are incontrovertible and, of course, demonstrably free of fraud. I believe that as long as you go around making statistical studies everywhere and on everybody, you are not likely to be convincing for a long time to come. (Margenau, 1982, 119)

Margenau went further than Feinberg to state what was only implied in Feinberg's criticism, the necessity of a theory of psi before the scientific community can take the concept seriously.

> Now, the second thing you need is theory. No amount of empirical evidence, no mere collection of facts, will convince all scientists of the veracity and the significance of your reports. You must provide some sort of model: you must advance bold constructs - constructs connected within a texture of rationality - in terms of which ESP can be theoretically understood. (Margenau, 1985, 120)

Their opinions are all the more relevant since both men are respected physicists. Neither could be considered a parapsychologist or paraphysicist.

The primary purpose of more than a half century of experimental research on psi is not to "prove" the existence of psi, as it may well seem to the non-parapsychologist or non-paraphysicist, but to ascertain the properties of a scientifically viable process in nature which has been dubbed psi. Nearly all researchers of psi seem to have forgotten this fact in their rush to overcome emotional and biased criticisms of the concept.

A property by any other name…

The actual properties of psi follow the same pattern of dualism that distinguishes the major categories of psi, the "mind to mind" and the "mind to matter" connections in nature. So, a theory of psi as well as a physics of psi must distinguish, as effectively and accurately as possible, which characteristics are physical, and which are biological and/or mental. Purely mental features of psi hinder the development of a physical theory, just as purely physical characteristics cloud an understanding of the mental models of psi. For example, the effect of personal or cultural biases on psi hits during experiments has no bearing on a physical theory of psi at the present level of understanding the process.

On the other hand, physics cannot ignore the mental or seemingly mental aspects of psi, but a physical theory does not need to include them directly. A physical theory of psi can be based upon the purely physical properties as long as there is a justification for doing so at a later period of development in the theory when all properties must be explained by a comprehensive theory of psi.

Carroll B. Nash, a biologist turned parapsychologist rather than a classically trained psychologist, is one of the few scientists to summarize and enumerate the various properties of psi, even among those who work in the field of parapsychology. He has concluded that …

> Psi appears to require life or mind; to be generally unconscious; to be unlimited by space, time, or physical causality; to act sometimes precisely and specifically and at other times without spatial and temporal precision; to lack reliability and durability; to act holistically; to be teleological and need fulfilling; and to be a two-stage process. (Nash, 1986, 213)

Determining properties for psi depends on an unwritten assumption that psi is a separate "thing" that can be characterized by specific properties. The situation is like that of the early development of a concept of space during the Middle Ages. The Aristotelian view that space was nothing, literally a 'nothing,' guaranteed that space could not be reduced mathematically for physical analysis. That viewpoint changed forever with the Newtonian concept of absolute space and again later with the Einsteinian view of relative space-time.

Newton's absolute space was essentially a 'nothing,' an empty container that is separate from the material bodies it contained, but it was not a 'no-thing' so it could have physical properties. These new views did not just appear at a specific point in time but were the crowning achievements of evolutionary processes in human thought. The same is true in the case of psi. If something has properties, then it cannot be a 'no-thing' and it must exist. Margenau and Feinberg understood this simple notion, at least at some very basic and probably unarticulated level in their own minds and consciousness. Establishing the properties of psi is more important than proving the existence of psi because the properties will lead to a tangible theory of psi which would not be possible unless there were something to the concept.

Harvey J. Irwin, a psychologist, has failed to find specific 'properties or characteristics for psi. Instead, Irwin describes the "character of performance" in an "experimental ESP task" which displays "certain consistencies in individuals" patterns of performance." (Irwin, 84) It would seem quite evident that he does not completely accept the notion of a separate thing called 'psi' which he would be forced to accept if he gave it properties. He strongly implies that the phenomena represent a specific and undetermined process or group of processes rather than a specific 'thing' called psi. His "patterns" include the bidirectionality of ESP or psi-missing, consistent missing, the focusing effect, position effects and decline effects, clustering, the differential effect, displacement, inherent response bias and variance effects.

These are some of the same patterns/effects by which Nash chose to characterize psi. But Nash sees these patterns as specific properties indicative of psi. Irwin further notes that PK is also bidirectional, demonstrates the focusing effect, differential effect, and displacement. Whether PK involves displacement and clustering is still an open question, (Irwin, 129-130) but not beyond possibility. Nash considers these same effects as characteristics of psi. He further finds enough similarities between the "patterns" or properties that he draws conclusions regarding the relationship between them, while Irwin seemingly refuses to draw any specific conclusions, especially any which would relegate these effects to anything but an independent existence.

Irwin does, however, note that there is a "reasonable degree of consistency and coherence in the process-oriented data" that he presents. (Irwin 105) Within certain constraints, including the possibility of cheating,

Irwin concludes that "the available data nevertheless might offer some encouragement to researchers who believe that the concept of ESP is viable, and certainly they provide foundations for theory building." (Irwin, 105) This statement is as close as Irwin will come to admitting that he is discussing specific characteristics and properties of psi and that a single theory to describe the phenomena could be possible.

Nash has also added psi masking, the relative weakness of psi, holism and the holistic shift, the teleological or goal-oriented nature of psi, diametric, memory relations and other characteristics to his list of psi properties. The list is fairly comprehensive. However, some of these features could be combined as different aspects or results of a single physical property. Holism implies a physical field effect, as does the goal-oriented nature of psi. In a normal physical situation, a process can be reduced to a cause-effect relationship between objects. According to Newton's third law of motion, there is an equal and opposite reaction to every action. But the field does not act in that manner.

Material objects do not come into physical contact within the field but interact over longer distances of physical separation via their fields. The third law is integrated into the field point of view such that the action/reaction sequence becomes the physical equivalent of an infinite regress and emerges as a mutual interaction between the fields emanating from two or more bodies. In holism (or wholism), an action/reaction sequence cannot be reduced, but must be considered as a whole, the complete and complicated group of interactions in an extended concept of field. The goal of any physical interaction process is related to the field from which it emerged, it is a part of the field because the action and reaction are not separate as in the strictly Newtonian viewpoint.

This is further related to the concept of a diametric whereby "the components constituting a complex target situation are apprehended by 'extrasensory perception,' the cognition is achieved in a single act rather than by a step-by-step process." (Thalbourne, 20) In other words, a target is sensed as a single whole not as a group of individual components. This property implies a holographic view of psi, which is again like a field.

Nash also lists another characteristic that he calls the 'gestalt.' The 'gestalt' resulting from the psi process is an alternate way of interpreting and expressing the field concept.

Nash's description of the gestalt as "an integrated whole that is greater than the sum of its parts" implies that psi is at least very field-like, if not directly a physical field, since a field is normally represented by field strengths at every point within its domain.

The field is greater than the sum of its parts, the individual points of space. While Nash's various properties are not strictly physical, they are mental properties which characterize how psi is cognized, they still imply that psi acts in at least a manner which is similar to the effect of a physical field. The two-stage process described by Nash is also quite informative. In the first stage the information is received or "becomes accessible without sensory mediation." The second step consists of cognizing the event by "normal mediation of the information ... through ordinary physiological machinery."

This property implies that the psi process is split into two distinct parts, the actual sensing or reception of the psi signal followed by a separate process by which the sensed signal is cognized. The two-stage process therefore offers a good approximation of the dichotomy between the physical and mental aspects of psi without actually requiring separate mechanisms for the transmission and interpretations of the psi signals. Sensing the effect of psi or detecting the signal could represent a completely physical/biological action while the second stage process of cognization would be strictly psychological.

For his own part, Irwin fully recognizes both the value and significance of a theory of psi as well as the separation of the detection and cognization processes. When he summarizes the various theories of psi, Irwin groups them as theories of psi mediation (based on the detection/transmission stage of psi), theories of the experiential phase of psi (based on the cognization stage of psi) and theories that encompass both stages. Quite simply, the various theories of psi reflect the characteristics and properties of psi that have been developed from experiment and observation. The significance of the viewpoints of these two scientists reaches several levels. The books in which their views are expressed are two of a very small number of textbooks that have been written for students of parapsychology. So, their opinions will help to influence and shape the opinions of future parapsychologists and scientists.

On still another level, the slight differences in their philosophical handling and presentation of the data reflect their training in two different fields of science, biology and psychology. The biologist Nash is more inclined to put his faith in properties and theories, while Irwin, the psychologist, is hesitant about committing himself to defining and listing specific properties of psi. However, Irwin deals with indefinable and nonspecific quantities such as mind, the subconscious and subliminal perception. On the other hand, Nash works with definable and quantifiable cells, organs, organisms and animate bodies.

In other words, the philosophical viewpoints of these two men reflect the very dichotomy that is inherent in the psi process, the mental/psychological versus the material/physical/biological. Yet these two men come to very nearly the same conclusions, in so far as it can be said that Irwin has reached specific conclusions about psi and the categorization of patterns/properties.

Both scientists unintentionally belittle the purely physical attributes and properties of psi. For his part, Nash lumps all the physical properties into a single property called "physicality," as if a living body itself is not physical.

In the sense of being independent of space, time, and physical causality, psi is non-physical. Physical causality presumes transmission of energy over time and space between the interacting bodies. Although control of psi has not reached the degree of precision that would permit determination of the duration of time between a psi cause and its effect, psi's apparent independence of physical causality suggests that, for it, cause and effect may be simultaneous. That psi is not a physical force in the classical sense is indicated by the failure of metal chambers and Faraday cages to prevent its occurrence, even though they are impervious to the transmission of most electromagnetic waves. (Nash, 1986, 185-186)

Nash certainly has a very narrow view of the limits of physics, and probably little depth of knowledge of the science itself. There is no evidence that psi is "independent of space, time, and physical causality," only that psi does not interact with space and time in a normal manner.

ESP and PK certainly exhibit physical properties that seriously challenge our present scientific views of space, time, and causality, but psi events occur in physical spacetime so they cannot be "independent" of spacetime. Nash also claims that psi has a property which he calls "physicality," but then claims that "psi is non-physical" in the very first sentence written to explain "physicality." This statement would seem to contradict his idea of "physicality." Nash's "physicality" represents a whole group of physical properties of psi rather than a single property or characteristic of psi and Nash belittles the role of physics in explaining psi by lumping the various physical properties together in this manner.

For his own part, Irwin does not list either physical properties or "patterns" of psi in his discussion of ESP and PK, even though PK is generally considered at least partially physical. The physical properties that Irwin considers are only listed as explanatory notes for the "mediation" theories that he summarizes. A "mediation" theory explains how information is "mediated between the environment and the individual" in either direction. (Irwin, 150) In this manner the process of mediation would seem to circumvent any need for a physical theory. However, there is a great gap in Irwin's logic on this point. His "mediation" process could possibly explain the connection between mind and environment, but it does not explain how the information travels through the environment extending between individuals, which would still be a physical process.

In fact, the theories, which Irwin misconceives as explaining his "mediation" process, are those theories that are strictly physical and explain how the information or psi signals travel through physical space and time. Irwin's decimation of a connection between physics and parapsychology is very nearly complete, but it leaves insurmountable theoretical, philosophical, and practical gaps in his overall concept of reality, whether psi is part of that reality or not.

The physics of psi fairs no better in Nash's view. He neither circumvents physics by inventing something new to replace physics nor does he cast it aside altogether.

Although psi may not be encompassed by the present boundaries of physics, changes in the concepts of the latter science may eventuate its inclusion of paranormal phenomena. Conversely, if mind proves to be basic to matter and physical phenomena are found to be only special manifestations of psychical activity, the science of psi will subsume physics rather than becoming encompassed by the latter. (Nash, 1986, 186)

Nash does not seem aware of the consequences of any possibility that "psi [would] subsume physics" if mind were proven to be basic to matter.

If mind were basic to all matter, then physics would explain mind, not the reverse. Even before that could happen, all matter would have to be proven to have life since psi is "life or mind requiring" even though there is no evidence that inanimate matter is animate. Again, Nash has not considered the consequences of his belittlement and neglect of the purely physical properties of psi by assuming a possibility that physics could become a branch of biology. This can be compared to Irwin's ignoring the physical properties altogether except for the fact that Irwin was obligated to include physical theories of psi in his textbook less it be considered incomplete.

Granted, the line of demarcation between the physical and mental aspects of psi is blurred at our present state of knowledge and might truly qualify as an artificial boundary erected by the human inability to comprehend an alternate view, but it is at least necessary to draw the line and make the distinction at this time to better understand psi. Other scientists and investigators do not run into this problem since they list the properties of psi relative to the various theories of psi that they are summarizing. Perhaps this very problem is the reason why more scholars do not just list all the properties of psi together without further explanation.

Physical properties at last!

Several authors have summarized the various theories of psi in comparative studies (Rush, 1986; Rao, 1978; Chari 1972, 1974, 1977), and each categorizes the theories as physical, mental, a combination of both, or a variation upon this same theme, as did Irwin. Grouping the theories along these lines provides at least circumstantial evidence that the distinction between the mental and physical aspects of psi is quite pervasive even though that distinction remains unrecognized by a portion of the parapsychological and scientific communities.

In fact, a *de facto* discipline of paraphysics has already emerged from psychic research even if it has been neither designated nor recognized as a distinct discipline of science. The nonrecognition of a physics of psi stems at least in part from the unorthodox nature of the physical properties that have so far been associated with psi. The physical properties of psi are far more clear-cut than their psychological/biological counterparts. That is the nature of the beast. They also have a much stronger experimental base than many of psi's mental features because they deal with quantities that are not as vaguely defined as mental properties. But the physical properties seem to fly in the face of standard scientific beliefs about the nature of space-time as described by physics. As best as can be determined, psi displays no attenuation, no exchange of energy, intervening bodies or physical fields do not block it; it seems to act without regard to the normal flow of time and defies common sense causality.

The fact that psi seems to defy so many commonly held beliefs about our normal physical reality has allowed scientists like Nash to declare that psi is probably nonphysical but still has a property of 'physicality.' Under these conditions, psi is either assumed to be either a nonphysical (mental) quantity like thought or our 'present physics cannot explain psi. Since our 'present physics cannot explain or otherwise accommodate psi, it is assumed that future advances in physics will expand the framework of space-time and our knowledge of physical reality so psi can finally be incorporated into a 'future physics.'

Nash's conclusion did not include all of the known physical properties of psi when he spoke of the 'physicality' of psi. Since the physical characteristics of psi go against the grain of space-time, Nash collapsed or combined them into the single statement that "In the sense of being independent of space, time, and physical causality, psi is non-physical." However, Nash was undoubtedly aware of the individual physical properties in order to come to that singular conclusion.

His own research on the attenuation of psi over distances was fundamental according to Scott Rogo.

The lack of attenuation of psi with distance is the most obvious of its attributes. Yet, it is only indirectly implied by experiment.

With attenuation, the signal intensity would normally decrease with distance because the energy of the signal spreads out as a spherical front emanating from the signal source. The energy at any point on the spherical front of radiation, or sectional area of that surface, would then decrease as the inverse square of the distance. The inverse square law is a fact of the geometric structure of normal space. On the other hand, low frequency induction fields present other possibilities.

Although the distances used in Nash's experiments were relatively small, the attenuation should have been significant. If the psi signal attenuated with distance, psi reception should have been ($1/10^2$) one-one-hundredth as strong at thirty meters as it was at three meters, a ten-fold increase in distance. Since it is assumed that the number of psi hits depends upon the strength of the signal, the number of psi hits would have a marked decline at the greater distance, but this did not occur. An alternative would seem to be that a point-to-point transmission (similar to a laser or maser beam) occurred, which is never assumed in psi research because of added difficulties in the possibility. Another alternative would be to assume that psi does not attenuate and therefore does not work in our three-dimensional space as do other influences and actions, *i.e.*, psi is not affected by the inverse square law.

And finally, there could be a small threshold energy which is necessary to initiate an action (such as conscious thought) in the brain for cognition and the energy did not decrease enough in the extra distance tested by Nash to fall below the threshold. This alternative would merely imply that greater distances be used to test the property of nonattenuation. There have been many other experiments that test for physical attenuation. Taken as a whole, these other experiments confirm that psi does not attenuate with distance. John Palmer gives an excellent summary of these experiments and has come to the same conclusion.

> Generally speaking, the experimental evidence indicates that ESP can occur at great distances and does not decline with distance. These findings do not fit well with most hypotheses that physical energies mediate the transmission of extrasensory information. Indeed, the information transmission model itself may be erroneous, a point I will return to at the end of the chapter. (Palmer, 77)

Still other tests of attenuation have been conducted.

Vasiliev carried out many such tests in Russia (Pratt, 1977, 890) and more recent tests were made with experiments in remote viewing. The remote viewing tests of attenuation were conducted by Puthoff, Targ and May at SRI as well as Jahn and Dunne at Princeton, among others. Their conclusions are unquestionably consistent in stating that psi does not display attenuation with increased distances. (Puthoff, Targ and May, 37, 76)

Palmer has taken the notion of a lack of attenuation to the next level and related it to the energy content of the psi signal. As Palmer points out, the idea of attenuation is intimately bound to the concept of energy and energy transfer between objects.

> If one accepts the possibility that a person can obtain information about distant events without use of the known sense, it is tempting to speculate that some transmission of physical; energy is involved, even though that energy may have yet to be discovered by scientists. One probable implication of most theories of ESP assumes such transmission is that receptivity should decrease as physical distance between the subject and the target stimulus increases. (Palmer in Krippner, 73)

Since there is no evidence of energy exchange during psi events, the very notion that 'physical energies' are somehow involved in the transmission of information as psi causes problems for the physics of the concept.

For a signal to 'travel' through space-time, from point to point or event to event, some form of energy must be associated with the signal. As has been noted, energy would 'dissipate' under normal physical conditions during the transmission process and the psi 'signal' would degrade. But this does not seem to occur with psi. "Communication, for example, requires energy; yet no known energy appears to account for ESP." (Rush, 281) One explanation for this anomaly would be that psi somehow utilizes energy from the immediate environment of the psi 'event,' but energy for ESP communication derived from the environment would break the second law of thermodynamics, it would be negentropic.

The problems are even greater for explaining PK phenomena. For an object to move under psychokinetic control it would require large amounts of energy and reported materializations would take insurmountable amounts of energy according to Einstein's formula for the energy equivalent of matter. Psi does not exhibit any energy transfer, nor have experiments or observations revealed any correlation between mass, weight and force (Rogo, 215) Margenau and others have pointed out that quantum mechanics and the concept of virtual particles can violate the conservation of energy at the submicroscopic level of the quantum (Margenau, 124) so there is some precedent for psi's violations of the physical rules governing energy, but that is a far cry from accounting for the large amounts of energy required for macroscopic manifestations of psi during psychokinetic events.

ESP could require smaller amounts of energy than originally thought given a quantum cascading process of firing neurons. Such was postulated by Sir John Eccles (Eccles, 1952), which would be helpful, but even accounting for that small an amount of energy remains troublesome for science. Both space and energy are intimately related to time. Space is coupled to time as an equal but semi-independent partner in the continuum while energy describes the time variance of momentum or simple matter in motion. Psi seems to defy our commonsense notion of both space and energy; therefore, the same must be true of time. But this is difficult to demonstrate experimentally, philosophically, or conceptually.

There are two aspects of time that are significant in the study of psi. One relates to space and the other to energy. In the first, the transmission time of a psi signal seems to be instantaneous. This fact corresponds to the non-attenuation of the signal as it travels through space. Attenuation assumes a finite or measurable time of travel as well as distance of travel.

This means that the signal has a definable and identifiable (at least in theory) instantaneous speed. Non-attenuation would be akin to an instantaneous transmission of the information carried by the psi signal, which leads to an infinite speed within the normal space-time framework.

The experimental evidence for the instantaneous transmission is much weaker than that for the other attributes of psi because of practical problems coordinating procedures to 'simultaneously' send and receive signals over long distances. The process of experimentally coordinating signals by their times is itself limited by the speed of light, even though psi may or may not be so limited, so evidence is only scant at best. Therefore, there is no definitive experiment that can test the amount of time it takes psi to travel between people or events, there are only suspicions based upon other facets of psi. However, in all probability the existence of precognition renders this a moot point since psi signals flow to and from the future equally in the case of precognition. It would certainly seem that psi is not time dependent if precognition is real, and there is enough scientific evidence to accept the reality of precognition.

It is with precognition that the second aspect of time comes to affect psi. The normal progression of time, from past to present to future, requires that an event is initialized before it occurs. This natural sequence in the temporal duration of an event is known as cause and effect, or causality. The initializing action is the cause, and the result is the effect. Normal Newtonian classical physics does not distinguish a preferred direction of time but does rely heavily on causality.

Time can be either positive or negative and not corrupt the Newtonian equations of motion. The only recognition of the forward progress of time in all of physics comes with the concept of entropy. Otherwise, the forward progression of time is an observed phenomenon only. Entropy naturally increases as time progresses forward, so entropy has come to be called 'time's arrow.' Nothing else in physics distinguishes this same progression of time toward the future. But entropy also entails the concept of energy.

Energy dissipates or spreads out in correspondence to increases in entropy. There are two ways to define entropy, one as the ratio of the energy change to the energy content of a system and the other as the increase in disorder of the system. So, time seems to be related to the natural tendency of energy to dissipate as in an attenuating wave front. If there is no attenuation, there is no natural energy dissipation; there is no time and no increase in disorder or entropy, at least as far as physics is concerned.

Since entropy is the only element in physics that distinguishes the forward progression of time, precognition could be possible in any nonentropic state and life is negentropic. However, precognitive knowledge of an event in the form of psychically received information about the event before it occurred places a conceptual burden upon causality that science cannot ignore. Precognition implies that causality can be violated at any level of physical reality, only it is more difficult at higher levels than the quantum because of the complicated entanglement of the larger number of quantum events in higher level systems. With precognition, information about an event can be gained before the event occurs which violates causality.

Information about an event can also be gained outside of the light cone (in a non-precognitive manner) of an event, but after or simultaneous to the event if the psi signal is instantaneous. In this case, causality is again violated. The physical paradoxes or anomalies that these acts represent could well be attributable to the lack of understanding of time by science, but still be psi related. Time is one of the great-unanswered questions of all 'time.'

Granted, there is a strange parallel between the possibility that psi is transmitted instantaneously, the lack of attenuation and the lack of any evidence of energy transfer during a psi event, but there is still more to be considered in the equation. These factors or properties are related to yet another strange property. Psi can neither be blocked nor diminished by intervening material objects or physical fields. In psi experiments, the separation of subjects is to be expected to circumvent cheating, many times by placing them in different rooms or locations. In such cases, walls do not seem to inhibit the reception of a psi signal.

Remote viewing has been successfully completed with objects or locations on the other side of the earth, so the intervention of the earth between the subject and object of the remote viewing does not seem to have interfered in any manner with the phenomenon. In the summer of 1976, specific experiments in remote viewing were successfully undertaken with the 'outbound experimenters' located in a cave at a depth of 200 feet. From this and similar experiments, it was concluded that

Remote viewing has been found to be neither spatially nor temporally dependent. Physical distances ranging up to thousands of kilometers do not "lessen the accuracy [or resolution] of the perception." (Puthoff and Targ, 1979, 45; Targ, Puthoff and May, 92,101) In all these cases, there are material objects and physical fields between the subject and object of the experiment. Yet there is no evidence that changes have occurred in the reception of psi, at least no regular pattern of variation has emerged due to the existence of intervening objects.

The principle of least action would seem to indicate that the psi signal follows the shortest path through space between the subject and object in any experiment, so it would be hard to justify a belief that the psi signal does not go through the earth during experiments with greatly separated subjects and objects. On the other hand, radio signals can follow the curvature of the earth to some degree and thus be detected at much greater distances because they are sometimes reflected between the ionosphere or clouds and the surface of the earth. Therefore, they can travel over greater distances past the horizon of the earth.

One of the earlier physical hypotheses or theories to explain psi simply postulated that electromagnetic signals were sent and received by the brain. Yet if psi were electromagnetic in nature, then psi could follow the curve of the earth in this manner rather than pass through the earth in a straight line. Yet many experiments have been conducted with the subject of the experiments placed in Faraday cages that would block most normal electromagnetic waves. In all these experiments, no appreciable differences were detected between receivers subjected to electromagnetic wave blockage and those not so subjected. (Puthoff, Targ and May, 75, 77; Rao, 1977, 248) So it is assumed that psi is neither electromagnetic nor blocked by intervening material bodies or fields.

There is a strange consistency in all of this. Since space, time, attenuation, energy, and causality are so intimately bound together, it would seem, or one would almost expect, that if one were violated all would be violated. This is the case with psi, and it causes grave problems for any physics of psi, especially as an experimental science. If psi were only independent of one of these aspects of physical reality, then science would immediately suspect that the fault lay in experimental procedures because each of these facets of reality are so intimately woven together.

But since psi violates all these facets of physical reality, a pattern emerges. This either means that psi does not exist (as the critics claim) or that psi is so fundamental that it is intimately interwoven into the very fabric of reality. It would either be more fundamental than the structure of space-time itself or the space-time framework that we accept in physics is woefully incomplete as it now stands. It is at this juncture of reasoning that psi enters physics on its own accord, naturally, since both experiments and observations are beginning to demonstrate that physics is woefully incomplete in many respects. It had long been thought that physics could not account for psi. However, according to Puthoff and Targ this trend is "not at all consistent with the framework of physics as currently understood.

In this emerging view, the often-held belief that observations of this type are incompatible with known laws in principle is erroneous, such a concept being based on the naive realism prevalent before the development of modern quantum theory and information theory." (Puthoff and Targ, 62) The most straightforward manner by which psi is entering physics is through the seeming necessity of a fundamental interaction between consciousness and matter when the wave packet of quantum mechanics 'collapses.'

Beyond this concept, experimental procedures indicate that consciousness can act non-locally which implies something very like psi is interacting with matter at the quantum level of reality.

Some physicists believe that the reconciliation of observed paranormal functioning with modern theory may take place at a more fundamental level - namely, at the level of the foundations of quantum theory. There is a continuing dialogue, for example, on the proper interpretation of the effect of an observer (consciousness) on experimental measurement, and there is considerable current interest in the implications for our notions of ordering in time and space brought on by the observation of nonlocal correlation or "quantum interconnectedness" (to use Bohm's term) of distant parts of quantum systems of microscopic dimensions. The latter, Bell's theorem, emphasizes that "no theory of reality compatible with quantum theory can

So, it seems that physics can now cope with psi and should be able to account for psi within its worldview.

However, physics has been unsuccessful in that attempt despite some very interesting and provocative hypotheses. Some minimal theories have been able to account for some types of psi functioning, but no single theory has been able to account for all the facets of psi and no theoretician has yet tried to tackle all the physical properties of psi so far enumerated within a single theoretical model. Since the preceding physical properties of psi have not yet resulted in a comprehensive theory of psi and cannot lead to a method of experimentally verifying psi, they imply the necessity of determining other physical properties to help isolate and define psi. On the other hand, if psi is so fundamental to nature and physics cannot accommodate psi, as physics is now understood, then physics must take psi into account in any future advances within its overall model of nature.

Rush has stated this paradox in another manner, relating it to the normal way in which experiments are conducted. A physical experiment would normally proceed with all the possible variables that can affect the outcome of the experiment fixed except the one that is being tested. But since no physical variables seem to affect psi, then science is unable to experimentally confirm the existence of psi even though the violation of all these various facets of physical reality is itself a pattern which could lead to a physical theory of psi.

As Rush states, "That familiar procedure has not worked for psi phenomena because of one stark fact. No physical variable has been shown to influence the scoring rate in psi experiments." (Rush, 281) Rush further attributes the skepticism of most scientists toward psi and parapsychological claims to the fact that "psi phenomena appear to be indifferent to physical constraints." (Rush, 281) Experimental verification of psi must therefore depend upon either a theory of psi developed to consider the stated physical properties or other evidence for psi linked to other properties than those already stated.

Footprints in the sand

Fortunately, psi leaves specific 'fingerprints' in physical bodies that cannot be attributed to other causes, or perhaps 'footprints' would be a more appropriate term. The term 'footprint' denotes both a signature-like uniqueness as well as a direction of movement toward a goal or endpoint, in this case a physical theory of psi. These 'footprints' occur for both ESP and PK phenomena, which also tends to confirm the basic assumption that a single underlying commonality such as psi exists for all the forms of paranormal phenomena being studied.

Various ESP phenomena are accompanied by physiological changes in human bodies, even when the psi signals are not mentally cognized. Not only do these physiological changes confirm a paranormal interaction or connection between the sender and a recipient, but they also indicate that the physical connection associated with psi is prior to the cognization of psi. They further confirm the two-step process described by Nash.

The other 'footprint' is quite clearly evident in the PK effect known as 'spoon' or 'metal bending.' 'Metal bending' and similar phenomena have been severely criticized by skeptics and other scientists as tricks played by magicians and hoaxers, but perhaps the greatest 'proof' of the reality of these phenomena comes in the post-event analysis of the psychokinetically deformed material bodies which cannot be replicated by magicians. However, submicroscopic analysis demonstrates abnormal features in the metal, which cannot be explained by normal physical processes. Given the results of the post-event analyses of psychokinetically altered metal, the explanation that seems to best fit 'metal bending' phenomena can only be that something has occurred that is beyond the normal physical limits of nature. Clearly, that physical process must be investigated.

The classic studies testing the body's physiological response to psi were made with a plethysmograph. This device measures changes in blood pressure (flow) or shifts of blood volume to various parts of the body. Changes in blood volume or flow can cause vasoconstrictions and vasodilation in the blood vessels to any part of a body, such as a finger. Emotion can also cause vasoconstrictions. S. Figar used this physiological response to emotion in the 1950's and later by E. Douglas Dean (Morris, 65-66) and Nash to test for emotional responses to psi signals. Emotionally laden events or pictures shown to one person caused unconscious physiological responses in the subjects who were connected to the plethysmograph. Some measured vasoconstrictions were quite strong (Rogo, 112) and clearly indicated the influence of psi. Since these early attempts were made, the experimental base has been expanded and includes other physiological responses, such as Electroencephalograph (EEG) readings and Galvanic Skin Response (GSR).

In summarizing and evaluating all these tests, Palmer has concluded that the EEG results were inconclusive, but "the presence of some positive findings should encourage further research." (Palmer, 197) The EEG results are not without inherent problems that skew the results. An EEG detects and records relationships between electrical activity in the cerebral cortex and cognitive processes, so there is no simple and conclusive way to differentiate between physiological responses to psi and normal, abnormal, or paranormal cognitive responses.

In other words, poor results with EEG experiments are quite likely due to psychological interference since there is no way to shield or otherwise eliminate the psychological effects of cognitive processes from the background of the recipient's inherent mental activities. Robert Morris has pointed out that "the EEG results considered as a whole are confusing," (Morris, 67) but given the type of responses elicited by the EEG this finding should not be unexpected. This problem does not exist with plethysmograph readings which measure completely autonomic responses which have very little to do with purely psychological responses.

GSR is also considered an autonomic response, but the results of skin resistance tests are 'less encouraging' than the plethysmograph tests. Palmer's conclusions on the combined results of both types of autonomic responses to psi were less than clear.

> Autonomic responses to remote stimuli of emotional salience to agent and/or percipient have been demonstrated with a fair degree of consistency in properly designed experiments when the measure is the plethysmograph, but not when it is the GSR. Although I am by no means an expert in autonomic physiology, I know of no physiological reason to expect different results with these two measures. (Palmer, 196)

But GSR is susceptible to a far greater amount of mental control and interference than blood volume changes, so it is harder to isolate psi effects from non-psi effects in the skin resistance tests than it is in blood flow tests. The quite significant result noted by Palmer is that emotionally salient information evoked a much greater psi related response. ESP phenomena leave a temporary emotional 'footprint' in the body of the person experiencing the paranormal event. In theory, there should be other methods of varying reliability that can be applied to investigate this 'footprint' in further detail as well as study its characteristics.

These three types of physiological response do not limit the possible unconscious responses that can be tested by scientists, but other types of tests and physiological procedures will not change the conclusion regarding the emotional nature of the responses. They will only confirm what is already known although they could refine our knowledge about the results. Similar tests have also been conducted on other animals and plants, but only in humans is there a chance to differentiate between cognitive and physiological responses with anything close to certainty, so the other tests are not pertinent at this time.

The major difference with the PK version of the 'footprint' is that it is a permanent alteration in the atomic structure of the inanimate 'recipient' of the psychokinetic action. 'Metal bending' and similar phenomena are the results of directed psi action on inanimate objects. The psi signals affect their targets by an unknown 'mechanism,' but the 'footprints' that they leave in their wake should demonstrate the characteristics of the 'mechanical' process of psi action. Use of the word 'mechanical' in this context is not meant to imply a Newtonian or any other theory-laden process. It is used merely for lack of a better word. All that is known about the 'mechanism' of PK is that it either does not utilize any normal energies or it uses energy in a manner that is wholly unknown and unsuspected in modern science.

Quite a few strange effects resulting from PK incidents of „metal bending" have been reported and investigated. In some cases, metal has not been so much bent as it had been "churned up," displaying extensions and contractions "at closely spaced intervals." The French researchers Grussard and Bouvaist have also reported that the structure of metal "is sometimes changed by metal benders even when no measurable plastic deformation has taken place" (Hasted, 105) demonstrating that atomic and molecular changes had occurred but had not been great enough to deform the piece of metal. In other cases, poly-crystalline metals have simply been fractured paranormally with no other dramatic alterations discovered. (Hasted, 108)

Wilbur Franklin conducted extensive microanalyses of pieces of metal bent by Uri Geller. Mr. Geller became quite famous for his paranormal abilities and underwent scientific testing during the early 1970s. His performance for scientists was neither spotless nor without controversies, but he did exhibit some scientifically viable results and he renewed scientific interest in what has become generally known as macro-PK phenomena. (Beloff, 199)

The detailed metallurgical analysis of three fracture surfaces in two metallic specimens broken by, or in the presence of, Mr. Uri Geller revealed two distinct types of fracture-surface microstructure in the SEM photographs. One type appeared quite similar to normal room-temperature ductile failure caused by mechanical loading, except for a viscous appearance at the bottom of a small lateral crack (see Plates 10 and 11).

In the second type of fracture surface, the predominant microstructures were not typical of ductile failure, fatigue, stress-corrosion, or shear failure, nor of room-temperature cleavage. In the platinum specimen, which exemplified the second type of fracture, localized regions of two types were observed on the same fracture surface only 0.02 cm apart. One region looked like ductile failure in an area that had been heated to the point of incipient melting (see Plate 15; the melting point of platinum is 1773 degrees C). The second region looked like low-temperature cleavage, with inclusions or vacancy clusters also appearing in the field of view (see Plates 16 and 17). These observations, which are not typical of SEM fractographs of failures by mechanical loading, indicate that the cause of fracture was not mechanical in nature nor was it a result of usual mechanical methods of fracture. In fact, the possible methods of, reconstruction of the fracture surface in the platinum ring by known techniques seem to require procedures such as partial cleavage at liquid nitrogen temperature (-195 degrees C) followed by ductile failure of the noncleaved portion and subsequent exposure of this portion to a small beam from a powerful laser in selected regions and a shear force in other regions. Such a project would not only be difficult to carry out, but could not, in fact, be conducted unless a number of people actually perpetrated fraud. (Franklin, 14)

Similar results have been found in the paranormal fracturing of crystals. Scientists have conducted electron microscope examinations of the cleavage surfaces of such crystals and discovered abnormal cavitation "as though some atoms have disappeared locally." (Hasted, 108) Hasted speculates that the PK effect involved in splitting crystals paranormally must include some type of atomic teleportation. There is a great deal of anecdotal and eyewitness evidence of teleportation, levitation, apportments and other mysterious appearances dating back to the days of mediumistic spiritualism, but never has such solid scientific evidence of teleportation been offered. These results are difficult for many scientists to accept and even some in the parapsychological community have mixed feelings about the authenticity and meaning of these phenomena.

On the other hand, 'teleportation' at the subatomic level is not completely unknown in modern physics. It is known as the tunneling effect in quantum mechanics, although it is considerably more difficult to explain the 'teleportation' or tunneling of whole atoms, which are groups of quantum mechanically entangled particles. Yet the 'footprint' left in the PK altered specimens offers a form of hard evidence of paranormal activity and possible tunneling initiated by psi, which is lacking elsewhere in psi research. There are also many other PK related phenomena that have garnered a great deal of attention by scientists and nonscientists alike. These include everything from paranormal healing to deflecting magnetic compass needles, influencing liquid crystal thermometers (Beloff, 208) and poltergeist activity.

But no other PK activities leave a permanent paranormal 'footprint' as do the 'metal bending' and related phenomena. Many of these phenomena are the hardest of all psi phenomena to accept because of the lack of any energy signature in the process. It would not be inappropriate to use the term 'fantastic' to describe these claims, but the microscopic evidence is difficult to refute. As Franklin has so aptly commented, the 'footprints' would be very difficult to manufacture by other than paranormal means and could only occur non-paranormally if "a number of people actually perpetrated fraud."

The picture that emerges regarding the psychokinetic 'footprint' is one of sheer structural randomness at the atomic level in fracturing, stretching, or bending material objects. Dislocation, the churning effect, and the structural differences between contiguous localized regions reported by Franklin all have an air of randomness rather than the application of a single overriding and overpowering physical process, as if anything necessary to reach the goal of altering the material was allowable at the atomic/molecular level of the material.

Hasted has also commented on the primary nature of dislocations resulting from psychokinetic intervention. "In these modified regions the crystals of metal show a large increase in the density of dislocations, and there are other effects as well. It seems that the occurrence of dislocations is a primary event brought about by the paranormal action." (Hasted, 106) In one case of 'metal bending,' a brittle metal alloy that could only be deformed (bent) under normal circumstances over a long period of time during which a smaller constant force is applied, in a process called 'creep,' was bent by PK. If a greater force were applied over any shorter length of time, the metal bar would simply break.

Yet the bars were bent paranormally without breaking in far less time than normal mechanical forces could have bent them. (Mishlove, 1993, 277-278) Since the brittleness of the bar is literally a function of the internal atomic/molecular structure of the metal, the bars could only be bent in the manner described by such a PK process involving alterations in the bars' structure at the atomic or subatomic level. These results indicate that PK acts at the submicroscopic level of material reality even though the PK events are perceived and interpreted at the macroscopic level of physical reality. This is the lesson that is taught by the PK 'footprints.' Such data supports the other properties of psi, indicating that psi operates at the most fundamental level of physical reality.

ESP has been described as a goal-oriented process that acts holistically, or in this case wholistically would be a more accurate descriptive term. It would seem from these various characteristics of psi, embodied in the 'footprints' left in both animate and inanimate material bodies, that PK is also goal-oriented and acts holistically. The mind of a person who wishes to affect a material object psychokinetically merely thinks of the goal of bending or stretching the object and the atomic structure of the object alters by some unknown 'mechanism' to accommodate that goal.

This action is holistic rather than reductionistic simply because the perpetrator of the act is not consciously aware of altering the internal atomic structure of the specified object. Indeed, consciously altering the object at the atomic level by sheer thought would require a far more acute and accurate knowledge of the nature of the atomic structure of the object than is presently known by the most advanced scientists.

The wholism of this action indicates that psi acts as, and probably is, a field phenomenon or process. Perhaps the word holism should be completely (everywhere in our language) replaced with a similar but lesser-known word having a stronger meaning, 'wholism.' The 'wholism' of PK also fits well with the wholistic emotional 'footprint' that characterizes subjects who have experienced ESP events and, of course, the goal-oriented and wholistic psychological characteristics of psi. All these properties indicate that psi is a field phenomenon while there is more direct evidence to support this conclusion.

The fact that Kulagina often misses the mark, so to speak, once again indicates that PK is not so much a mental force but some sort of field. Forwald has suggested this in his PK work, and the experiments with Stella C. and Ingo Swann also support this theory. Lastly, poltergeists seem to be shifted about by a moving field. In Kulagina's PK, photographic paper placed under the target objects have often been found clouded as though some force field had affected it. The field theory of PK was further substantiated by experiments with an American counterpart to Nina Kulagina, Felicia Parise. (Rogo, 223)

Notwithstanding Rogo's elevation of the field effect or property of psi to a theory of psi, his narration still indicates the validity of viewing psi as a field phenomenon. He further explains that Miss Parise was able to deflect a compass needle. When the compass was taken away and then returned to the same spot, it was again deflected. It was as if she either established an independent field or disrupted the normal magnetic field at the location in space where the compass had originally stood.

Photographic film placed at different locations around Miss Parise were also fogged, with the greater amount of fogging appearing the closer the photographic paper was placed to Miss Parise. (Rogo, 223) These results further indicate the field like nature of psi. Graham and Anita Watkins also performed tests on anesthetized mice that were prematurely awakened by ESP suggestion from gifted agents. These experiments confirmed a 'linger effect' that has also been attributed to the field nature of psi. (Rogo, 224) When all is taken into consideration, the field nature of psi seems to be well established by experiment given this work.

The strangest facts one could imagine

From this study of psi phenomena, a specific set of physical properties has emerged. While these properties are non-physical in a sense, they are not affected by the same physical restraints as most other properties of nature, they are consistent and seem to be intimately interwoven. In other words, when these properties are taken together as one interacting and interdependent whole, they suggest a number of different patterns of action that should lead to a viable physical theory of psi.

The physical properties of psi can be enumerated quite thoroughly and simply as follows:

1. Psi is not restricted by the normal limits of time.

2. Psi is not restricted by the normal limits of space.

3. Psi can act non-causally.

4. Psi cannot be blocked by matter or other physical fields.

5. Psi does not attenuate with distance from a source.

6. Psi does not carry, nor does it display any exchange of energy with its environment.

7. Psi is a goal-oriented process.

8. Psi acts holistically.

9. **Psi acts in the manner of a field.**

These properties clearly suggest that psi acts at the most fundamental levels of physical reality, so psi should not be dismissed as a myth or superstition.

In essence, the fact that psi even has physical properties demonstrates the reality of psi and the fact that psi acts at the most fundamental level of reality means it is something that cannot be so easily grasped, realized, reduced, or understood. Psi is among the subtlest of physical actions, so it only rarely manifests itself at the macroscopic level of reality and thus seems to act randomly and unpredictably.

It is only as physics itself reaches into the most fundamental aspects of reality that it finally comes to the point in its own evolution where it can deal with the reality of psi. This idea has been expressed by others, although never as forcefully and fully as the idea has been expressed here.

> Modern physics has reached the stage where the clear outlines of a new world-view and methodology are visible, and these provide an "integral understanding" of the world and its psychological and mystical background. With wider acceptance of such world-view and methodology, new fields of research may be expected to open out, and parapsychology may be considered to have become an established science. (Whiteman in Wolman, 751)

While reduction is the way and method of science, it ultimately leads to wholism. A fact that has not yet been incorporated not science.

To better understand psi, it must be reduced if only momentarily and the physical and mental aspects explained separately before they can be understood together as a whole. Parapsychology has been unsuccessful in explaining psi because it has traditionally worked at cross-purposes with the physics of psi. It is generally understood that "We have no comprehensive theory accounting for all the reported physical, physiological, and therapeutic effects of psi." (Chari in Wolman, 807)

However, the reason for this lack of a theory has never been adequately defined or explored. All things taken into consideration, it would certainly seem that the failure to separate the physical and mental properties of psi has at the very least unnecessarily complicated the search for a theory of psi if not prevented the development of a theory. Once the physical and mental aspects of psi have been separated and individually considered the science of psi will begin to progress at a much more rapid rate than ever before and a complete theory of psi will emerge.

CHAPTER 3
The History of Future Science

In the Beginning

As this century comes to a close, and the third millennium of the Common Era comes closer to reality, scientists and scholars will begin a long process of analyzing the events and scientific accomplishments of the last century. This process offers both strange parallels and deep contrasts with similar analyses that came at the end of the nineteenth century.

Although both the nineteenth and twentieth-centuries experienced unprecedented growth and scientific advances, the overall outlook of the scientific community at the end of the two periods is vastly different. The nineteenth century ended with many scientists reveling in the success of Newtonian physics and believing that science had only to fill in a few gaps to complete their explanation of nature. Yet by 1901, a new and totally unexpected revolution in science had already been initiated. On the other hand, this century will end with many scientists expecting another revolution in thought in the very near future. It is to this end that we must now turn.

The nineteenth century came shortly after the beginning of the industrial revolution. With this came a new emphasis on the science of energy and power as well as subsequent shifts in political attitudes and cultural mores. Both the technological and scientific advances of that age can be characterized by the development of two devices, the steam engine and the electric battery. While these devices were physical in character, corresponding developments in chemistry and the atomic worldview were also significant. The science related to steam power formed the basis of the new science of thermodynamics and the kinetic theory of matter over the next several decades. While these were primarily extensions of the Newtonian worldview, they did introduce statistics into physics in the name of the advancement of knowledge.

The electric battery led to the birth of electromagnetism and the new concept of fields of potential. While electromagnetism was decidedly non-Newtonian, the first interpretations of electromagnetism were undoubtedly Newtonian. Indeed, these were the very scientific advances that led scientists and scholars to the belief that they were on the verge of a total understanding of nature by the end of the nineteenth century.

These and other scientific advances also had unintended consequences which are normally ignored by scholars, historians, and scientists. The successes of science were a primary factor in the philosophical debates of the late nineteenth century that began to question what it was that science was investigating. Did the laws of nature describe the reality of the world or were they just our common human perception of that reality? This philosophical movement was antithetical but intimately related to the popular movement of spiritualism.

On the one hand, the overwhelming success of science and technology encouraged an interest in them by the common, and many times, uneducated and inadequately educated masses. Segments of this cultural movement combined with popular forms of spiritualism and religion to give birth to what has become known as 'modern spiritualism.' Yet the success of science offered a unique opportunity for scientists, scholars, and philosophers to once again address issues that had been popular a century earlier but largely ignored as science built upon its own foundations. These controversial issues included questions regarding the nature of life and the relation between living and dead matter, as well as the mind/body paradox.

The issues were not just the results of advances in physics, but were also the stepchildren of advances in biology, chemistry, and geology as well as the advent of the Darwinian theory of evolution. Such issues developed the collective consciousness of the whole scientific and scholarly communities. Out of this mixture of questions on mind, body, life, spirit and the human perception of reality emerged the new science of psychology. Even today, very few scientists and especially psychologists would admit that psychology's parents were physics and spiritualism, and its closest sibling is parapsychology, but that fact is nonetheless true.

When all these notions are considered, it is easy to see why so many believed that science was nearly complete at the close of the nineteenth century. Yet, at the same time, it is easy for modern scholars and historians to find the first signs of the new revolution in science within the confines of this grand milieu of ideas. The new revolution that evolved out of this period was non- Newtonian even though it was rooted within the greatest successes of the Newtonian worldview. One early crack in the Newtonian worldview came with the discovery of non-Euclidean geometries at the outset of the nineteenth century. Newtonian physics and its corresponding worldview were based squarely upon Euclidean geometry. There was no other geometry to be considered until the non-Euclidean geometries were discovered and the Euclidean correspondence to Newtonian physics was taken completely for granted by scientists and scholars alike.

The importance of Euclidean geometry to the Newtonian worldview is clearly emphasized by the fact that first discoveries of the non-Euclidean geometry were held in secret by J.K.F. Gauss who feared the "clamor of the Boeotians," a vague reference to the uproar the discovery would make given the Newtonian view of reality. When the non-Euclidean geometries finally became known within the academic community during the second half of the nineteenth century, some scientists immediately applied the new geometries to physics and some astronomers began to search for observational evidence that the universe as a complete whole was non-Euclidean. (Beichler, 1988, 1996)

Scientific and philosophical debates ensued, and the issue was not settled by the end of the nineteenth century. All that was known was that space might be Euclidean but matter itself could not be proven Euclidean. If space in the large scale was non-Euclidean, it was so close to Euclidean that astronomical observations could not distinguish the exact geometrical nature of space. (Robert Ball, 1881, 518-519)

The case for a non-Euclidean or hyper-dimensional physical space was not helped by the fact that some spiritualists and religious philosophers attempted to locate ghosts, spirits, angels and even heaven in higher dimensions of space that angered the scientific community. This fact was an important factor leading to a building backlash tot eh latest advances in Newtonian and non-Newtonian science. Other, more serious cracks in the Newtonian picture came during the last two decades of the century. Ernst Mach made the first successful argument that Newtonian absolute space was not necessary in physics, although his argument had no practical significance to physics until the twentieth century. The earlier attempts to explain electromagnetic theory in terms of Newtonian physics resulted in the development of the concept of a luminiferous aether.

This aether was a non-inertial or massless substance whose sole significance lay in the fact that it lent to absolute space enough substantiality to allow for the transmission of electromagnetic waves through the vacuum of space between the stars. The luminiferous aether had no other physical characteristics that affected mechanics. Michelson and later Michelson and Morley proved rather decisively that no aether existed. Although their experimental work provided a believable coffin in which to place the aether theory and that specific Newtonian interpretation of electromagnetic theory, it did not provide an immediate burial of the aether concept. Scientists realized the importance of the Michelson-Morley experiments almost immediately, but still attempted to place bandages over the wounds that it caused in the Newtonian worldview.

And finally, the discovery of radioactivity just before the close of the nineteenth century opened a whole new vista of research and physical concepts that had not been dreamed of before. Coming very near the end of the century, researchers would not have been familiar enough with the characteristics of radioactivity to even guess the problems that its existence posed for the Newtonian worldview and all of physics. Nature does not follow the man-made calendar and has no psychological stake in the turning of a century or millennium. It is only coincidental that these changes came at the turn of the century, which is no more than a psychological marker for human culture and science. The turn of a century is only a convenient psychological moment for humankind to assess its past and attempt to foresee its future.

At the turn of the century, psychology was first beginning to emerge from its roots in the mind/body problem in physics and philosophy as well as its paranormal roots. This last break is best illustrated by the work of William James. In a sense, his work offers a cusp or watershed that is useful when gauging the emergence of psychology as a science according to its own merits. James believed in psychical or paranormal phenomena. Not spiritualism, but ESP and similar phenomena. He developed the concept of 'subliminal perception' to explain ESP. That fact would come as a shock to many modern psychologists who see the concept as an important factor in any psychological research, devoid of all psychic or paranormal content.

Yet in its modem form, 'subliminal perception' is an important and valid concept within standard psychology and is completely dissociated with the paranormal. 'Subliminal perception' marks the limit of one's psychological awareness of his or her surroundings while psychic phenomena are past that limit in modern psychological thought. James emphasized the concept of consciousness as well as the psychic aspects of psychology. The concept of consciousness has been ignored by psychologists for all those decades since it was first emphasized by James, just as the paranormal was effectively stripped from the study of mind and psychology. However, a new emphasis has been placed upon the direct study of consciousness, just within the past two decades.

Many have attributed the long gap in the history of this subject as a result of an early takeover of psychology by the 'behaviorists.' Behavioral psychology has recently been toppled from its high perch as psychologists begin to look more closely at environmental factors in human psychology. In a very true sense, behavioral psychology emphasized a mind/mind or human-to-human approach to psychology, while subtly maintaining the mind/matter split of the Newtonian and Cartesian worldviews, even during this century when the Newtonian worldview has been de-emphasized in physics and the physical sciences.

The dominant view of the behaviorists grew at the expense of the mind/body or human/environment emphasis in which physicists and other scientists participated. This marked the modern breach between psychology and the other sciences that has only been healed within the past two decades. But the growth of the behavioral school coincides directly with the growth of the non-Newtonian concepts in the physical sciences that have marked the Second Scientific Revolution of this century. The new emphasis on the study of „consciousness‟, which has developed in the scientific community within the past two decades, is indicative of the recent changes in the scientific view toward modern physics that has evolved concurrently within the past two decades.

The new revolution in physics during this last century evolved from the problems inherent in unifying the electromagnetic theory with Newtonian physics. This unification came within a scientific climate clouded by the discovery and explanation of radioactivity. Electromagnetic theory alone was successful in describing how electricity and magnetism worked independently as well as together, especially in the transmission of electromagnetic waves through space.

But it was not so successful in explaining the minutest interactions of matter and electromagnetic waves. Max Planck's explanation of blackbody radiation, Albert Einstein's explanation of the photoelectric effect and Bohr's model of the electronic shells of Hydrogen and other atoms all came as answers to questions raised concerning the interaction of matter and electromagnetic waves at the atomic and subatomic levels of physical reality. These and subsequent developments in quantum physics formed a very serious blow to the time-honored Newtonian concepts of nature and the world. They introduced the new ideas of probability, chance, indeterminism, and the wave/particle paradox into modern physics. During the previous century, statistics were introduced into physics via the study of thermodynamics. Even though statistical analyses gave the correct mathematical models for large ensembles of particles, the action of individual molecules and particles of matter were still thought to act causally and deterministically in a Newtonian manner.

Statistics provided a stopgap method to describe large ensembles of motion in cases where the individual Newtonian motions could not be practically measured and accounted for. Statistical analysis of such large ensembles mimicked measurable and discernable macroscopic quantities that resulted from microscopic physical interactions. On the other hand, radioactivity elevated statistical methods to probabilities and demonstrated that chance and randomness were essential features of nature at the most fundamental level. This fact opened the door for the incursion of indeterminism into the newly forming worldview after the turn of the century. This indeterminism was institutionalized in physics by the Heisenberg uncertainty principle.

These various conceptual changes took place over the first three decades of the twentieth century and are regarded as the "thirty years that shook physics." These were the revolutionary years in physics during which quantum theory came to form the most important basis of science, although not the only basis of modern physics. Einstein and others, as expressed in the null experiment of Michelson and Morley, solved the other side of the electromagnetism/Newtonian mechanics problem, with the development of the theories of relativity. Special relativity came first. Although it is now considered in its more mechanistic manifestations of time dilation, energy/mass equivalence, Lorentz contraction and explosive increase of inertial mass at near light speeds, it originated as a theory of electrodynamics, an attempt to reconcile the Newtonian and Maxwellian theories.

With the main emphasis of the purely mechanistic properties of special relativity, its electrodynamic origins have almost been forgotten while the resulting misinterpretation of special relativity abounds in common culture as well as science in general. It is often thought that special relativity forbade the existence of absolute space and the luminiferous aether, but it only rendered them superfluous as unnecessary concepts for use in modern physics.

This was only a *de facto* banishment of the concepts and general relativity has occasionally been accused of rendering a new absolute spacetime. Special relativity was itself limited in scope from its very inception. It covered only systems moving at relative constant speeds, so Einstein generalized his theory of relativity to include accelerated systems leading to the general theory of relativity. The special theory itself was revolutionary in that it destroyed the Newtonian concepts of constant mass and absolute time, unifying space and time within a single structure as well as deriving the basic relationship between mass and energy. But general relativity went further in adopting a non-Euclidean geometry as the fundamental geometry of the universe.

There are three general interpretations of the fundamental principle of general relativity, the curvature of the space-time continuum. (Graves, 182-183) Einstein espoused each of these interpretations at different times during his career. In the first interpretation, mass distorts space-time to cause the curvature, so mass represents the reality. In the second, mass distorts spacetime and changes in curvature cause the motion of matter. Matter and curvature are equally real.

And finally, the curvature itself is the fundamental reality and matter is just a manifestation of curvature within the continuum. Although the mathematical model of spacetime represented by general relativity does not distinguish between these three interpretations, the distinctions made in these interpretations lead to different consequences in advancing field theory to include other fields such as the electromagnetic field.

Einstein was not the first to equate space curvature with matter and ripples in space curvature as the motion of matter. William Kingdom Clifford first proposed this conceptual advance four decades before Einstein's theory. (Clifford, 1870) Although Clifford's theory was fundamentally electromagnetic, he planned to include all the forces of nature in his theory including gravity at a later time (Clifford, 1887), while Einstein's was a theory of gravitation alone and his later search for a unified field theory was to include electromagnetism into his space-time model. Einstein dealt with a space-time continuum rather than a separate space and time, as did Clifford. His theory was also successful in describing the finer points of gravitation as opposed to Clifford's theory, which was never completely enunciated before his untimely death. Even so, Clifford's theory is closer to the later attempts to derive a unified field theory than it is to Einstein's original version of general relativity as most scholars claim today.

Einstein's development of both theories of relativity was concurrent with the development of quantum theory. In fact, Einstein's physics can be found at the most basic levels of development in both branches of physics in the new scientific revolution, relativity, and the quantum. It is something of a historical paradox that Einstein's work was crucial in the early development of the quantum theory because Einstein never accepted the probabilistic interpretation of quantum mechanics as expressed in the Copenhagen Interpretation.

The different strands of thought that represented quantum theory jelled into a single coherent body of knowledge at the 1927 Solvay Conference against Einstein's arguments and objections. What evolved at that time has come to be called the Copenhagen Interpretation of quantum mechanics. In this interpretation, quantum theory, in the form of the Heisenberg uncertainty principle, placed a precise limit on human knowledge of the world and physical reality. Since momentum and position could not be known simultaneously with any degree of accuracy, the fundamental basis of physical reality was thought to exist (or perhaps not exist) beyond all human attempts to observe and probe that reality, for now and all the future of science.

When placed in terms of wave mechanics, the Schrödinger wave equation was given a probabilistic interpretation such that the square of the wave function became the probability of locating the wave, as an individual material particle, at any point in space-time. Before the act of measurement, as initiated in the laboratory, the wave representing a material particle had some probability of being located at any position in spacetime. The act of measurement localizes the particle-wave to a single position in space-time, an action that is known as the 'collapse of the wave function.' According to the Copenhagen Interpretation, the particle does not physically exist prior to this 'collapse' and therefore has no physical reality until the interaction that caused the 'collapse' in the first place.

In the most extreme version of the Copenhagen Interpretation, there is no physical reality until our consciousness interacts with the wave function so science cannot probe beyond this point. Therefore, the Heisenberg uncertainty principle marks a strict and unbreakable limit to our knowledge of physical reality. It is this interpretation of quantum theory which Einstein so sternly and steadfastly objected to, but it became the primary interpretation of quantum mechanics after the Solvay Conference of 1927. Einstein's refusal to accept this interpretation branded him something of a pariah from this time forward and he spent the rest of his life in the development of a unified field theory from which he dreamed that the quantum of action would evolve from the field equations.

While this brief explanation does touch on the major contributing events in the conceptual developments of the Second Scientific Revolution, as well as summarize the evolving trends, it is far from complete. What is important is the fact that these new physical concepts were alien to the Newtonian worldview and thus represented a fundamental conceptual shift in scientific thought and attitude. The worldview of the scientific community thus began to follow a new evolutionary track that sought the incorporation and elaboration of these new ideas. The success of the quantum theory under the influence of the Copenhagen Interpretation was both phenomenal and immediate, as opposed to the general theory of relativity, which languished (by comparison) over the next four decades. Until the 1960s, there was little practical application of general relativity, so most physicists chose to work in the areas of physics dominated by the quantum mechanics.

The Copenhagen Interpretation was a philosophical force to be reckoned with within the physics community. During this period, it once again came to seem that science was nearing a state of completion and the quantum theory would eventually explain all of nature. A state of scientific complacency and smugness emerged toward new ideas which were at odds with the Copenhagen Interpretation and science seemed only to increase the accuracy of measurements, dot their i's and cross their t's, while the last vestiges of a quantum world were being discovered.

Success followed success in the application of' quantum concepts to nature, leading to the development of quantum electrodynamics and 'quantum field theory.' Yet no one objected that the concept of a 'quantum field theory' is itself an illogical proposition. The phrase is an oxymoron. The quantum of action represents the discrete and particulate in nature, while the field represents that which is physically continuous. Since quantum and field represent opposing physical realities, the concept of a 'quantum field theory' is ridiculous as it stands. The major objective in 'quantum field theory' has been to reduce all the forces of nature, which act in the manner of fields, to the exchange of material particles such as gravitons, photons, and virtual particles.

The field concept would thus disappear at the expense of the quantum of action. No one has sufficiently addressed this contradiction in terms with respect to developing a theory or explanation of the difference between the quantum and field. The prevailing view of those who adhere strictly to the Copenhagen Interpretation and believe that the Heisenberg uncertainty principal limits how closely and accurately we can observe reality has been largely accepted without challenge to its basic assumption of a discrete physical reality. Yet this is only one of the philosophical problems that beset the standard quantum conception of nature.

Parallel developments at the fringe

While these ongoing events dominated the world and attitudes of science, another silent revolution has been approaching at a much slower pace. Like the Second Scientific Revolution, its roots can also be found deep within the science of the nineteenth century. However, unlike the events leading to the last revolution in science, this revolution is still in its developmental stage and has not yet come full circle to burst upon the scientific community.

As has been explained, the scientific community became confident enough in the Newtonian worldview and physics by the middle of the nineteenth century to again seek answers to earlier philosophical questions that had been ignored during the developmental phase of Newtonian physics. These questions concerned the role of 'spirit', in both of its references to our sense of spirituality and our afterlife spirt body, in science, the mind/body problem and the physics of life.

When combined with Darwinism, other historical factors and trends in thought, modern spiritualism, the precursor to modern parapsychology, resulted. Between the turn of the century and about 1930, the movement in modern spiritualism languished. An extremely important factor in the partial collapse of this movement was the development of behavioral psychology. The human/human interaction emphasized by behavioral psychology and the de-emphasis of the mind/body or human/environmental interactions cut physics out of the parapsychological equation. This trend was further exacerbated by political events as well as the new emphasis on the quantum, which absorbed ail of the physicists" thoughts and scientific efforts.

By 1930, an intellectual opening appeared. Modern parapsychology was born shortly thereafter under the direction of J.B. Rhine. Quantum physics had reached a specific plateau of sophistication as its earlier developmental stage reached completion and behavioral psychology was well entrenched in the scientific world. The newly found influence of statistical methods and probability in physics and science in general formed a new basis for interpreting the results of lower-level nonrepeatable psychological events, such as ESP hits with Zener cards. The emphasis of behavioral psychology heavily influenced lab standards and provided research procedures that proved fruitful in the earlier stages of parapsychological research.

But parapsychology had been severed from its physical basis to a very large degree. Despite the obvious physical interactions by which ESP and PK events are characterized, only two physicists, Robert A. McConnell, and Joseph H. Rush, made continuing contributions to the field of parapsychology prior to the 1960s. In this respect, early parapsychology followed the behaviorist model in psychology. This trend was not bad for the development of parapsychology and may very well have been necessary to reduce and define ESP and PK phenomena for study. But physics was effectively and unnaturally severed from the early evolution of parapsychology despite the early coexistence of the two sciences. In the long term, physics must again be considered for the study of psi phenomena to reach any degree of completeness.

That part of physics that deals with the same events or phenomena as parapsychology, now termed paraphysics, has been de-emphasized within the whole scientific community, not just the larger part of it, ever since mind and body were separated within physics. Physics and questions concerning life, mind and spirit have been intimately connected for as long as science has existed in the modern sense of the term. For the first Scientific Revolution to progress in its reductionist manner, such questions were necessarily deemphasized. Evidence of this is quite clear given Rene Descartes' dictum separating mind and body (or matter). Yet there has always been a strong undercurrent of connection between mind and body, even though such ideas have been expressed under several different guises. It is this undercurrent that forms the historical basis of modern paraphysics, since the ideas and concepts expressed therein go 'beyond' the normal physics depicting all of reality as 'matter in motion.'

During the past two decades, the development of science has come to a point where the evolutionary path of paraphysics has crossed the path of normal science, thus completing a circle that began more than a century ago at the height of the modern spiritualism movement. It is the long and slow evolution of paraphysics that forms part of the backbone of the silent revolution that will soon burst onto the stage of the scientific world. The split between mind and body was first institutionalized in science by Newton.

Newton's *Principia* culminated the opening phase of the Scientific Revolution and served science very well since that time by acting as a conduit for the separation of mind and matter. After it was first published in 1687, the *Principia* was severely criticized for leaving God out of science. Although this criticism reflects the important influence of religion as a cultural force during Newton's Day, it also reflects the extent to which mind (or in this guise spirit) and matter were severed during the Scientific Revolution.

Although this criticism reflects the important influence of religion as a cultural force during Newton's Day, it also reflects the extent to which mind (or in this guise spirit) and matter were severed during the Scientific Revolution. Newton answered such criticisms by including a "General Scholium" in the second edition of the *Principia*. There he equated the "Sensorium of God" to his concept of "absolute space." But God, like absolute space, mind and spirit had no practical use in the new physics, so the Newtonian worldview developed independent of mind, spirit, and religious beliefs, to the dismay of the clergy and religious philosophers.

Gottfried von Leibniz, an early critic of Newton as well as cofounder of the calculus, attempted to develop a theory of matter based upon 'monads.' He combined mind (or spirit) with physical matter at the lowest and most fundamental level of existence in an atom like structure. However, his theory of 'monads' was never more than philosophical even though it formed a starting point for later connections between mind (or spirit) and matter. Leibniz' ideas did influence later philosophers who in turn influenced the Romantic Movement. Romanticism was a reaction to the mechanistic worldview and the social philosophies that it engendered during the Enlightenment by Newtonianism. Newton's physics had been so successful that philosophers concluded that society itself must follow similar natural laws and thus begot the Enlightenment.

It was out of Romanticism and the Romantic era that the 'occult' again became popular in deference to the successes of Newtonian mechanism. Romanticism was a cultural backlash against the strictures of Enlightenment thought and Newtonianism. Spirit, mind, free thought, the macabre and an elemental view of nature as a living organism, strictly non-Newtonian concepts one and all, were stressed during the Romantic period. In turn, these Romantic notions indirectly influenced the modern spiritualism movement several decades later.

Modern spiritualism differed from earlier forms of spiritualism in that it was scientific to some degree. On the one hand, legitimate scientists studied the psychic phenomena associated with modern spiritualism, a practice that began the march toward the development of parapsychology. On the other hand, the general popularity of science among the common uneducated masses led many non-trained pseudo-scientists to claim that modern spiritualism had a strictly scientific basis. However, their scientific claims for spiritualism were often misguided and many times questionable. Whichever the case may be, it is certainly true that the advent of modern spiritualism marked the first widespread scientific attempt to unite mind, as portrayed by spirit, and matter.

According to Frank Podmore, an important member of the Society for Psychical Research, the roots of modern spiritualism could be found in witchcraft and the "animal magnetism cult" which began in Mesmerism. (Podmore, II, 34) While his view is certainly accurate it is nonetheless incomplete. Podmore was a product of the same cultural environment from which modern spiritualism developed, so his opinion does not include historical trends that have only recently become evident from our later, less biased perspective.

Mesmerism and phrenology were undoubtedly important trends that fed and enhanced the development of modern spiritualism, but alone they cannot account for the breadth of ideas and strength of conviction that characterized modern spiritualism. The significance of these developments for paraphysics can be found in the fact that mesmerism was originally a physical theory of 'magnetism,' although today we associate Mesmerism with hypnosis and altered states of consciousness. This finding clearly indicates that the forerunners of paraphysics were evident in one form or another in the pre-developmental period of modem spiritualism and thus parapsychology.

When these and all factors are considered, several distinct periods of development are found to have emerged from the historical record, offering a convenient method of categorizing the development of paraphysics. These periods parallel developments in both physics and parapsychology although they may not always conform to standard developmental periods in the history of science. These periods of parallel developments in both standard physics and parapsychology although they may not always conform to standard developmental periods in the history of science. They can be designated as the prescientific era from earliest times to 1850; the early scientific from 1850 to 1930; the middle scientific from 1930 to 1970; and the late scientific from 1970 to the present. Since no comprehensive theory of psi, either physical, mental or both, has yet been offered in the world of science, all these periods comprise the developmental, preparadigmatic era of paraphysics.

The pre-scientific period can be characterized by the early separation of science and superstition as well as mind and matter, the spread of the Newtonian paradigm, early (nonscientific) spiritualism and the cultural reactions of the Enlightenment and Romanticism. The early scientific period envelops both the rise of modern spiritualism and its decline after the turn of this century as well as the developments thermodynamics, electromagnetic theory, relativity, and quantum mechanics in physics. A neo-Romantic cultural movement also occurred in the late nineteenth century, coinciding with the heyday of modem spiritualism.

The middle scientific period began roughly with Rhine's early scientific work to develop parapsychology as a legitimate science and lasted until the acceptance of the Parapsychological Association as a member of the American Association for the Advancement of Science. During this period, the Copenhagen Interpretation of quantum theory reigned supreme in the world of physics. Physical theories of psi were put forward mainly by non-physicists while parapsychologists spent most of their time trying to prove the existence of psi against the arguments of the skeptics.

The late scientific period includes the present day. Starting about 1970, many physicists have entered the field of parapsychology and the word paraphysics has increased in popular use. The stranglehold of' the Copenhagen Interpretation on quantum theory has been challenged and the general theory of relativity has gained in stature due in part to technological advances in astrophysics as well as the space program.

Since physicists have now entered the field of' study, physical theories of psi have been far more likely to be suggested by physicists than non-physicists. More scientists have become willing to accept psi as a reality so less emphasis has been placed on proving its existence although the skeptics remain quite vocal in their denunciations.

Meanwhile, other physicists and physical scientists, while not directly advocating research in psi phenomena, have been more open to accepting the possibility of psi. Challenges to the Copenhagen Interpretation have also led to a wedding of physics and consciousness within the physics community itself; so, some physicists have entered paraphysics via a different route than the direct study of psi phenomena.

When the physics of psi, consciousness, mind, spirit and life finally come together, the Third Scientific Revolution will have begun. And this will only happen when a fully unified field theory has brought the relativity and quantum paradigms together under one theoretical model of reality. Many scientists already suspect that the life sciences, if not a combination of physics and the life sciences, will dominate science in the twenty-first century, just as physics has dominated all of science during the twentieth century.

The pre-scientific period

It is difficult to speak of theories of psi before the present era since the concept of psi, a collective term representing the common element of action in all ESP and PK phenomena, was only expressed for the first time in l946. (Berger and Berger, 338,436) But ideas related to what we now call psi, or rather phenomena that seem to demonstrate a psi factor, are evident in physical explanations and theories dating to the time of Newton and before. The term 'theory' is also used rather loosely and perhaps the term 'hypotheses, or even 'suggestions' would better fit many of the historical situations. But the term 'theory' will be used just the same for reasons of consistency.

The parapsychologist William C. Roll has traced physical explanations of paranormal phenomena to an earlier era when John Webster first associated 'corporeal beams' with the influencing of objects external to the organism. (Roll, 38) Otherwise, as early as 1671, Sir Henry More related the existence of spirits to a hyper-dimensional space in his book *Enchiridion Metaphysicum*. This case is especially interesting since mathematicians had not yet considered hyperspatial geometries. His work is therefore cited in many histories of geometry that have nothing to do with parapsychology or paraphysics. For the purpose of paraphysics, his suggestion was the first in a long line of theories suggesting that geometries other than a Euclidean three-dimensional geometry could be related to paranormal phenomena.

More, a Neo-Platonist philosopher at Cambridge University in England proclaimed that our three-dimensional bodies extend into a fourth dimension that exists beyond our normal senses. This external extension of living matter was made substantial in the form of a quantity that More called 'spissitude.' More came up with the suggestion that spirits should be ur-dimensional. He phrased this in terms of an occult quality he called 'spissitude,' meaning something like 'denseness of substance.' His idea seems to have been that the differences between the physically identical bodies of a dead person and a living person would be that the living body has more spissitude, and that spissitude is physically unobservable because it corresponds to a certain hyper-thickness in the direction of the fourth dimension. (Rucker, 53-54)

For all intents and purposes, 'spissitude' was a material substance only after a fashion since it could not be measured by normal Newtonian mechanical means. For all intents and purposes, 'spissitude' was a material substance only after a fashion, since it could not be measured by normal Newtonian mechanical means. More's model is even more interesting because it came a decade before Newton published his *Principia* (1687), the foundation upon which Newtonian mechanics arose and more than a century before multidimensional geometries began their own long road of development.

As a basis for the science that was evolving, the German philosopher Leibniz took a path different from Newton. He thought that science should be based upon an atomic theory. However, he did not follow the notion of separating mind and body to derive his theory. Leibniz developed the theory of monads during this same time frame. When matter is reduced to its smallest and most fundamental constituents, mind in the form of spirit and matter were unified as the monad. The common factor of the two was a flexible boundary that could exist between the natural and the supernatural as the monad.

Such a boundary could be extended whenever necessary by any new developments in science. In his *Monadology*, Leibniz stated that …

The passing condition that involves and represents multiplicity in the unity, or in the simple substance, is nothing else than what is called Perception. This should be carefully distinguished from Aperception or Consciousness, as will appear in what follows. In this matter the Cartesians have fallen into serious error, in that they treat as nonexistent those perceptions of which we are not conscious. It is this also which has led them to believe that spirits alone are monads and that there are no souls of animals or other Entelechies, and it has led them to make the common confusion between a protracted period of actual unconsciousness and actual death. They have thus adopted the Scholastic error that souls can exist entirely separated from bodies, and have even confirmed ill-balanced minds in the belief that souls are immortal. (Leibniz, 253)

To be more precise, Leibniz did not postulate an interaction between spirit and body. Instead, there was a "pre-established harmony" between the two.

The 'substance' to which Leibniz referred was the substance of the scholastics. Leibniz thus introduced continuity between the world of spirit and that of matter which was lacking in the Cartesian view and thereby attacked Cartesian doctrine. According to Leibniz, soul and body in the monad were inseparable but also autonomous from each other. This continuity between the world of mind and matter, reached by the reduction of matter to the atom-like monads, was no different from the later spiritualist views of a single world consisting of a material segment continuous with the spiritual.

Two great cultural and philosophical movements, the Enlightenment and Romanticism, separated Leibnizian philosophy and modern spiritualism historically. The rationalism usually associated with the Enlightenment of the eighteenth was objective. It was an attempt to explain the world as mechanistic by reducing of the world to a combination of matter and forces. In the extremes of Enlightenment philosophy, the separation of mind and matter was complete. If then, Romanticism was a complete rejection of that rationalism, it would be completely subjective, and not bother with the reduction of the world to matter and forces. But Romanticism as such is misunderstood and is not completely subjective.

There can be no doubt that the Romantic philosophies were a rejection of the mechanistic reductionism, but at the same time Romanticism was also a continuation of that rationale. Those scholars who were the Romantics realized that the reduction of the world to a mechanism, a simple automaton explained by Newtonian physics, was not enough to completely explain the world of life and nature. Leibniz was aware of this shortcoming of science a century before the Romantic period sought a solution to the problem in his association of spirit and soul with the monad, a physical entity. The monad itself was a reduction of our physical world, but it was also a unification of the physical with the mental, mind with matter, objective with subjective. Yet Leibniz" ideas did not directly influence the Romantic Movement and thus affect modem spiritualism indirectly.

Herder, a follower of Leibniz, discarded the autonomy of the soul and body in the Leibnizian monad and made them interactive; this change gave the monad a unified substance that strove to realize itself, a 'spirit, a 'kraft' or force. Herder influenced Romanticism through his extension and amplification of Leibniz' ideas, so Leibniz did not influence Romanticism directly. Herder's filtered view of Leibniz' monads directly influenced the Romantic ideals of the organic view of nature as well as an attitude of the unity of forces and nature which was the cornerstone of the German *Naturphilosophie* movement.

Many other traits of Romanticism are also evident in the smaller movement of mesmerism. In fact, "French romantic writing is full of electric shocks, occult forces, and ghosts" (Darnton, 150) which may have been adopted from mesmerism and "Mesmerism provided [Alexander] Dumas and other romantic writers with the material they wanted ... 'the fantastic, the mysterious, the occult, the inexplicable'." (Darnton, l50-l51)

Mesmerism was developed shortly before the French Revolution and first took root in Paris. Just as Leibniz" monads went beyond Newtonian physics, and were thus an early form of paraphysics, Mesmer tried to unify the early physics of magnetism with life, again an early paraphysical endeavor. The 'animal magnetism' portion of Mesmer's theories was crucial to later developments in the sciences of the paranormal. For theoretical and perhaps financial purposes, Mesmer made use of a common scientific method of explaining phenomena outside the realm of Newtonian mechanism in terms of a non-mechanical, unspecified, and massless fluid. He then considered the possibility that an imbalance of his hypothetical fluid in the human body caused illnesses and speculated that his fluid was magnetic in nature.

Mesmer's manipulations of the fluid in a sickened body, either through massage or by waving magnets over the body, were thought to heal a person's illnesses and diseases.

Within the mesmerism movement there were also suggestions that other psychic phenomena, such as telepathy, were observed.

Mesmerism eventually dissociated itself with the use of magnets and developed into a purely mental discipline before merging with phrenology and modern spiritualism. This evolutionary path, parting ways with its roots in physics, was forced on the movement as science learned more about electricity and magnetism and better explained these and their associated phenomena. There were fewer unknowns for the Mesmerists to utilize for their own non-mechanistic and non-Newtonian ends. Mesmerism was forced to shed its physical baggage and evolve into a new mental discipline, thus emphasizing other facets of its doctrine such as telepathy and strictly psychic phenomena.

It would be difficult to demonstrate that specific theories evolved during this pre-scientific period of paraphysics, since the 'psychic phenomena' which we associate with psi today were ill-defined at the time, if they were defined at all. What we now call psi phenomena, or psi events, would have been cast in a supernatural, occult or religious light during this era. What really emerged from this period was the beginning of a separation between superstition and fact due to the advent of Newtonian physics as well as general ideas about the paranormal which would later be more properly and scientifically addressed when science itself had reached a higher level of sophistication.

Perhaps one of the best examples of the mutual influence between Romanticism and the early precursors of paraphysics can be found in Mary Shelley's popular book *Frankenstein*, published in 1818. The Frankenstein monster was a hodge-podge of inanimate body parts from different sources, made whole by Dr. Frankenstein, but still inanimate. The whole but lifeless body represented the Newtonian view of a living body, a mechanism that is the sum of its parts. In the view of the Romantics, there is more than to life than could be explained by Newtonian physics and the body remained inanimate until the spark of life was given to it by nature, via lightning and electrical stimulation.

The equating of life itself with something beyond Newtonian mechanisms, something extra-mechanical like electricity, was important in itself. This related the Frankenstein monster to the 'animal magnetism' of mesmerism as well as more recent physical theories and advances in electrical and magnetic theory. It should be evident that two separate and central ideas emerged from the pre-scientific period. First, that something extra which science saw as mind, life, or spirit, should be equated to something extra in physical reality, or rather, beyond the mechanics of physics. And secondly, that the something extra which was needed could take the form of dimensions of space beyond the normal three of human perception, or the forces of electricity and magnetism which had not been considered in Newton's *Principia* or a hypothetical and non-substantial fluid.

Newton had suggested that something extra was needed in this regard by equating the "Sensorium of God" to his own concept of absolute space and so set a precedent for others to follow when considering new phenomena outside the scope of the Newtonian worldview. Even today, these three categories of theories rule the paraphysics of psi.

The early scientific period

Scientific attitudes toward the paranormal after 1850 cover the whole gamut of opinions. Michael Faraday, the founder of electromagnetic theory, basically set the tone for most later physicists. He felt that science was justified in taking a stance on spiritualism and testing the veracity of reported psychic and spiritual phenomena. In his own experiments on the subject, Faraday disproved the spirit phenomena that he witnessed. Yet other well-known scientists, including several physicists, were more open to modern spiritualism. Among the physicists so inclined were William F.

Barrett, Sir William Crookes, and Sir Oliver Lodge, all of England, the astrophysicist J.K.F. Zöllner of Germany, the French astronomer Camille Flammarion, and the Italian astronomer Schiaparelli.

The opinions of these and other scientists who supported psychic research were quite varied as to which phenomena they accepted as real and their opinions and beliefs were not static, they changed over the decades. But in general, they remained convinced that something was happening that required the attention of science even though they could not quite isolate and explain what was happening.

The closest that any scientist has come to developing a complete physical theory was the non-Euclidean theory advanced by Zöllner. Non-Euclidean geometries were first discovered at the beginning of the second quarter of the nineteenth century. However, they remained fairly unknown until about 1860. After that date, a great deal of speculation on the geometric nature of our physical space came out of the scientific community. Following this trend, Zöllner felt that a non-Euclidean hyperspace could account for the existence of spirits as well as other paranormal phenomena.

In the early 1870s, Zöllner visited England where he was influenced by Crookes' work on psychic phenomena. Tapping on his previous flirtations with the non-Euclidean geometries, he found parallels between the purported physical properties of four-dimensional geometries and various psychic phenomena. Mathematicians had already demonstrated that motion into and out of a three-dimensional enclosure could be completed without going through the walls of the enclosure if one traveled through a fourth dimension. Furthermore, three-dimensional knots, no matter how complicated, were not knots when investigated using four-dimensional geometries. This mathematical property led Zöllner to propose that psychic phenomena could be explained by the existence of a fourth-dimensional component of our normal space.

Zöllner theorized that the fourth dimension and the Kantian concept of an "absolute area" were associated. He further equated Kant's 'thing-in-itself' with a four-dimensional material object whereby the three-dimensional part of the object exists in our three-dimensional space as a projection of the complete four-dimensional object. In this theoretical development, Zöllner clearly considered the spiritualist phenomena as legitimate scientific subjects. (Hermann, 82) He felt that his theory was founded upon solid philosophical and scientific grounds, and it may have been better if he had left it so, but he extended his work into the experimental arena by studying and interpreting the tricks of the American magician Henry Slade within his theoretical framework.

Zöllner saw in Slade's stage tricks, the untying of complicated knots without apparent human contact or intervention and the appearance of writing inside encased and thus inaccessible chalk slates, as evidence of the existence of beings in the fourth dimension. Slade's tricks were thought to represent a manipulation of three-dimensional objects via the four-dimensional space.

> In these phenomena, or some of them, Zoellner found experimental confirmation of his hypothesis of a fourth dimension of space, - a dimension which should stand to the known dimensions of cubic space, height, length, and breadth, in the same relation which height now bears to the two dimensions of plane space. Given the fourth dimension, the existence of which is mathematically foreshadowed, Zoellner pointed out that to a man or a spirit endowed with the capacity of dealing with it, the abstraction of objects from a closed box, the knotting of an endless cord, or the removal into invisibility of a solid object would be tasks of no special difficulty. (Podmore, 15-16)

Purely mathematical models of geometry seemed to support the fact that four-dimensional beings could accomplish the tricks that Slade demonstrated before assembled groups of scientists lending credence to Zöllner's physical theory.

Slade was eventually exposed as a charlatan and fake, discrediting Zöllner within the scientific community. In a scientific backlash to Zöllner's theoretical work, Ernst Mach attacked Zöllner's ideas (without naming Zöllner) in his book the *Science of Mechanics*. Mach castigated anyone who would believe that the tricks of "prestidigitateurs" and magicians could represent real scientific phenomena. (Mach, 589-59l) He stated that religionists had mistakenly found in the fourth dimension a convenient place to put hell while mediums could locate their spirits there. Mach could not accept the four-dimensional hypothesis himself until objects began to appear out of nowhere, literally pop into and out of space, a possibility that he did not believe would ever occur.

Mach did not blame the mathematicians who had developed the non-Euclidean geometries and whose work had been usurped by the spiritualists for these purposes. For his own part, Mach refused to believe in the physical reality of non-Euclidean geometry as well as anything else that could not be normally perceived and held those beliefs until his death during the second decade of the twentieth century. The breadth of Mach's statements would seem to indicate that there was far more to the use of the hyperspace hypothesis to explain spiritualistic phenomena than was evident in Zöllner's work alone. His statements on these relationships also raised questions pertinent to the present use of the hyperspace hypothesis with psi. Zöllner's work influenced many people outside of his native Germany, although he was academically discredited.

C.C. Massey translated an abridged version of Zöllner's book *Transcendental Physiks* into English for the Society of Psychical Research (the SPR) in 1880. Between this book and various popular press reports regarding Zöllner's work with Slade, English and American spiritualists had ample opportunities to learn of Zöllner's work. Yet hyperspatial concepts had become far more popular than was evident from Zöllner's notoriety and publications alone. The concepts and ideas of non-Euclidean geometries became quite popular during this period among scientists and philosophers in their purely mathematical form. In other words, even in a form dissociated from spiritualism and the paranormal, the knowledge of non-Euclidean geometries filtered through to the general populace in America, Britain, Europe, and the rest of the civilized world.

English mathematicians such as Clifford, J.J. Sylvester, and scientists such as Sir Robert S. Ball supported the possibility that physical space was non-Euclidean and four-dimensional rather than Euclidean and three-dimensional. (Beichler, 1996) Clifford developed what would today be called a 'unified field theory' based upon space curvature in a higher dimension. Sadly, his theory was left incomplete with his untimely death at the age of thirty-four, but others carried on his work. But as Clifford left the stage of history, a new 'force' pushing the idea of hyperspace on the public emerged. Charles Howard Hinton wrote several popular books and essays on the mathematical subject of curved hyperspaces. Although he did not associate his ideas with the spiritualism movement or with ghosts and the supernatural in general, there was such a hint in his popular expositions of the new geometries. Nor did he cite Clifford's work even though the telltale signature of Clifford's work can be found in Hinton's publications.

English mathematicians such as Clifford, J.J. Sylvester, and scientists such as Sir Robert S. Ball supported the possibility that physical space was non-Euclidean and four-dimensional rather than Euclidean and three-dimensional. (Beichler, 1996) Clifford developed what would today be called a 'unified field theory' based upon space curvature in a higher dimension. Sadly, his theory was left incomplete with his untimely death at the age of thirty-four, but others carried on his work.

As Clifford left the stage of history, a new 'force' pushing the idea of hyperspace on the public emerged. Charles Howard Hinton wrote several popular books and essays on the mathematical subject of curved hyperspaces. Although he did not associate his ideas with the spiritualism movement or with ghosts and the supernatural in general, there was such a hint in his popular expositions of the new geometries. Nor did he cite Clifford's work even though the telltale signature of Clifford's work can be found in Hinton's publications.

Clifford's philosophical nature allowed him to tackle the bigger issues of possible links between the more general concepts of linking mind and matter, although his research in non-Euclidean spaces was purely mathematical and physical. He even developed the concept of mind-stuff to explain the interactions between mind and matter in nature, but he did not directly equate it to his hyperdimensional research in physics. Although these men did nothing to directly support or further the connections between their own scientific research and spiritualism, others were free to both interpret and misinterpret their work in the light of spiritualism.

In the meantime, the popular expositions of hypothetical four-dimensional spaces did imply the existence of beings in this new extension of space to anyone not familiar with the scientific method of explanation of a difficult subject by analogy. Meanwhile, in order to explain the physical consequences of a real fourth dimension of space in common terms, other popular writers imagined fictional worlds peopled by beings that realized higher dimensional spaces for the first time. The short book *Flatland*, published in 1884 by Edwin Abbott, related the story of two-dimensional beings that encountered our three-dimensional space for the first time.

This book was a purely fictional allegory meant to explain a difficult mathematical subject without mathematical language, but spiritualists who could imply that such beings could exist could easily have misinterpreted it. Even Frank Podmore added a footnote to his brief explanation of Zöllner's theory that referred the reader to *Flatland*. "The idea of space of various dimensions has been well worked out in an amusing little book called *Flatland*, " (Podmore, 16)

References to connections between ghosts, spirits, and psychic phenomena, on the one hand, and the fourth dimension on the other hand, also appeared throughout the popular scientific and mathematical literature on non-Euclidean geometries throughout this historical period. Simon Newcomb, the American astronomer and one of the first American scientists with a worldwide reputation, made indirect contributions to the association of non-Euclidean geometries and the paranormal. He made early contributions to the mathematical theory of non-Euclidean spaces, but later implied connections between ghosts and the higher-dimensional spaces necessary for the real existence of space curvature.

Newcomb publicly used the phrase 'fairyland of geometry' on several occasions during the period of 1890 to 1910 to describe the non-Euclidean geometries and hyperspaces. (Newcomb, 1902) His use of this phrase has been interpreted by historians and other scholars as implying that the existence of a real four-dimensional space was fictional, the stuff of dreams and fantasies, but Newcomb did believe that our physical space was non-Euclidean and exhibited real curvature in a higher dimension.

The word 'fairyland' only referred to the ephemeral nature of physical space itself. It had nothing to do with a connection to psychic studies, although some people who were unfamiliar with either the abstractions of science and mathematics or the humor of men of science may have thought that the phrase implied Newcomb's personal belief in spirits or ghosts. On the other hand, they may have thought that he and others used the word fairyland' because they did not take the possibility of a real physical hyperspace seriously, when in fact Newcomb did take it quite seriously.

On the other hand, they may have thought that he and others used the word 'fairyland' because they did not take the possibility of a real physical hyperspace seriously, when in fact Newcomb did take it quite seriously. The hypothesis that a fourth dimension could be used to explain psychic phenomena also appeared in several short stories and books published by H.G. Wells during this same period, thus demonstrating the lasting popularity of such hypotheses. Even as late as 1909, nearly on the eve of general relativity, Crookes wrote to Lodge that he wondered if spiritual beings might not reside in a four-dimensional space. (Quoted in Oppenheim, 351) (Quoted in Oppenheim, 351)

Further evidence that non-scientists had tied the popular conceptions of non-Euclidean geometries to paranormal phenomena can be found in a simple statement that the mathematician J.J. Sylvester (teaching at Johns Hopkins in Baltimore at the time) wrote to his friend and colleague, Arthur Cayley. (Sylvester, 1866) In a short postscript, Sylvester noted that the Americans were using hyperspaces to find spirits, a fact that seemed to amuse Sylvester by the tone and context of his notation.

Both mathematicians had been Clifford's teachers, although they later became his students in a more general sense when his mathematical research surpassed their own. They sincerely believed that Clifford was correct in assuming that physical space was four-dimensional. This belief was held by some scientists, mathematicians, and scholars in England and elsewhere, but little became of their convictions after Clifford's death except for some simple theoretical speculations.

Electromagnetic explanations of psi were also in vogue following the recent successes of Faraday and James Clerk Maxwell's electromagnetic theory. As early as 1871, Crookes came to believe that the "obscure phenomena" demonstrated by the spiritualists were "objectively true." At first, he proposed "a new force, or a new form of a known force" to explain the phenomena. At first, he proposed "a new force, or a new form of a known force" in order to explain the phenomena. (Quoted in Oppenheim, 348)

Crookes had come to the simple conclusion that "normal physics" could not account for the observed "obscure" phenomena associated with spiritualism and acted accordingly. More than two decades later, he still held this opinion, but he had become even more resolute that the force could not be accounted for within the framework of physics, as he understood the subject. Like Zöllner, he turned to a popular showman to provide experimental evidence for these phenomena. He used the famous medium D.D. Home as a subject of scientific investigation.

He never ceased believing that psychical research was revealing the operations of a new force, but he was no longer certain that scientists could assign it a place on the map of modern physics. He had attempted to locate psychic force on that map during his tests with Home: He had placed the medium "in a helix of insulated wire through which electric currents of different intensities were passed"; he had brought strong magnets near to Home and to the objects that moved in home's presence; he had illuminated the experiments with different colored lights - all to see what effect the known agencies of physics might have on the medium's manifestations. Crookes had noticed no effect whatsoever. (Oppenheim, 349)

Still, he could find no explanation of the phenomena that Home displayed. Unlike Slade, Home was never caught cheating and departed the stage of history leaving his spiritual and psychic accomplishments unexplained.

Crookes later studied the purely mental (psychic) phenomenon of telepathy. Assured of the existence of telepathy, he employed the newly discovered x-rays as a possible conveyance of telepathy as he explained in his 1897 presidential address to the SPR.

That in these rays we may have a possible mode of transmitting intelligence, which with a few reasonable postulates, may supply a key to much that is obscure in psychical research. Let it be assumed that these rays, or rays even of higher frequency, can pass into the brain and act on some nervous centre there. Let it be conceived that the brain contains a centre which uses these rays as the vocal cords use sound vibrations (both being under the command of intelligence), and sends them out, with the velocity of light, to impinge on the receiving ganglion of another brain. In this way some, at least, of the phenomena of telepathy, and the transmission of intelligence from one sensitive to another through long distances, seem to come into the domain of law, and can be grasped. (Quoted in Oppenheim, 349-350)

But he still came no closer to a theory than these simple analogies.

Barrett's approach to the "obscure" phenomena of modern spiritualism was somewhat different. He immediately showed an interest in phenomena such as clairvoyance and thought transference as displayed by some mediums while in a state of trance and shied away from the more controversial and purely spiritualistic phenomena such as communication with the dead.

By 1876, he attempted to develop a theory of thought transference based on the known properties of electromagnetism.

Barrett was working toward a theory of thought transference, "the action of one mind upon another, across space, without the intervention of the senses." He was certain that much more was involved than the normal processes of suggestion between two people - more even than heightened sensitivity and abnormally acute perceptive powers on the part of his subjects. What was involved he did not pretend to know, but he proposed a tentative hypothesis based on the model of electrical induction, or influence, across space. Defining thought as a form of nervous action. Barrett inquired at Glasgow: "May not nerve energy, whatever be its nature, also act by influence as well as conduction?" If the nerve force were a radiant energy of some kind, "might it nor be capable of throwing the nerve tissue of passive, receptive individuals into states of activity corresponding to the states existing in an active adjoining mind?" (Oppenheim, 357)

Several years later, he still spoke of "nervous induction" and made analogies between electrical and magnetic induction and telepathy, but he only suggested a connection without deriving any specific explanatory hypotheses. Barrett never went beyond analogies and suggested courses of action to develop a theory of psi, yet he fully believed in the reality of the mental phenomena he studied and hoped that they would eventually be explained within the context of physics.

Oliver Lodge's theoretical approach was somewhat different from the approaches of his friends and colleagues. Lodge was an early investigator in the transmission and detection of radio signals and became a lifelong believer in the existence of a luminiferous aether, which he thought necessary to transmit electromagnetic waves across space. He invented the basic circuit (LRC) that became the tuner for all radio transmissions and receptions, so is considered the inventor of the radio. So, Lodge based his own suggestions and hypotheses for an explanation of spirits and psychic phenomena on the existence of this aether. Continuity was an important property of this aether while the general concept of continuity also had spiritual connotations for Lodge, who, unlike other physicists, fully accepted the possibility of the existence of spirits.

Lodge reasoned that the essence of an individual human did not exist wholly in the physical body but formed an "aetheric body" through which psychic powers acted. Such "aetheric" bodies had been suggested by mediums that had been known to eject a sort of ectoplasmic substance during séances. He explained that …

> What we have learnt physically is that the ether can act on matter through electric and magnetic properties: we also know that mind can somehow act on matter, though probably indirectly. Our assumption is that we possess an ether body or animated structure of modified ether here and now, that life or mind is closely in touch with the ether body, and that through its action on this at present imperceptible body it is able to exert an action on the familiar material body. To assume that mind acts on ether and that ether acts on matter is I hope an assumption in the direction of truth: and it appears to be justified by psychical facts, which show that the action of mind can be independent of matter. (Lodge quoted in Oppenheim, 384-385)

The "aetheric body" was the intermediary between spirit and the physical body. Lodge held this view and fully supported his theory of the "aetheric body" as late as 1930, despite the new advances in physics of the twentieth century.

In this manner, Lodge became the philosophical heir of those who sought to explain psi and other paranormal phenomena by the adoption of a hypothetical fluid. Flammarion offered a similar suggestion in his own dealings with spiritual mediums. In his 1907 book Mysterious Psychic Forces, Flammarion argued for the existence of a "magic substance" which could explain the phenomena that he had witnessed.

This "magic substance" was unique.

> … unknowable in its essence. We see and touch only its condensations, its aggregations, its arrangements; that is to say, forms produced by movement. Matter, force, life, thought, arc all one. In reality, there is only one principle in the universe and it is at once intelligence, force and matter, embracing all that is and all that possibly can be. That which we call matter is only a

Flammarion's conclusion was like that of other scientists in that he could not explain the phenomena in terms of the physics of his era, so he resorted to some imaginary medium or substance. Flammarion's explanation coincided with the opening phase of the new revolution in science.

Podmore best expressed the final verdict of the nineteenth century scientists on the scientific worth of spiritualism. Even though he wrote his opinion in 1902, it aptly describes the state of the science of the paranormal at the beginning of the scientific period in 1930. The intervention of World War I, recent technological developments and the Second Scientific Revolution left little room for new research in the paranormal until things settled down during the 1920s.

Podmore was never trained as a scientist, but he was still an acute and discriminating observer of the phenomena associated with spiritualism. In his capacity as an official of the SPR, he concluded that various spiritualistic phenomena had no scientific foundation, but he could find no fault in the various forms of telepathic communication, which were purely psychic phenomena and had nothing to do with spirits.

From this brief review of the evidence - experimental and spontaneous - for telepathic communication, many topics of interest have necessarily been excluded. In the present chapter the examples have been selected mainly from the class of visual hallucinations, because these phenomena are in themselves more impressive and explanation by chance coincidence is more obviously precluded. But the narratives here quoted, though they represent the evidence, either as regards its amount or its variety, very imperfectly, are sufficient to afford some idea of the character and importance of the problems to be solved. First amongst these problems is the nature of the agency by which the results are brought about. On this question there has been speculation enough, from the first crude analogy of two tuning-forks sounding in unison, to elaborate theories, with experimental demonstration, of radiant neuric force, or a comfortable belief in the omnipotence of the ether. But in truth we know neither the medium by which the telepathic impulse is conveyed, nor the organ by which the impulse is originated or received. By some, indeed, it is held that telepathy is but one of a group of transcendent faculties, which point to a world beyond the world of sight and touch: the germ of powers which cannot reach their full growth until man has ceased to be man. Such a view is perhaps little more than the expression of the difficulties involved in any physical explanation. That mind should reach to mind over miles of intervening space without discoverable apparatus may, indeed, appear to call for supernatural means. But so, to the peasant might appear the discovery of rayless stars, the analysis of the sun's photosphere, or the familiar miracles of the electric current. The properties of the ether and the mechanism and functions of the nervous system, it may be suggested, are still imperfectly explored and it would be rash to assert that the nerve changes which are the presumed accompaniment of thought could not be conveyed by ethereal undulations to a kindred brain over distances at least as great as those which are indicated by some of our thought-transference experiments. Even the greatest distance vouched for in the spontaneous cases of death-apparitions - even the whole diameter of the earth - would be an insignificant fraction of the distance traversed by the waves of ether which strike upon our retina the image of a star. (Podmore, 266-267)

The attitudes of those open-minded scientists who accepted at least the possibility of paranormal abilities tended to dismiss the spiritualistic phenomena to center on the related psychic phenomena after most of the mediums were exposed as fakes.

The rest of the mediums were assumed to be fakes, charlatans, and con artists. But spontaneous telepathy and other psychic phenomena could not be so easily discredited or discounted. So, this assessment of the phenomena essentially formed the basis upon which scientific investigation of the paranormal was to continue during the next period of scientific development. Until the new ideas in physics trickled down to the study of psychic phenomena, the older Victorian views dominated the study of the paranormal. In other words, the scientific revolution to develop parapsychology did not come until nearly 1930. The inauguration of this revolution is generally attributed to the work of Rhine, but the date also reflects the important influence of physics for the growth of parapsychology as well as basic changes in overall scientific attitudes resulting from the revolution in physics.

Since the study of psychic phenomena remained constant until the late 1920s when the new concepts of physics coalesced into a coherent system and began to influence other areas of academic study and culture in general, Frank Podmore's 1897 assessment of the state of psychic research remains valid for all the work between 1850 and 1930.

The middle scientific period and parapsychology

The historical evidence that has thus far emerged indicates that there are several categories that could be useful in classifying the different types of psi theories. Specifically, three major categories have evolved by which physical theories of psi can be easily classified: (1) Theories which are hyperspatial and/or non-Euclidean in nature and (2) field theories which are essentially electromagnetic. These two categories overlap to a small degree and will eventually merge when a unified field theory is developed.

The work of More and Mesmer preceded them, respectively. Another category would seem to flow from (3) the adoption of a hypothetical fluid form of matter or other hypothetical particulate matter. Mesmer's concepts of a vitalistic fluid as well as Lodge's aether serve as examples in this category even though Mesmer's concept of animal magnetism would more properly be placed within the electromagnetic theories category.

And finally, a new category of physical theories emerged after this scientific period. It is based upon the more recent developments in quantum mechanics, as will soon be demonstrated. Many scholars who analyze questions of this type refer to the quantum-based theories as 'observational' theories since an observer is necessary to collapse the wave packet or otherwise interact with the wave function to create physically measurable quantities such as position and momentum. However, these theories will simply be classified as quantum theories of psi since the observer also interacts in an intimate way with the observed in all field theories including the relativity theories which are essentially field theories and thus nonquantum mechanical.

As before, there is quite a bit of overlap between the quantum theories of psi and theories in the other categories. For example, a few modem theories of psi postulate hypothetical entities such as virtual particles or tachyons as signal carriers for ESP. These hypothetical particles may be the result of speculations in either quantum theory of relativity, but the theories of psi that have developed from them more closely resemble the fluid/hypothetical particles theories than the quantum or field theories and are thus classified in the fluid category.

The quantum theory classification was neither established nor defined during the middle scientific period, although quantum ideas surely influenced other theories between the 1930s and 1970. This stems from the fact that general applications of the quantum theory in other areas of science did not occur until well after 1930 when the ideas and concepts of the quantum theory had disseminated throughout the scientific and academic communities to a high degree. Also, most of the physical theories of psi that were proposed during this period were developed by non-physicists who did not necessarily have an adequate understanding of quantum principles and their physical consequences. There was nearly a thirty-year lag between the latest changes in physics and their applications to psi phenomena.

With this structure of classification in mind, the continuing evolution of ideas concerning psi makes far more sense. The possibility that psi could be electromagnetic in nature has always been readily accepted, but as easily refuted. The modem era of electromagnetic explanations precedes the beginning of the middle scientific period by several years, but most of the work done on these new views occurred within the middle period.

Modern electromagnetic theories of psi begin with the research of F. Cazzamali in the early 1920s. During his research, Cazzamali detected electromagnetic impulses of about one centimeter in length within the human brain. Based on this discovery, Cazzamali designed an ultra-high frequency apparatus to test for human telepathy. (Beal, 429) Expanding upon this work, Hans Berger further discovered that the impulses were regular or 'rhythmic' in 1928 and redeveloped the electroencephalograph to study these impulses.

The 'rhythmic' nature of the impulses led Berger to conclude that electromagnetic waves might act to carry signals extrasensorally as in ESP phenomena, but his claims were never substantiated. Berger worked on this hypothesis in one form or another for the next three decades. The author Upton Sinclair popularized the idea of such a *"Mental Radio"* in 1930 in a book of the same title. Sinclair's book presented a popular description of experiments that he had conducted himself and included an introduction by Albert Einstein.

Berger later noted that the detected brain impulses are far too weak to send or receive signals electromagnetically, so he suggested that a „psychic medium" of transmission must exist. This psychic medium could interact with the physical processes of the brain, even at so small an energy threshold as he detected. Yet he still needed and so postulated a new 'psychic energy' in nature to accomplish this goal. This new unknown energy could span the distances of space without attenuation or obstruction by physical objects as well as interact with the human brain at extremely low energies.

Bechterev in the Soviet Union also conducted fundamental research concerning the relationship between electromagnetism and psi during the 1920s. Bechterev believed that "the emanation of electromagnetic waves would provide the best working hypothesis to account for telepathic phenomena." (Quoted in Burt, 89) The Russian scientist Vasiliev continued Bechterev's research and performed numerous experiments on the relationship between electromagnetism and psi between 1927 and 1963. Vasiliev's physical experiments only tended to prove the idea that the electromagnetic transmission of waves resulting in ESP or other psi phenomena was untenable.

From the physiological point of view, Vasiliev also noted "brains and nerves are surrounded by fibers and liquids which possess greater electromagnetic conductivity than the nerve tissues themselves. This makes any meaningful transmission of electrical potentials from brain to brain farfetched." (Rao, 142) The major obstacles to any electromagnetic theory of psi are the lack of attenuation and absorption. The necessary energies are so small that activation in the human brain would seem unlikely if not physically impossible. So, all the electromagnetic theories of psi have faced very real and serious physical limitations. Despite these limitations, they remain popular to some small degree.

On the other hand, the hyperspatial theories have proven more resilient over the years. Hyperspace theories of any type have become more favorable since the development of Einstein's theories of relativity. Although neither special relativity nor general relativity use an extra dimension, special relativity can be represented as a Lobachevskian space-time and the general theory a Riemannian geometry. The successful application of non-

Euclidean geometries to physics in this manner suggests the possibility of a real higher-dimensional embedding geometry for the space-time continuum.

The first person to utilize this idea within the context of psi phenomena was J.W. Dunne. Dunne's 1927 theory was based on the concept of a fourth-dimensional time. Our consciousness travels in this temporal dimension in such a manner that we seem to experience a motion in time in our three-dimensional world. "Real" time is then the fifth dimension in which second consciousness moves, and this process continues in still higher dimensions. Precognition becomes possible when the second consciousness moves about in our waking conscious time of four dimensions. (Rao, 166-167)

C.D. Broad proposed a similar theory in 1967. It is also based on a second dimension of time but lacks Dunne's infinite regress of higher time dimensions. In Broad's theory, precognition is explained in approximately the same manner as Dunne's explanation with a secondary consciousness associated with a second time dimension. Precognition in our awareness occurs in the past of the second time dimension since the second time dimension precedes normal time. (Rao, 167-168)

Both ideas seem quite fantastic, indeed "A topological generalization of these theories seems to be possible, but no coherent interpretation of the amended theory can be provided without the most sweeping changes in all our causal conceptions of the passage of events." (Chari, 1977, 816) Both of these theories are based upon the relativistic view of a four-dimensional space-time continuum. While they are decidedly five-dimensional, incorporating a second time dimension, they are loosely based upon the five-dimensional theory of Theodor Kaluza.

In 1921, Kaluza developed a unified field theory in which a fifth spatial dimension was used to account for electromagnetic phenomena. The fifth dimension, or rather the fifth component in the metrical field, had no physical significance; Kaluza merely used it as a mathematical artifice. A few years later, Oskar Klein extended Kaluza's theory to include quantum effects, while Einstein and other scientists continued to develop variations of the theory throughout the 1930s and 1940s.

Einstein gave up the pursuit of a five-dimensional theory in 1943 declaring that the five-dimensional approach could not be considered valid until a proper explanation of why the extra dimension could not be sensed was given. Since Kaluza attached no special physical significance to his fifth coordinate or dimension, it could be considered legitimate to use it as an extra time dimension. At the very least, Kaluza's theory set a precedent for developing a five-dimensional theory of psi.

Carroll B. Nash made use of the four-dimensional space of Hermann Minkowski and Einstein in yet another way. A particle or material body prescribes a „world-line" in the four-dimensional space-time continuum, a line that stretches from past to future tracing out the history and future of an event or object, when it is represented as a point in three dimensions. Nash has given Minkowski's world-line a physical reality that it does not have within ordinary relativity theory.

In physics, the world-line has no more physical presence than a dotted line representing the parabolic trajectory of a thrown ball on a student's homework paper. But according to Nash, these world-lines act as connections between our brains and remote events in time and space. The world-lines themselves react with our brains within the limits of the uncertainty principle. This interaction gives rise to ESP, precognition, and PK. (Roll, 1966, 39)

Science recognizes the concept of world-lines in Minkowskian spacetime diagrams, but they are not physically real strands which stretch through space and time. Nash's ideas may be mentally appealing, but too much is left to the imagination for his theory to hold any merit. At the very least, a more precise mechanism is needed to explain how his version of the world-lines can interact with the human brain, if not a better explanation of how these world-lines can have a real physical existence.

Still other multi-dimensional theories of psi have been proposed. In 1965 H.N. Hart proposed a theory like Nash's. In his theory, our material universe is only a four-dimensional cross section of a five-dimensional manifold. (Hart, 16-22) Events take place by moving along a four-dimensional timeline with each observer "reading his past, but sometimes looking ahead non-inferentially into the future." (Chari, 1977, 815)

Precognition thus becomes a valid consequence of the space-time geometry. In 1967, J.R. Smythies" thoughts ranged toward a three-dimensional physical space and a three-n-dimensional psychic space in which ESP and PK are exchanges between the two spaces. His theory was based upon the concept of a non-Cartesian dualism. In other words, "non-Cartesian dualism suggests that the world consists of the physical universe extended in physical space and a number of substantive minds extended each in a space of its own." (Smythies, 6)

In the case of Smythies" theory, there is a constant flow of information between the individual's mind in its own space and the individual's brain in physical space. This interaction can be envisioned if "the total human organism is extended in an n-dimensional space." (Smythies, 7) This model allows the mind-brain interaction to be treated as a part of normal science and offers "possibilities for deductive development by topologists, geometers and physicists." (Smythies, 9)

Some unspecified type of signal coming from another individual's mind-space can interrupt the normal connection between one individual's mind and brain. This interruption explains ESP. Since the signal is outside of the normal physical space-time continuum of the brain, the normal laws of physics, such as the attenuation of waves with distance, do not apply. In applications to psi, any theory of hyperspace would assume the higher dimensions to be real. This idea makes any theory of psi based on hyperspaces untenable with normal perception, a fact that also forms the main argument against unified field theories based on a five-dimensional hypothesis.

Before Einstein's development of relativity theory, mathematicians and scholars discussed the possibility of four-dimensional physical spaces with a separate time, but after Einstein these concepts evolved into a four-dimensional space-time continuum embedded or curved in an external fifth dimension. After development of special relativity in 1905, the background for nature became a four-dimensional space-time continuum and the previous work on four-dimensional spaces was either forgotten or inadvertently purged from science.

Modern field theories as well as the hyper-dimensional theories of psi have been dissociated from Zöllner's and similar concepts in their past by Einstein's omission. We could say that the scientific revolution wiped the slate clean and left the field unbiased toward non-Euclidean applications in physics. Although Einstein used Riemannian geometry to explain gravitation, it has never been decided whether a fifth dimension was needed to explain gravitational forces.

In mathematical language, general relativity utilizes an 'intrinsic' space-time curvature, so a higher-dimensional embedding space is not necessary. Nearly everyone who studies Einstein's theory assumes a four-dimensional space-time structure. By utilizing the analytical methods of Riemannian geometry, this intrinsic model is made plausible without the need to embed the four-dimensional space-time continuum in a fifth dimension.

On the other hand, an 'extrinsic' geometry whereby a fifth dimension is needed for curvature has never been ruled out, either by requirements of the mathematical structure or by philosophical arguments. This fact allowed Kaluza to develop his five-dimensional unification of gravity and electromagnetism. So, a fifth dimension in the form of an extra space or time dimension could be adopted without changing or otherwise altering our normal laws of nature. However, scientists would never adopt a fifth dimension unless there was a compelling reason to do so, and psi phenomena alone do not constitute the necessary compelling argument.

The hyperspatial theories of psi suffer from many problems, the least of which is the inability to consciously sense the extra dimension or dimensions. But this same characteristic also forms a chief advantage for such theories. Since we do not normally sense the fifth dimension, any real phenomena associated with it would necessarily have unique properties that are very difficult to isolate or detect.

This would give a new meaning to the term 'extra' sensory in ESP. 'Extra' sensory would be equivalent to sensing the 'extra' dimensionality of space-time. Among these problems can be included the concepts of free will and consciousness traveling in other time dimensions. Hinton in the late 1800s and Robert Browne in the 1920s discussed similar issues in the popular literature, so they are not unknown to the general community of scholars.

If world-lines and time dimensions are accepted as pre-existent, such that human consciousness travels along a trajectory in the time dimension then the concept of free will fails. The psychological sense of time would be reduced to the consciousness moving along a pre-existent world-line and the idea that individuals choose their own future from moment to moment would disappear. So, the possibility exists that acceptance of a hyperspace theory of psi might result in the abandonment of our present concept of free will. This possibility seems to be the only way that precognition can be explained using an extra dimension of time. However, this problem is a general problem of physics, not just parapsychology and paraphysics. From the first moment that philosophers began discussing the consequences of the physics of a four-dimensional spacetime continuum, they have discussed its effect on the concept of 'free will,' and in most cases this discussion has had nothing to do with ESP, precognition, or psi phenomena.

The main advantage of both the electromagnetic and hyperspatial theories is derived from the fact that they utilize pre-existing structures (*i.e.*, theories, concepts, and models) in science. Maxwell's electromagnetic theory is basic to modem science while the hyperspatial and non-Euclidean geometries are well documented within mathematics. In general, there are two alternatives for developing a physical theory of psi. The first, and by far the most accepted, would be to extend pre-existing structures (as above) to cover psi phenomena by tweaking them in just the right way. The difficulty is finding how to tweak them. But too many concerned scientists and scholars believe that science in general and physics are just not up to the task of explaining psi at the present time, either in principle or in practice.

So, the other alternative would be to invent new structures in science, in the form of newly defined measurable quantities, new concepts or wholly new physical entities, to explain psi. This has been the general case for what have been termed fluid theories. They are termed fluid theories because they are wholly new entities as were the original fluids utilized to explain everything from heat (caloric) to the transmission of light waves (the luminiferous aether) in centuries past. Such fluids have always been characterized by the fact that they are non-detectable except through the phenomena that they have been invented to explain. The fluid theories of psi do not necessarily posit such hypothetical fluids to explain psi but are so classified because they posit new hypothetical entities to explain what many scientists regard as hypothetical phenomena.

Science does occasionally advance by the development of new entities filling these criteria, but usually only by small incremental changes unless there is overwhelmingly strong evidence that indicates gross changes and new structures for science. Such is not the case for psi. Mach who wrote that he would not accept the reality of a non-Euclidean space until objects began to pop into and out of our three-dimensional space suggested similar circumstances. Such 'overwhelmingly strong evidence' would certainly necessitate the development of new structures in science.

Coincidentally, Mach made this comment in association with his rejection of the use of non-Euclidean geometries by Zöllner to explain spiritualistic and psychic phenomena. Mach was finally proven wrong in this regard, since Einstein successfully used a Riemannian geometry to explain gravitation. Ironically, Mach never fully accepted Einstein's concepts of relativity even though Einstein's ideas were inspired by Mach's arguments and philosophical views.

The hyperspatial theories are, by definition, field theories. They are either derived from or otherwise associated with relativity, which is a field theory. However, other field theories of psi, which have nothing to do with electromagnetic fields, gravitational fields, or non-Euclidean geometries, have also been proposed. Many such theories were developed during the middle scientific period. They were pure inventions by their developers, not associated with pre-existing scientific structures, so they fall under the banner of fluid theories.

In the first such case, Wasserman developed a system of four new physical fields to explain psi. These were the p-field, M-field, B-field, and the psi-field. The M- or morphogenic-field acts to steer a living organism's development while human behavior is a result of the mutual action between an individual's B-field and matter field. Both of these two fields were specializations of the p-fields, which are inherent in all forms of matter. When these fields interact with matter according to quantum mechanical transition rules, ESP and other psi phenomena result. This proliferation of fields is purely „ad hoc" and has no scientific value beyond their value as an "interesting speculation." (Rao, 169)

William G. Roll borrowed from Wasserman's ideas and developed his own brand of a psi-field, halfway between Wasserman's psi and p-fields. Roll's psi-field was defined as "the region in space in which psi phenomena are detectable." (Roll, 1966, 47) His psi-fields were associated with both animate and inanimate bits of matter. Interactions between individual psi-fields tended to leave a 'psi-trace' that would in turn be imparted to an individual as a psychic event.

Roll further specified sets of rules, conditions, and postulates (all rather ad hoc) governing the interactions of the psi-fields. Roll's model of psi-fields also shares some characteristics in common with Ninian Marshall's 1960 model of 'eidopoic influences.' Although Marshall's model is only semi-physical at best, it is related to other physical theories and thus invites consideration as a physical theory of psi. The 'eidopoic influence' creates resonances between various complex systems, such as human brains, in such a manner that human brains regardless of the spatiotemporal distances involved can detect very faint signals.

Although the strength of the 'eidopoic influences' is related to quantum probabilities, this is not a quantum theory-in-itself. Since the 'eidopoic influence' acts like a physical cause without exhibiting the properties of a physical cause, it is not restricted by ordinary physical limitations. These properties render the concept quite revolutionary, but far more difficult to accept. One final theory is related to these fluid theories. Rush, who groups it with Wasserman and Roll's field theories even though it is quantum mechanical by design, has classified H.A.C. Dobbs' 1965 theory of psi-trons as a "quasi-physical" field theory. (Rush, 1986, 281)

In 1953, the British neurophysiologist Sir John Eccles offered a theory on how a person's will could affect the activation of a single neuron without attaining the necessary energy to fire the neuron. This theory of 'subtle influences' depended upon a quantum cascading of energies. The theory had nothing to do with psi, but Eccles could not help but mention possible connections to ESP and PK phenomena in the closing sections of his book. (Eccles, 278-279, 284)

Dobbs later suggested that Eccles' 'subtle influences,' as perceived in the brain, could have the form of mathematically 'imaginary' energies. These, he argued, are not without precedent since physicists use the concept of virtual particles, which carry mathematically imaginary energy and moments, to describe the fields around real particles. These virtual particles are only inferred from experimental results and not observable as real 'things-in-themselves.'

Dobbs hypothesized a two-dimensional theory of time to account for his 'imaginary influences.' The first dimension of time corresponds to the normal perceived time of physics and psychology and is represented as the real axis in a complex number system. The second time dimension, "in which the objective probabilities of different possible outcomes of events are ordered," (Dobbs, 250) is represented by the imaginary axis.

Different probabilities in this second virtual time dimension can be co-present and may be interpreted as Karl Popper's objective dispositions or 'propensities.' These are related to 'precasts' of events that are highly probable but not certain to happen in the future. The 'precasts' spread out in a quasi-temporal order in all directions around physical events. When the mind perceives these 'precasts,' precognition occurs. The 'precasts' are analogous to the virtual particles of quantum theory.

Dobbs further postulated the existence of particles called psi-trons, which are the carriers of psi information. Psi-trons "register the probabilities in the second time dimension and contribute to the EEG alpha rhythms of the brain." (Chari, 1977, 814) The same arguments made against the physical theories of virtual particles, two-dimensional time and 'propensities,' can be made against Dobbs' theory.

Although an important part of quantum theory, virtual particles are completely undetectable. Dobbs' psi-trons are also undetectable, as would be expected since they have only mathematical and not physical significance. Dobbs' psi-trons are also undetectable, as would be expected, since they have only mathematical and not physical significance. They are imaginary in the mathematical sense of the term, which does not mean to imply that they are physically imaginary. Perhaps this is why Chari claims "the 'psi-tron' theory of the late H.A.C. Dobbs can claim no unequivocal support from the logic of quantum mechanics or from the mathematics of EEG rhythms." (Chari, 1974, 3)

Dobbs" quantum mechanical theory of complex time and energy can also be used to explain general ESP. It cannot be classified as a quantum theory, as explained above, because it is not 'observational.' It is a fluid theory because Dobbs has invented a new entity or structure to explain psi. Unfortunately, Dobbs' theory suffers from the fact that it "offers no good experiments, nor does it provide for clairvoyance or psychokinesis." (Rush, 284)

There is still another aspect of modern physics that has proven advantageous for the development of a theory of psi, the concept of plasma. Plasma theories of psi can also be categorized as fluid theories because their originators invent specialized plasmas whose only purpose is to account for psi. Andrija Puharich proposed one such theory, which is rather unique, in his book *Beyond Telepathy*, published in 1962.

Puharich suggested that a psi-plasma be adopted to explain paranormal phenomena. The psi-plasma can be attributed with form since form "is perhaps the only level at which we have any idea as to the nature of the "psi-plasma." (Puharich, 72) In normal physics, plasma is the fourth state of matter. The three normal states of matter are solid, liquid and gas. Each of these states is defined by its physical properties and charact

eristics. However, several decades ago superheated gases were found to have specific properties that marked them as different from normal gases. So, plasma physics, the study of matter under these special conditions, was born. By analogy, a psi plasma would be a special state of matter which is characterized by the ability to interact psychically with other bodies of matter. The psi plasma is characterized by a pilot wave, a concept which is well known in quantum physics, and must therefore always have a velocity greater than that of light. Puharich was very well aware that anything moving faster than light could not interact with our slower than light physical world and thus developed a complicated structure of pseudospheres of plasma, with parallel pseudospheres spinning in opposite directions.

As these pseudospheres spin in vortical motion they will draw the other pseudospheres into them, creating spherical bubbles, which will spin at right angles to the axis of the pseudospheres. The four upper spheres are the electron, proton, neutron and neutrino. Opposite spheres spin in opposite directions, resulting in giving rise to positive and negative electrical charges. There are also four spheres on the underside of the rim of the pseudosphere. These are the positron, antiproton, anti-neutron, and anti-neutrino. The upper four spheres or particles represent the first domain or positive energy, and the lower ones represent the negative energy of the second domain.

The pseudosphere itself is made up of the psi-plasma, the fourth domain. According to Puharich, "all exchanges of energy in the physical world are governed by the psi plasma properties ... and ... the psi-plasma records and in a sense remembers every such transaction." (Puharich, 179) Further, a gravitational force couples the psi-plasma and physical fields. Once his physical structure had been developed, it only remained to relate the structure to ESP.

Any event, physical or mental, causes a perturbation in the psi-plasma field due to changes of state in the particles involved. This perturbation travels as a pilot wave, at a velocity greater than light, to the observer's brain where a similar particle is set in motion by the wave. Thus, a standing wave is set up which can act as a channel for ESP while the observer's brain is stimulated, and the ESP message is cognized.

In order to present this theory, Puharich has not only invented a new entity, the psi-plasma, but he has also introduced a very complicated and unorthodox structure of space and matter, which is quite untenable with modern physics. A theory such as this seems to call for a complete d comprehensive worldview on almost no evidence. Science and culture would not just throw away their cherished although well-worn worldviews of reality so easily.

Only newly discovered and persistent physical phenomena would force science to look for a completely new structure of physical reality, although small chips in the wall of science's worldview could cause some scientists to begin searching for an alternate view of physical reality. Puharich's theory is not justified given the present state of science. There are mysteries and paradoxes in the world of physics, but there is nothing so radically different as to warrant the type of worldview change that Puharich has proposed.

Perhaps that is why there seems to be no mention of his theory in any publications rather than his own. There seems to be no secondary literature regarding Puharich's theory, so it is questionable how seriously the parapsychological community let alone the general scientific community have received his theory. The scope of Puharich's theory just goes too far beyond that of explaining psi.

It would be unthinkable to even suggest that all of physics should be rewritten to accommodate a theory of psi. However, the opposite is highly possible. If physicists were to develop new physical theories, in essence rewrite physics to incorporate new advances in science, then it might be possible to incorporate psi into the overall structure of nature.

In yet another plasma/fluid type of theory, the age-old concept of a vitalistic fluid that supplements mechanical matter to account for life has taken a new form. This notion of theoretical reasoning has led to the adoption of a biophysical plasma or similar device to explain psi. Newton once suggested that one form of the all-pervading 'aether,' like the form involved in gravitation, "also pervades 'animal juices' and controls the movement and even procreation of living things." (Roll, 1966, 37)

Such speculation as this has led to other theories concerning biogravitation, bioplasmas, and various other biophysical fields and entities. Early attempts to explain life and specific properties of animate matter were made by Mead with 'animal gravitation' and Mesmer with 'animal magnetism.' Biological organisms do, of course, exhibit electrical and magnetic properties, and this has introduced field theories into the realm of biology. Various plasma theories are also heirs to these early theoretical hypothetical structures and are related in some ways to the field concepts.

The field concept has seemed more likely an explanatory tool in theoretical biology only when the patterns, which guide biological functions such as organization, growth and development, were not found to be part of any single part of an organism. This belief follows a very old tradition that the mechanistic worldview has never been able to account for life and living processes. It does not matter that we now have a quantum mechanistic worldview rather than a Newtonian mechanistic worldview; still more is needed to explain life and the processes associated with living organisms.

In 1935, Burt and Northrop proposed that bioelectric fields could influence the pattern and development of organisms. However, no proof has ever been developed to substantiate such claims despite the known presence of electrical fields within organisms. W. Elsasser has postulated 'biotonic laws' inherent in living beings "and drawing upon accumulated quantum mechanical and information-theoretic uncertainties." (Chari, 1972, 203)

As recently as 1968, Inyushin and his colleagues have worked with a "bioplasmic interaction" to explain psi. They consider the "bioplasm" to be a "fourth state of matter" and claim that it is responsible for the effect detected in Kirlian photography. They further claim that "All living things - plants, animals and human - not only have a physical body made of atoms and molecules, but also a counterpart body of energy." (Schroeder and Ostrander, 217) C.T.K. Chari criticized all such speculations and hypotheses by stating that a "'bioplasm' or 'psycho-plasm' with totally unknown properties cannot claim to be a rightful link between physics and psi phenomena." (Chari, 1977, 812) Yet the issue of a link between physics, biology and psi is of growing concern and, as many scientists believe, it may be an inevitable outcome of future research.

Since many scientists and scholars see the twenty-first century as the century of biology in much the same manner that the twentieth century has been dominated by physics, the merging of biology, psi and physics would not seem to be so improbable. However, even this speculation on the future of science does not render the development of new entities in physics such as bioplasmas more valid.

One last theory developed during this period is worthy of mention even though it is different from all the others in that it defies classification, but it is an extension of preexisting structures in science. Haakon Forwald proposed a new physical theory of psi, which utilizes gravity as a means of accomplishing PK phenomena and thus shares no characteristics with the other theories. Forwald's theory hypothesizes no new structures, nor does it tend to be anything but physical. Forwald's theory is derived directly from the experimental results of mental influence on rolling cubes. His experimental findings seem to indicate that observed PK phenomena are of a gravitational nature resulting from a mental influence on the atomic nuclei of the material making up the cubes.

The energy which causes the displacement seems to come from the transformation of some of the cube's mass to energy according to the formula $E=mc^2$. The energy is derived from the nucleus in the form of gravitational potential differences by a yet unknown process. If it is assumed that some unknown physical force acted on the cubes, Forwald used instruments to test for all types of forces (electrostatic, magnetic, nuclear, etc.), but could find no forces acting on the rolling cubes.

Thus, he arrived at a gravitational theory of PK by a process of elimination. He justifies the choice of gravitation in various ways including "what seems a highly dubious argument from his position effect, the parabolic U-curve." (Rush, 225) The position effect is a graphical representation of the probability and change of position, which yields a figure that closely approximates a parabola. Considering that other psi phenomena also seem to display parabolic functions, he compares this to the fact that an object thrown within a gravitational field, travels along a parabolic curve.

This argument is tenuous and specious. The analogy is not completely valid and any "correspondence between the position effect in those experiments and the gravitational trajectory of a moving mass seems purely formal." (Rush, 22) There is no precedent nor is there any compelling reason to assume that any phenomena which can be graphically represented by the same geometrical curves are themselves related. There are, however, other arguments that could be made to support a gravitational theory. Forwald points out that "it would be more correct to use the expression 'psi is nonenergetic' instead of 'psi is non-physical'" (Forwald, 66) and the psyche also seems to be structural (nonenergetic) rather than energetic, giving a correlation between mind and gravity on a formal level.

Psi phenomena have been shown to be unable to shield and they seem to present no measurable energy loss (attenuation). Gravity also meets these same requirements whereas other physical forces do not. Gravitational effects are transferred non-energetically and gravity cannot be shielded. Forwald's hypothesis is also open to experimental verification. "The loss in mass is far too small to be measured as the difference in the weight of a cube before and after an experimental series.

While psi phenomena cannot be shielded or blacked in any way, gravity also acts the same way whereas other physical forces do not. Gravitational effects are transferred non-energetically and gravity cannot be shielded. Forwald's hypothesis is also open to experimental verification. "The loss in mass is far too small to be measured as the difference in the weight of a cube before and after an experimental series.

But it should in principle be possible to determine by mass spectrometry whether the isotope composition of the cube material changed during the experiment. A possible result in such an investigation might give a direct proof of the existence of psychokinesis." (Forwald, 6) But once again, Forwald has overstepped the bounds of credulity. According to general relativity, gravity is not a force, but rather an effect due to the curvature of space-time. To say that psi affects the curvature of space-time, which is certainly implied in Forwald's theory, is quite different from what his intent seems to be. Forwald's thinking in this matter is far too Newtonian which does not measure up to modern standards in physics.

As in other attempts to derive a physical theory of psi, Forwald's hypothesis only covers one type of phenomena from among the many different kinds, which seem to be exhibited by psi. He does try, albeit indirectly, to apply his theory to other psi phenomena when he says that "the gravitational field ... should have had a guiding influence on the psyche. This would mean that the field should to a certain extent have limited the freedom of the psyche to act deliberately." (Forwald, 19)

However, to assert that gravity affects the psyche and 'thinking' is a bold assumption and which can easily be tested as space travel takes us further from the gravitational field of the earth. But so far, traveling and living in the weaker gravitational fields of space have shown no effect on the psyche and thought patterns of astronauts and cosmonauts. Of course, if Forwald's hypothesis could be proven true, then it must also have a real affect for all psi processes. Those theories that are classified as fluid theories seem to have dominated the middle scientific period. Such a historical observation can be easily explained by several other historical facts. Most of the physical theories during this period were developed by non-physicists including scientists with limited understanding of the principles of physics.

Jumping to a quick conclusion and inventing new entities just to explain psi is a characteristic of someone who does not have a complete knowledge of the present state of physics. In the case of the theories developed during the middle scientific period of paraphysics, parapsychologists developed theories and physical models of psi that used words and ideas that sounded to them like physics but were merely hypothetical ramblings.

At the time, many parapsychologists believed, and still believe to this day, that psi is a purely mental phenomenon and physics has nothing to do with psi. In other words, even some of the best and most qualified parapsychologists have ignored any possibility of a physical theory of psi, surrendering the development of a physical theory to an even smaller number of parapsychologists and scientists than even their own ranks could muster.

On the other hand, physics has traditionally been considered inadequate to cope with psi phenomena, so developing hypothetical new quantities or entities to explain psi would seem both easy and necessary for any scientist interested in a physical theory of psi, let alone one who was not up to par on his knowledge of physics. Since physicists were more involved in other matters, they did not care about psi phenomena and surrendered the discipline to the parapsychologists. Physicists were too busy applying quantum theory to physical systems, such as those studied in atomic and nuclear physics, which seemed by far the more pressing concern for science during the three or four decades after the flowering of quantum mechanics of the 1920s.

At the time, many parapsychologists believed, and still believe to this day, that psi is a purely mental phenomenon and physics has nothing to do with psi. In other words, even some of the best and most qualified parapsychologists have ignored any possibility of a physical theory of psi, surrendering the development of a physical theory to an even smaller number of parapsychologists and scientists than even their own ranks could muster. On the other hand, physics has traditionally been considered inadequate to cope with psi phenomena.

So, developing hypothetical new quantities or entities to explain psi would seem both easy and necessary for any scientist interested in a physical theory of psi, let alone one who was not up to par on his knowledge of physics. Since physicists were more involved in other matters, they did not care about psi phenomena and surrendered the discipline to the parapsychologists. Physicists were too busy applying quantum theory to physical systems, such as those studied in atomic and nuclear physics, which seemed by far the more pressing concern for science during the three or four decades after the flowering of quantum mechanics of the 1920s.

Interest in developing a physical theory of psi did grow, if only slowly, throughout the period from 1930 to 1970. More physical theories of greater variety emerged in the 1960s than ever before. In 1969, the Parapsychological Association was admitted to the American Association for the Advancement of Science. Although some scientists fought their admission and still fight for the removal of the Parapsychological Association, their new elevated status gave this science of psi a new legitimacy that it had never before enjoyed before. Admission to the AAAS also brought more non-parapsychologists into contact with the parasciences as well as providing an important new forum for established parapsychologists to present their evidence for the reality of psi.

The late scientific period and paraphysics

After 1970, paraphysics finally came into its own. The word paraphysics began appearing in a few specialized books and scientific articles. The appearance of the word in literature marked a turning point in the evolution of this science and the beginning of a return of physicists to the study of psi phenomena. The factors influencing this return are numerous and mark a specific shift in attitude of the whole physics community.

One aspect of this change can be found in the first development of 'observational' theories of psi, as others call them. The 'observational' theories of psi demonstrate the introduction of a new level of competence in physics into psi research. Quantum theory had been incorporated in some manner in several earlier theories of psi, but not at the same level of sophistication as became evident with the „observational" quantum theories of psi. The earlier theories had used quantum effects to explain psi, but the new 'observational' theories were based directly upon the philosophical foundations of quantum theory, right to the point where the observer and the observed quantum system interact.

Rosalind Heywood documented these changing attitudes in 1967. However, she spoke and referred to changes in parapsychology rather than paraphysics, even though her arguments were equally valid for paraphysics.

> One key to this beginning of a change in the mental climate in relation to ESP may perhaps lie in the astonishing revolution in physics, which is now going on, and which has already shown that the classical notions about the nature of time, space, and matter are not all-embracing when it comes to the realms of the very large and the very small. ...
>
> The revolution in physics, of course, has been going on for over half a century, but new ideas need a lot of time to seep through and modify the general outlook, and the most fundamental change of all - to a concept of „"solid" matter as being convertible into elusive „"intangible" energy - is only now beginning to exert its influence in the outside world. In the early days of psychical research men could say that no mystery remained in the universe and that the subject of physics was exhausted. Now a pioneer of the new physics can write that „"the physics of the future a few centuries hence could well be as different from the physics of today as the latter is from the physics of Aristotle", and, again, that "we must never forget how limited our knowledge must always be". [The quote is from Louis de Broglie, Nouvelles Littéraires, 2 March 1950] (Heywood, 60)

Her notion that new concepts take a long time to 'seep' into the mainstream of science and "modify the general outlook" is a general feature of the evolution of science, but her conclusion was not wholly correct.

She was wrong in limiting the more recent changes in attitude toward psi research to the original events of the Second Scientific Revolution alone. She missed the point that radical changes in attitude toward both quantum theory and general relativity had occurred since 1930 and these changes were having a profound effect on psi research as well as the growing relationship between physics and psi on the one hand, and physics and science in general on the other hand.

After 1950 Physicists attitudes toward the Copenhagen Interpretation of quantum mechanics began to change quite radically, although slowly. This change in attitude had (and even now has) a far greater effect on physical theories of psi and the growth of paraphysics than the slow 'seep' or dissemination of quantum mechanical ideas and concepts into sciences other than physics. In essence, serious challenges to the standard Copenhagen Interpretation of quantum mechanics have been gaining momentum since 1960.

The standard challenge to the Copenhagen Interpretation was voiced and put to print by Einstein, Boris Podolsky and Nathan Rosen. This challenge came in a published paper that has become known only as EPR. This document expressed Einstein's and the other scientists' concerns with the probabilistic interpretation of quantum and wave mechanics. In essence, they felt that quantum mechanics was 'incomplete' since it limited our view of reality and knowledge of the physical world of nature. The EPR viewpoint implied that underlying factors, which could give a more precise and accurate view of physical reality, existed beyond the limits set by the Heisenberg uncertainty principle.

Although the ensuing debate was well publicized and generally acknowledged throughout the scientific and general academic communities, the true significance of EPR is only now becoming known and then only after many other events have taken place. In 1951, David Bohm reopened the debate with a philosophical argument leading to the concept of 'hidden variables.' He reasoned that some type of unknown physical quantities, 'hidden variables,' were implied by the EPR argument. The eventual discovery and identification of these 'hidden variables could render quantum mechanics 'complete' in Einstein's sense of the term.

Then, in 1964, John S. Bell offered a possible solution to the debate over the existence of 'hidden variables' that is today known as Bell's theorem. Bell distinguished between the 'local' and 'non-local' properties of physical interactions. Since these philosophical arguments were made, a great deal of literature on these and related subjects has appeared and directly affected experimental physics.

The EPR argument and its consequences had earlier been debated by either philosophers or scientists in philosophical terms since the outcome of the debate was not expected to affect the experimental results of the quantum theory which are always the primary interest of physicists. Yet the philosophical arguments regarding Bell's theorem have resulted in new experiments that the physicists had not considered. This reintroduced philosophy into physics in a very concrete manner. All these conceptual changes in Physics came despite the clear successes of quantum electrodynamics and other early versions of a quantum field theory, or perhaps because of them.

After the 1930s, an emphasis on the unification of all the forces of nature within the quantum model of reality evolved just as there was a movement for unification under the auspices of pure field theory within the relativity camp. But the development and success of quantum electrodynamics was not followed by a corresponding success in unifying gravitation and the other forces of nature under the single banner of the quantum. The proposed graviton particles necessary to explain gravity were not to be found. Gravitons were not detected at the time and have never been detected, even to this day, yet the existence of gravitons is implied and accepted within the quantum field theory.

The historical picture is further clouded by the mutual interactions of so many different factors and ideas that it is difficult to determine the relative position and relevance of many of the various factors, let alone to identify them all. However, it can be determined that just as the quantum theory was reaching its greatest pinnacle of success, there was an explosion in the number of' elementary particles. The rapidly increasing number of 'elementary' particles began in the early 1970s. The increasing numbers of these fundamental constituents of matter challenged the very notion that they were fundamental or 'elementary.' The development of' Quantum Chromodynamics with the mathematical necessity of quarks as the new fundamental constituents of matter seemed to quell these problems for a while.

Yet quarks have never been observed or detected as individual particles existing outside of their combined state within the older elementary particles. Nor can quarks, as they are presently viewed, account for all the 'elementary' particles that have been observed, so their fundamental nature is also questionable. It would certainly seem that every answer which quantum theory can develop in its search for the ultimate nature of material reality runs into roadblocks. The successes of quantum theory only raise more problems and related questions – and the naming of more particles to fill gaps and unverified theoretical speculations. This situation is at least disconcerting for some scientists, if not fundamentally and philosophically unsettling, but it continues unabated.

In the meantime, both the common and scientific worldviews were expanded by humankind's first ventures beyond the atmospheric envelope of our home planet. This venture represented both a widening and a broadening experience for the human ego and cannot be discounted as indirectly influencing our concept of who and what we are and thus acceptance of the Parapsychological Association by the AAAS.

Early space exploration could never have occurred without specific technological advances, just as continued space exploration influenced the rapid advancement in those and other technologies. Both the direct and indirect influences of the space program on the study of psi cannot be underestimated. Its direct affect was clearly recognized by Heywood and depicted as the "changing mental climate" toward psi research during the mid-1960s.

In the 1950s then, the situation looked like a stalemate. ESP still could not be harnessed to order, and merely to demonstrate its overwhelming probability did not satisfy critics who hated and subconsciously feared its heretical implications and past association with the supernatural. But now, in the sixties, a softer wind is beginning to blow. One reason for this is practical: the idea that telepathy might be a means of communication in space flight. (Heywood, 57) While she directly attributed the space program as an influence on the changes in attitude, the indirect influences could not have been known to Heywood because she was too close to the events.

Other technological advances such as computers (with artificial intelligence), the new technology extending astronomical and astrophysical observations, and communications technology played a significant role in the continuing enlargement of the perspective of the human species. From the founding of general relativity until about 1960, there had been little scientific interest in general relativity. Very few of the phenomena explained by general relativity directly affected our normal lives.

But the new scientific and technological advances, such as those in the space program or related to the space program, placed a new emphasis on relativity and this all but neglected area of advanced physics experienced a revival after 1960. The renewed interest in relativity theory was as much due to these events as it was to the success of a simple quantum field theory or its failures, all of which seemed to emphasize a field theoretic view of nature rather than the discrete view of nature that is inherent to the quantum theory.

Nor can strictly cultural influences be ignored. During and after the 1960s, sociopolitical and geopolitical factors induced changes in the worldview of western culture as a distinct whole. A holistic view of life became popular as opposed to the reductionist views of western science. Eastern philosophies such as Taoism and Buddhism became popular in the United States and Europe and Americans began to learn about Chi, Ki, Qi, Karma, and Prana. These new philosophies, or rather old philosophies that were new to western culture, embodied a holistic view of nature and reality. All the life sciences were affected, as were medicine and psychology. Evidence of the effect of these philosophies on physics can be found in the publication and immediate success of Fritjof Capra's book, *The Tao of Physics*, in 1974.

It will take decades for scholars to discover and interpret all the factors influencing the rise of paraphysics and come to understand the vast complex of influences that they have had on one another, but there can be no doubt that the holistic view of nature was growing within western culture and this pattern of growth corresponded to an upsurge in the field theoretical view of reality. By the late 1970s, Kaluza's unification of general relativity and the electromagnetic field was again revived and the unification of the fundamental forces of nature began to move in a new direction.

On the other hand, the more classical electromagnetic theories of psi had neither died nor faded away during the middle scientific period of paraphysics but continued to develop despite their inherent shortcomings. These theories were inherited by the new generation of psi researchers and theoreticians and modified accordingly. The new attempts to resolve old problems with the electromagnetic theories were quite unique, reflecting the latest advances in science.

I.M. Kogan of Russia combined electromagnetic theory with information theory. Even though electromagnetic waves lose strength as they travel through space, the old problem of attenuation, no information is lost as the strength decreases. Even the weakest signal would contain the same information as the strongest. So Kogan suggested "with a suitably devised formula we can posit a hypothetical wave yielding a 'telepathemic bit' (*i.e.*, a binary psi unit) of information of specified duration at a specified temperature." (Chari, 1977, 811)

In 1979, Michael Persinger added another new element to the electromagnetic controversy by relating the information transmitted as psi to naturally occurring ELF waves. ELF waves are „extremely low frequency" electromagnetic waves with just the right wavelength to interact with the human brain. It was discovered that ELF waves are emitted by geophysical, atmospheric, and other disturbances in nature, so Persinger proposed that these naturally occurring ELF waves might imprint information on human brains which could be interpreted by those people who are especially sensitive to them.

Both Kogan and Persinger's suggestions imported new elements into the physics of psi from other branches of science, which reflected the latest advances and discoveries in those other branches of science. Computer sciences had not existed several decades earlier, and geophysics went through a revolution with the introduction of plate tectonics during the 1960s. And yet classical electromagnetic theory yielded still other gems for psi researchers.

It has been known by physicists for a long time that solutions to the standard electromagnetic wave equations allow for the transmission of waves both backward and forward in time. But scientists had generally ignored the reversed time solutions. As attitudes changed toward the acceptance of more abstract ideas, Richard Feynman and J.A. Wheeler took a new look at the backward traveling waves and developed an 'absorber' theory to physically account for the destruction of that portion of the waves.

As bizarre as this idea may seem, they were acting well within the accepted scope of normal physics. Given all these new ideas and concepts being considered, it would clearly seem that physics itself was beginning to approach something that looked more paraphysical then just physical, or more like psi activity. Gerald Feinberg and I.J. Good (Good, 152) later suggested that the "advanced" absorbers, waves traveling back in time, are not all canceled or otherwise lost and could act as the carriers of information during precognitive psi events. This type of absorber theory was mentioned by Chari as early as 1974.

As bizarre as this idea may seem, they were acting well within the accepted scope of normal physics. Given all these new ideas and concepts being considered, it would seem that physics itself was beginning to approach something that looked more paraphysical then just physical, or more like psi activity. Gerald Feinberg and I.J. Good (Good, 152) later suggested that the "advanced" absorbers, waves traveling back in time, are not all canceled or otherwise lost and could act as the carriers of information during precognitive psi events. This type of absorber theory was mentioned by Chari as early as 1974.

The physicists Harold Puthoff and Russell Targ suggested a similar theory early in the 1970s, but their theory was based upon the space-time structure of special relativity. They proposed that "significant events create a perturbation in the space-time in which they occur, and this disturbance propagates forward and, to some small degree, backward in time." (Puthoff and Targ, 526) Since the wave propagates both forward and backward in time, the theory is like Feynman's 'absorber' theory as well as good's theory of precognition. However, Puthoff and Targ's model is relativistic within a spacetime framework rather than quantum mechanical. An event will not be a completely instantaneous occurrence in time but will extend backward and forward from the moment of occurrence in the direction of both negative and positive time.

If an observer is close enough in time to the event, approaching the event which will occur in his or her immediate future, then that observer might become aware of the negative time extension of the event and thus precognize the event. This backward time extension of the event would be stronger the larger the „magnitude" of the event for the observer and fall off as the temporal distance from the event increases. Even though this theory is based upon a commonly accepted space-time framework, it is still classified as electromagnetic since the special theory of relativity is an electrodynamical theory as originally developed.

In 1973, M. Ruderfer announced an explanation of precognition within the framework of electromagnetism that was qualitatively different from other theories and hypotheses. He suggested that neutrinos and tachyons (particles which travel faster than the speed of light) could interact with normal electromagnetic waves in such a manner to carry psi signals across the barrier established by the speed of light.

Igor Shishkin, a physicist in the Soviet Union, in the late 1960s, advanced similar ideas. Ruderfer hypothesized a 'phasor-neutrino derivation' to explain ESP and PK and a 'tachyon-neutrino-tardyon' interaction to account for precognition. Neutrinos are normally so difficult to detect that their properties have not been completely determined and a debate concerning the possibility that they could or could not have some small mass is currently being waged.

The neutrinos elusiveness makes it highly unlikely that neutrinos could interact with normal matter in any manner which would allow them to pass information to the brain for psi stimulation, but so little is known about them that psi-like properties can be attributed to them without altering their known physical attributes. There is no real evidence that neutrinos or any other material particle can ever cross the 'barrier of the future' (travel faster than light) while tachyons exist only on a speculative basis. So, this theory is extremely tentative. Any "neutrino theory of ESP cannot be positively falsified or verified at present, [so] we must entertain many misgivings about them." (Chari, 1977, 811)

Whereas new variations of the electromagnetic theories of psi hold minimal appeal and seem to be fighting a rearguard action in overcoming problems diagnosed many decades earlier, the hyperspatial theories are riding into the future on a groundswell of popularity. In 1972, Gertrude Schmeidler, who is not a physicist, offered the first hyperspatial theory of psi of this period based upon embedded space-time geometry. She used a model of "topological folding" which contended that our four-dimensional spacetime continuum was somehow folded on itself.

Distant or non-local points representing events in space-time could be quite near to one another for any signals that could travel out of our physical space-time envelope, from one single fold to another. If psi could travel unhindered between folds, then ESP and other such phenomena could be easily explained. This assumption may not be as far-fetched as one might imagine considering the latest theoretical advances in physics.

Einstein abandoned physical theories utilizing the concept of a higher-dimensional embedding manifold during the 1940s, but other scientists continued to develop the notion. In the late 1970s, the Kaluza-Klein five-dimensional theory again gained popularity and now many physicists theorize that as many as ten (or even twenty-six) dimensions are needed to explain the various normal physical phenomena that are evident on the quantum scale.

Within this context, Schmeidler's notions are not as radical as they might seem at first glance. However, she provided no explanations, in the form of mechanisms, of how signals could travel between different points in space-time beyond exiting and reentering different folds. The geometry of 'folded space-time' goes beyond the presently accepted Riemannian structure of space-time used in general relativity. So, Schmeidler's ideas seem at odds with general relativity when in fact general relativity offers the only evidence that physical space-time is characterized by curvature.

The primary difficulty with any hyperspace theory, whether of pure physics or psi, is our inability to sense or otherwise detect the extra dimension or dimensions. Current theories postulate that the fifth dimension is compactified or curled up in a cylinder of nearly infinitesimal width that circumvents any normal physical detection. The overall extension in the fifth and higher dimensions shrunk to such small sizes during the 'Big Bang,' giving our four-dimensional space-time its present characteristics while guaranteeing that the higher dimensions can only be detected at extremely high energy levels. The enormous amounts of energy needed to detect the higher compacted dimensions are at present and for the foreseeable future unobtainable.

Despite such difficulties, the hyperspatial/hyperdimensional theories do have a clear potential for explaining psi. The most obvious advantage comes by way of the fact that these theories utilize a space-time model in which time acts as an extended dimension similar, after a fashion, to our normal spatial dimensions. This model thus allows an easy explanation of precognition in that precognitive events reduce to the same explanation as ESP phenomena. They are simply a signal passing between two points in space-time rather than a signal passing between two points in space. This advantage of the spacetime model has already been used by other theorists of psi such as Dunne, as well as physicists and philosophers within a purely physical context.

Karl Brunstein has given the arguments supporting a hyperspatial theory of psi. In 1979 he published the book *Beyond the Four Dimensions: Reconciling Physics, Parapsychology and UFOs* that essentially offered philosophical support for a five-dimensional explanation of psi and linked psi to the observation of UFOs. In the case of psi alone, having nothing to do with UFOs, he states "ESP phenomena can today be viewed as quite firmly, if not yet comfortably, scientifically established." (Brunstein, 150) He further believes that paranormal phenomena are "indicative" of an "extradimensional force" which can "relativize and distort space and time" in a manner that could be expected to account for psi phenomena. A physical force is needed to bind our normal three-dimensional space with the time dimension. Brunstein identifies this binding force as electromagnetism.

By analogy then, a fifth dimension would necessitate another force to bind it to our normal four-dimensional space-time continuum.

> What is connoted is that if this fifth dimension is to be lacked onto the other four, orthogonally or otherwise, it – like time – requires some kind of associated force, or effect, to go along with it, for basic definitive, dimensional, and normalization purpose. ... For want of a better term for it, and for largely historical reasons, we dub this unparalleled kind of force and its effects "extrasensory." (Brunstein, 150)

This new "force" is structural, and it is first identified as "extrasensory." We only sense physical events within our space-time continuum. Therefore, anything that exists outside of that continuum is beyond our normal ability to sense it, and thus "extrasensory" by definition.

Brunstein qualifies his use of the term "force" and eventually identifies his new "force" with consciousness itself.

In man, we found ultimately the manifestation of another natural force, a new force – an unconformable kind of force. We call this new force finally "consciousness," or "intelligence." Just as with the four physical forces, the new force springs directly from matter. It springs from discrete lumps of it we call "human." This new force, as the others, manifests itself by effecting physical changes within reach of the material lumps that form its seat. Sometimes these changes can be vast. Again, this is just as with the other forces.

> We say the new force is "unconformable" because it does not fit the accepted definition of a physical force; that is, it cannot be equated to a product of mass times acceleration. In this sense, it is in the physicist's neat world most definitely not a force. (Maybe force is not the best word to describe it, but the parallel

The identification of consciousness with the fifth dimension is unique, but not unprecedented. Hinton, Browne, and several others have associated consciousness with the fifth dimension as long as a century ago and more. Brunstein's concept is unique and different from others in that he quantifies consciousness as a binding "force" of sorts and thus relates it to physics. In other words, Brunstein more 'forcefully' binds (please forgive the obvious pun) the concept of consciousness and thus psi to physics.

While Brunstein's model seems a bit more ambiguous, Saul Paul Sirag has proposed a more precise mathematical model relating consciousness to hyperspace derivation is philosophical and qualitative, a problem facing nearly all the physical theories of psi. Sirag proposes that our physical reality is only a "sub-realm of a larger reality" in which all physical forces are unified within a hyperdimensional structure. Using Wigner's interpretation of quantum mechanics whereby consciousness "projects the state vector onto the eigenvector" during the act of measurement, the non-locality expressed in Bell's theorem suggests the existence of a "universal consciousness." (Sirag, 329)

This consciousness can be represented by a mathematical structure known as a "reflection space." While mathematical reasoning cannot choose between the many different reflection spaces that can be used to represent the "universal consciousness," physics can. The specific reflection space that Sirag envisages "exists in the intersection of the Mackay group algebra and the Lie algebra." (Sirag, 341) In fact, its role is to mediate between these two algebras and is therefore identified with the "universal consciousness." Unlike most other theories, Sirag's theory is represented by a clear and precise, although quite complex, mathematical model.

Bob Toben and Alan Wolf have taken yet another path to relating hyperspace and consciousness. If anything, their theory is disjointed as presented in the 1979 book of hand drawings entitled *Space-Time and Beyond*. The theory is more a list of statements than a coherent and logical explanation of either psi or consciousness. In their own manner and after their own style, they state that consciousness is hyperspatial, beyond normal space-time, while normal space-time follows the physical model presented in general relativity.

The four-dimensional spacetime continuum is itself constructed of quantum foam that is linked to consciousness via gravity (curvature in a higher dimension) while gravity is the "master field." "Vibrations of thought patterns in specific harmonics structure all 'matter' and light as we experience it" (Toben, 61) and the human mind acts as a "filter" which focuses our thoughts on specific physical events. (Toben, 62) Psychokinesis is an effect of consciousness influencing energy/matter fields (Toben, 67) while ESP or telepathy occurs when signals travel through "wormholes" in "the sea of space." (Toben, 78)

Living beings generate "biogravitational" fields, which allow them to manipulate space-time curvature to various ends. It is through this manipulation, carried out by consciousness, that other paranormal and psi phenomena are manifested. Therefore, the higher a being's consciousness, the better the being can utilize consciousness to affect psi. Toben further postulates that many higher levels of consciousness than humans have yet imagined or detected may exist.

These ideas were supported and amplified by Jack Sarfatti in the "Scientific Commentaries" portion of the book. He relates Toben's ideas to the previous work of other philosophers and scientists and describes the psi processes in more detail, although they are still not rendered in a quantitative model. No real mechanisms or physical processes have been offered to support Toben's theory. Despite some provocative and promising points made by Toben, Wolf and Sarfatti, the ideas presented have not been developed into any type of a coherent theory of consciousness, psi or even physics.

Two other hyper-dimensional theories are also worthy of note. Elizabeth Rauscher in 1977 and Puthoff, Targ and Edwin May proposed these in 1979. In these cases, all the conspirators are physicists. Rauscher's theory is simple and elegant, as well as mathematically supported. She contends that our four-dimensional. Rauscher contends that our four-dimensional space-time is only a portion of a larger eight-dimensional spacetime.

Each of the four dimensions in normal space-time can be represented by a complex number having a real and an imaginary part. Therefore, an imaginary four-dimensional space-time exists alongside of our normal four-dimensional space-time. This imaginary space-time acts as a realm for the transfer of signals during psi processes. An individual's consciousness "is free to access information in the entire complex space," (Rauscher, 69) so mind is not limited to the four-dimensional space-time continuum as are the brain and matter in general.

This model addresses an important problem that plagues many other theories. In the other theories, it has never been adequately explained how energy transfer between remote positions in space-time might occur during the psi process, but this is not a problem for an imaginary space-time.

> Instead of hypothesizing a model which involves energy transmission, and associated problems of energy conservation, we chose to develop a model in which remote information is accessed in four-space as though it was not remote in a higher dimensional geometry. The relativity theory formally describes the relationship of macroscopic events in space-time and, in particular, their causal connection is well specified. Higher dimensional geometries appear to reconcile precognition and causality and define a formalism in which the special and temporal separation of events (in four space) appear to be in juxtaposition in the higher dimensional geometry. (Rauscher, 56)

Within the context of this imaginary space, Rauscher can postulate that there is no distance between psi events, at least in so far as distance is measured between points in normal space-time.

So, the normal laws of physics, such as the limit to velocities imposed by special relativity, need be of no concern in this paraphysical model. Therefore, there is no energy transfer between points in space-time during the psi process. Some type of least action principle would also be necessary to govern signal transfer through imaginary space-time. (Rauscher, 69-70)

The application of a least-action principle would allow a mathematical model to be developed.

Rauscher's model thus offers the possibility of developing a mathematical formalism, which is missing from other theories. Other areas of physics have successfully used complex numbers to explain physical phenomena, so Rauscher's method is not without precedent, the care, which Rauscher has taken in relating psi theories to the theories of normal physics, seems to be unique in comparison to other approaches to psi.

A short theory offered by Puthoff, Targ and May also utilized an eight-dimensional manifold that is quite like Rauscher's model. The similarity should be expected since Rauscher acted as a consultant at Stanford Research Institute where Puthoff, Targ and May developed their theory. They acknowledge this fact and further stated that their theory was developed "in conjunction with Gerald Feinberg." (Targ, Puthoff and May, 100) Their theory was only directed toward explaining remote viewing, but it does have the potential to account for other psi-related phenomena.

They hypothesized an eight-dimensional model of space-time using complex numbers to represent our normal four-dimensional space-time. This model could be represented by a proper time which would differ from the normal proper time of special relativity by the addition of two new terms, such that:

$$\Delta s^2 = 2\Delta s \Delta s^0 = \Delta x^2 + \Delta x'^2 + c^2 \Delta t^2 + c^2 \Delta t'^2$$

The primed terms represent measurements in the imaginary part of space-time. These new elements add a special distance factor of $\Delta s0$ to the formulation. Remote viewing would occur when the element Δt goes to zero and the other terms are fixed so that Δs also goes to zero. In this case there would be neither temporal nor spatial distance between remote physical events.

Since the primed terms are only accessible to consciousness when the separation is not limited by the properties of physical space-time, there is a clear separation between physical space-time and mental space-time, yet the two are intimately connected under proper conditions that are theoretically testable. This model could be verified by determining the physical conditions that would give rise to this situation. In other words, determining the conditions for optimum success at remote viewing could possibly verify (or not) this theoretical structure.

The model provides a "geometrical interpretation of the quantum interconnectedness principle, by which events remote in space-time are nonetheless connected by nonlocal correlations or, in this interpretation, by the nature of the space-time fabric itself." (Targ, Puthoff and May, l00-101) Modern physics is dominated by either field or quantum theories, which generally reflect different sides of the ancient debate of whether reality (and/or matter) is continuous or discrete. In turn, this debate is intimately related to the mind-matter question.

Quantum theory has so thoroughly dominated physical and theoretical research in the recent past that almost all attempts at unification were made by explaining fields in terms of the quantum theory, yielding quantum field theories. But renewed interest in relativity theory, the icon of field theories, has reversed that trend and fostered attempts to base quantum theory on the field perspective. This was a fortunate happenstance for psi research since many experimental results as well as spontaneous events indicate that psi is a field phenomenon.

This fact also bodes well for hyperspatial theories of psi that are ultimately founded upon one or another form of general relativity, which has been unified with electromagnetism and the other fundamental forces in several recent theories. These last theories in part reflect this change in attitude, but the change in attitude is still more far reaching. Some scientists believe that new advances in physics, especially those dealing with these new unified field concepts, will eventually precipitate a paradigm shift if not the next revolution in physics.

Any such revolution in thought would surely be significant for psi research since some form of consciousness would have to play a central role in any new theories of physics. While consciousness seems to be necessary in most interpretations of quantum mechanics, the role of consciousness in relativity-based theories has not been so clear-cut. So hyperspatial and hyperdimensional theories that account for consciousness are certainly far more progressive than those that do not.

The fact that consciousness must take a role in future theories of psi is far easier to demonstrate in the case of quantum theories. The already established role of consciousness in quantum mechanics has been an opportune happenstance for quantum theories of psi. Evan Harris Walker in 1974, Helmut Schmidt in 1974 and Richard Mattuck in 1977 developed independent theories exhibiting the quantum relationship between consciousness and psi.

Robert G. Jahn and Dunne offered further insights along this line during the 1980s. These theories carry more weight in the scientific community because they are based directly on experiment and have themselves led to the further development of more refined experiments. They have thus proven to be quite popular among non-physicist researchers of psi. There is also a good deal of literature on both the experimental and theoretical development associated with these 'observational' quantum theories.

Walker claims "consciousness is a nonphysical, but real, entity." However, "physical reality is connected to the consciousness by means of a single physically fundamental quantity." (Walker, 547) He further notes, "There exists at least one physical quantity that connects the consciousness to the physically real world." This connecting physical quantity cannot be a gravitational field, nuclear forces, or weak interactions because their magnitudes in the brain are too small.

By eliminating the other known physical forces, this leaves only electromagnetic forces as possible links between the physical world and consciousness. Yet the electric currents in the brain are extremely localizable and brain waves fail as a source of psi because superposition of the waves would destroy any coherency. With electromagnetic forces likewise ruled out as connectors, only Schrödinger's wave equation remains to explain the connection between the physical world and consciousness.

Walker has determined that "consciousness is a nonphysical entity connected to the physical world by means of the state vector for a quantum mechanical process linking the synapses of the brain." (Walker, 549) But Walker's process of elimination does not put forward the best face of science. Science does not progress forward by eliminating all the possibilities until only one candidate to answer a problem is left over. Science is not a game of musical chairs. There must be a sound physical reason to use a concept to explain an event or phenomenon other than all the other explanations do not work.

Sticking strictly to the Copenhagen Interpretation of quantum mechanics, Walker introduces a new variable whose purpose is to make the system causal thereby representing the consciousness of the observer. The variable is based upon Bohm's original concept of a 'hidden variable.' It has two properties. The 'hidden variable' must only interact physically by means of the measurement process, and secondly, Bell has shown that it must act non-locally. Non-locality means that the 'hidden variable' is independent of time and space and is thus not dependent on other physical processes, yet it still affects the 'collapse of the wave packet.' The purported properties of the 'hidden variable' have far reaching consequences because they lead to an explanation of psi in conjunction with consciousness, if for no other reason. However, they were not introduced only to explain psi and consciousness. The possible existence of 'hidden variables' has been debated quite effectively without ever introducing psi.

Both the 'sender' and the 'receiver' that constrains the wave packet to 'collapse' into a single state can explain telepathy as an intersubjective agreement in consciousness or rather a mental act. Under these circumstances, all that is necessary for a telepathic event is two conscious beings. Since no message is sent in the traditional sense of the concept, but a future state is being agreed upon at a mental level, no sender is really needed. The event just occurs, and no psi 'signal' is needed in the classical sense of the word.

Clairvoyance is reduced to another form of telepathy combined with precognition. PK involves the same basic process as clairvoyance, only the selection no longer occurs as a quantum mechanical brain process. Instead, it is an accompanying divergent effect of the relevant physical or energetic system. All the different aspects and forms of psi are related to the properties of the 'hidden variable' which was first introduced into pure physics to explain things other than psi. The 'hidden variable' model of psi cannot be classified as a fluid theory because the concept of a 'hidden variable' was not invented to explain psi. It was developed within and is a part of normal physics, which is its main advantage.

Schmidt developed a novel mathematical model of psi, "which permits a logically consistent discussion of a world with psi" (Schmidt, 302) rather than establishing a physical theory explaining the nature of psi. By focusing his attention on the environment rather than the observer, the medium through which psi propagates so to speak, Schmidt was able to utilize a break between psi and physics due to the goal-oriented character of psi in PK tests. Schmidt has attempted the duplication of a psi-like process independent of true mental intervention to establish the physical ground rules for psi communication in the presence of mental activity.

Random number generators and computers were used to replace the experimenter and psi sources. The computers were programmed to mimic the properties specified by a 'psi axiom,' allowing Schmidt to electronically duplicate some aspects of PK and ESP. The 'psi axiom' was based upon the notion the "Psi sources are, generally, devices with a signal input through which the source can be stimulated." Schmidt further refined the 'psi axiom' to the mathematical equation $p'/q' = \theta(p/q)$, which represents the source of psi at time t. (Schmidt, 306-307) The ratio of probabilities of a psi 'hit,' or an electronically successful telepathic message, is expressed in the ratio p'/q'. It should be the same as the ratio of the probabilities of the message being sent by the psi source, p/q, except by a factor of the messages" relative strength θ.

For strictly experimental purposes, the mock 'psi-source,' the random number generator, has been set up to act in a way like a PK subject. The 'psi axiom' ensures that the experiment simulate space-time independence since it contains no factor for attenuation. It is the key to understanding psi. The different forms of psi appear as logical consequences of this single 'psi axiom' within the experiment.

Schmidt's model enables the generation of "a large number of testable implications and may serve as a useful basis for future theoretical and experimental studies." (Schmidt, 301) His electronic systems do not actually send or receive ESP messages, just as they fail to move objects by PK, but they simulate psi events in a manner that leads to the construction of a mathematical language to describe psi. In other words, by mimicking psi events with a computer, a standard by which real psi phenomena can be judged is established.

Basically, Schmidt assumed a physical interaction between the brain and a material body during PK. Since a signal must pass between the two, it must conform to certain conditions and criteria independent of the brain (transmitter) and the body (receiver). Therefore, a mechanical/electronic device should be able to transmit signals that would act exactly as the psi signal acts during transmission. These falsely generated signals would mimic psi and thus allow Schmidt to study the properties and characteristics of psi.

The Danish physicist Mattuck has taken another approach, although his theory is dependent upon Walker's work. Mattuck's theory differs from

Walker's in that he emphasizes the sources of energy needed to accomplish the psi act.

Since PK phenomena generally resemble those produced by ordinary forces, most of the attempts to explain them have postulated some new type of force coming from the mind. This paper presents a different approach based on a proposal by E.H. Walker. In this proposal, mind makes use of the energy which is already present in matter in the form of random fluctuations or noise, reorganizing this energy in such a way as to achieve the desired PK effect. In this work, I will not only discuss thermal noises, but the general ideas apply to any kind of random fluctuations. (Mattuck, 191)

His theory is based on the notion that "consciousness (or mind) can directly influence a physical system by utilizing the quantum mechanical fluctuations in the system properties." (Mattuck and Walker, 112)

Mattuck's calculations indicate that enough 'local' energy is available to accomplish PK phenomena on the macroscopic scale, (Mattuck, 194) not just the sub-microscopic level as is the case in Walker's theory. Since the energy in quantum fluctuations is 'uncertain' before the wave packet is collapsed, all that mind or consciousness must do is 'select' or 'reorganize' the already existing quantum "fluctuations in a non-random way so as to produce the PK effect." (Mattuck and Walker, 114)

By utilizing the energy that is available locally, there is no need for an extraneous 'force' to transmit energy from the subject initiating the PK event to the object acted upon under the influence of psi, thus circumventing one of the major arguments against the existence of PK as well as preserving the conservation of energy.

W. von Lucadou and K. Kornwachs of Germany have also suggested a quantum model based upon an interaction of consciousness and the wave function. Their theory has not yet been finalized but has at least reached a stage whereby it can be described. To the normal ψ function of quantum mechanics, which "describes the condition, and the development of a quantum mechanical process," (Lucadou and Kornwachs, 187) they have added a new probabilistic ϕ function that describes complex quantum Systems.

The quantity $|\phi|^2$ represents the probability of the appearance or transfer of information during a physical event. ϕ itself would represent a complex material system such as the human brain and its value would depend upon the complexity of the system it represents. Any physical event described by quantum mechanics would therefore need to consider both ψ and ϕ and could thus be represented mathematically as $\Psi = a\psi + b\phi$. Ψ is the normal state vector of wave mechanics and a and b are simple constants. ψ and ϕ represent both the wave packet and the conscious act of collapsing the wave packet, respectively.

In the case of psi, the quantum mechanical wave function Ψ would merely be constructed to represent the psi event. Otherwise, their model is more general and can be used to describe any physical phenomenon by the correct choice of Ψ. The problem is that ψ and ϕ are mutually dependent, which makes them difficult to specify. The theory will not be complete until they are specified.

This model bears some resemblance to Jahn and Dunne's approach to explaining PK. In the 1987 book *Margins of Reality*, they announced their conclusion that the human mind can psychokinetically affect the operation of machines. This finding, reached through extensive research, formed the basis of their theoretical work. Jahn and Dunne do not believe that subatomic particles are real until consciousness interacts with the 'environment.'

In this respect, they reflect a common viewpoint within the scientific community that is a fundamental aspect of the Copenhagen Interpretation of quantum theory and mechanics. However, they part with most scientists in their definition of 'consciousness.' They believe that „consciousness" is "anything capable of generating, receiving, or utilizing information." (Talbot, 146) By defining 'consciousness' so broadly, neither human consciousness nor any intelligent consciousness is necessary to establish either our physical environment or physical reality itself.

They reasoned that if consciousness "finds wave mechanics a useful complement to particulate physics" then "consciousness may also find a wave-mechanical metaphor to be conceptually and functionally useful for representing itself." So Jahn and Dunne postulated a "probability-of-experience" wave to be associated with consciousness. (Jahn and Dunne, 219) Such "consciousness waves" range freely over space and time, but "if a particular consciousness wave is confined to some sort of 'container' or 'potential well,' representative of the environment in which that consciousness is immersed.

Characteristic patterns of standing waves, or eigenfunctions, will be established that represent the experiences of that consciousness in that situation." (Jahn and Dunne, 242) These waves could then escape the potential well in which they are bound, just as ordinary matter waves escape potential wells in quantum physics. The potential well in which they are bound in this case is the human brain. When the "consciousness waves" escape a particular brain, they could interact with either the environment or another human brain, resulting in a psi event.

PK and remote viewing, the psi events with which Jahn and Dunne were primarily concerned, could thus be considered resonances between the mental and physical natures of consciousness and matter. There is absolutely no need to rely on an active 'force' or a 'transfer of energy,' both views of which lead to serious problems in physics. Although their theory is independent of other theories, there are a number or similarities and correspondences with other theories. More specifically, their idea that consciousness must be represented in a manner like matter within the quantum theory, *i.e.*, by its own wave function, sounds vaguely like von Lucadou and Kornwachs" quantum model.

Another possibility of applying new revelations from the physics of the quantum comes from a suggestion made by Richard Feynman in 1949. He noticed that a positron was mathematically equivalent to an electron traveling backwards in time, so he suggested that a positron was just that, an electron moving back in time. There is no physical law or principle that would prohibit this assumption on the sub-atomic level of reality.

However, its mathematical counterpart, as derived by P.A.M. Dirac, can be interpreted as displaying a negative energy just as Feynman's positron hypothetically travels backwards in time. Both ideas could provide and have provided explanations for various aspects of psi. Upon this basis, Pearson derived a hypothesis whereby "energy densities of brain chemicals are supposed to be high enough to permit some anomalous manifestations (telepathy, clairvoyance, and precognition) of the 'non-zero properties' of the 'perfect vacuum'." (Chari, 1977, 815) Others have also postulated negentropic explanations of psi events.

The French physicist O. Costa de Beauregard has come to a similar conclusion through an entirely different evolution of thought. His own personal changes in attitude began as early as 1951, but he felt it unwise to voice his "rational conversion" to a belief in psi until 1975. Once again, this reflects the changing attitudes of the scientific community in general and the physics community in particular prior to and during the decade of the 1970s. He contends that human "conscious awareness" has "two faces."

Or rather our conscious knowledge of reality springs from two procedures, "decoding a message (that is, an ordered structure) and emitting a message (that is, producing order) by means of one's information…" These quantities (or qualities) are commonly known as cognition and volition (or will), respectively. Consciousness, an "attention to life," is comprised of both cognition and volition, in a broad sense, which is "tightly bound to corresponding symmetries in the real world." (Costa de Beauregard, 177-178)

The two faces of human consciousness are cognition and will and it would be expected that both should show up during the process of quantum measurement. The transfer of information, which is the key to knowledge, is essential to quantum measurement. In these processes, the flow of information is actually "negentropic." In physics, entropy commonly represents a measure of disorder of a system. The thermodynamical fact that entropy naturally increases in a system renders the concept of what is commonly called "time's arrow." Alternately, negentropy could then be thought to represent an increase in order, which could or could not be related to time reversal.

On the one hand, biological systems exhibit negentropy because they order matter into a complex structure. Since probability is the "hinge about which mind and matter" interact, biological systems are probability decreasing processes. Advanced wave processes are also probability-decreasing processes. So, the advanced wave process can provide a physical model of the ordering process or negentropic aspects of biological systems and the transfer of information from the future to the present can account for psi phenomena.

In fact, "under appropriate conditions, information as an organizing power should act as a sink of advanced waves, just as information as a gain of knowledge acts as a source of retarded waves." (Costa de Beauregard, 182) Thus, the whole process is quite natural and a consequence of accepted physical theories and laws. Psi phenomena should therefore be viewed as a "very rational" consequence of relativistic quantum mechanics rather than an "irrational" speculation. (Costa de Beauregard, 186)

In still another view based on quantum principles, Sarfatti favors a concept of "back action" at the quantum level to describe both consciousness and non-local interactions as well as psi. Sarfatti's model follows from the most basic concepts of physics. According to Newton's third law of motion, every action generates an equal and opposite reaction.

> We now come to "back-action" which is the main idea of this paper. The Origin of this idea is Newton's third law that for every action there is an equal and opposite reaction. We now know that this is a consequence of translational symmetry in physical space.... It can also be shown that quantum spontaneous emission of real radiation by virtual zero -point vacuum fluctuations can be explained as advanced wave effects from the future that are classically associated with radiation resistance. Feynman also used the term "backaction" to explain the generation of quantized vortices in superfluid helium. (Sarfatti, 2 of 35)

And,

> By the term "back-action" I mean that the quantum wave field is "directly affected by the conditions of the particles". It is qualitatively obvious that such a direct dependence is the counterforce or reaction to the quantum force. The combination of the quantum force of wave on particle with the counter-force or backaction of particle on wave forms a feed-back control loop which is able to control the formerly uncontrollable guidance of the particle by its wave. This results in a distortion of the statistical patterns of orthodox quantum mechanics. This is the mechanism of intent or free will. This is the mechanism of intent or free will. (Sarfatti, 4 of 35)

In other words, Sarfatti agrees that consciousness collapses the wave packet to realize or materialize a solid particle in accordance with the quantum mechanical model, but he has extended that idea by adding that the particle reacts equally and oppositely to the collapsing wave in the form of a "back action." Just as consciousness acts physically through an exchange of information which causes the wave collapse, the particle reacts through its "back action" to pass information on the state of the collapse back to consciousness. Therefore, the solid particle must affect the consciousness that collapsed the wave to create it.

Sarfatti implies that his theory of "back action" can explain psi effects by a mutual exchange of information at the quantum level rather than a one-way exchange of information from consciousness to wave function. He ultimately associates his "back action" with perception and consciousness, as well as intent, free will (as described above) and the elan vital or life force upon which scholars speculated over two centuries ago. (Sarfatti, 8 of 35)

He even associates his back action with the Chinese concept of Chi and the Japanese Ki. (Sarfatti, 18 of 35) And finally, he associates "back action" with several other characteristics, quantities or qualities of cognitive thought: Purpose, meaning, intention, and intuition. (Sarfatti, 13 of 35) These associations may well prove true, if not for Sarfatti's "back action" then for some other physical explanation of psi since all these quantities are either associated with consciousness or some level of thought itself. However, Sarfatti gives no further physical mechanisms that can help to distinguish between these various types of mental activity. It was perhaps unwise for Sarfatti to have associated his "back action" with so many different things before he has more fully developed his theory and provided some further verification of his ideas.

Contrary to the progress in the other classes of theoretical work, the fluid theories have not fared so well during this latest period of paraphysical work, with just one new contender for a physical theory of psi. The fluid theories seem to have evolved into field theories, or at least they have been associated with field theories as of late. With all the recent advances in quantum and field physics, there is no longer a need or an incentive to invent new structures to explain psi.

Yet a few such theories persist. Toben utilized a biogravitational field in conjunction with his hyperspatial model of psi and consciousness, but his was not the only utilization of this or similar ideas. Alexander P. Dubrov, a biophysicist, as a "field-energy system", first suggested Biogravity in 1973. (Dubrov, 231) It could be transformed into any type of field or energy, a feature that he developed as the cornerstone of any unified field theory of the future. This field was thought to exist in all living beings, including humans. It was called biogravitational because its properties are akin to both living organisms and gravitation. Dubrov suspected the field played a role in such biological processes as cell mitosis as well as physical processes such as gravitational waves. Between these two extremes came biological effects of the whole organism such as "psi photography and levitation." (Dubrov, 234)

Dubrov spelled out the properties of the biogravitational field rather explicitly. They represented a combination of biological and gravitational properties. (Dubrov, 234) However, the concept of biogravitation still suffers from the same severe problems that plague other fluid theories. Despite this detailed list of properties, the theory is admittedly qualitative and lacks mechanisms or processes to accomplish the stated results. These properties simply beg for some kind of mechanical or other explanation and further clarification before any use can be made of this theory.

Like the earlier fluid theories of bioplasm, biomagnetism, animal magnetism and a psychic aether, biogravitation is a quantity that was invented merely to describe psi and other paranormal phenomena. It is always dangerous to invent hypothetical quantities to explain hypothetical circumstances and events. One valid complaint against such theories is that they "seem to be cast in a holistic framework (the anomalous effect shows not in the individual living cells, but in the organism as a whole) and remain obscure for even the limited purposes of experimental testing and replication." (Chari, 1977, 807) In other words, it is quite unlikely that Dubrov's biogravitational theory can never be tested let alone confirmed.

The popularity of fluid theories seems to be in decline with very few new versions added to the list in the past few decades. Quite simply, the fluid theories seem to have run their course. There may be no need for such unique new structures as more and more concerned scientists have concluded that physics is either presently able to deal with psi or will soon be able to explain psi. The notion that and accepted modern concepts and theories in physics can account for psi is rapidly gaining popularity and this would represent a completely new situation in the study of psi. Previous generations of scientists who worked on psi or psi related subjects have believed that an explanation of psi was beyond the physics of their day.

Electromagnetic theories also seem to be going in the same direction. This trend would seem to be a natural occurrence in the evolution of physics since any unification of physics toward which many scientists now move, either under the banner of the quantum or a field theory seems closer at hand. Any unification of the fundamental forces of nature will naturally reduce all electromagnetic theories to the single field or a similar structure by design.

A theory of psi based on a unified field theory would be an electromagnetic theory by default and vice versa, so there will soon be no need to distinguish an individual class of electromagnetic theories of psi. Fluid theories are no longer needed just as electromagnetism is very possibly merging into the unified field. The burning questions for the immediate as well as distant future are: What form will the unification take? And then, will psi emerge as a consequence, or side effect of the unification? Or will the unification need to be modified to incorporate psi?

When trends collide

A closer study of the progression of physical theories of psi throughout history in conjunction with recent developments in theoretical physics strongly implies the evolution of a new and significant trend in science. It is not a trend in the science of psi alone, but a trend in physics that directly impacts the study of psi. This trend is marked by the acceptance of consciousness as a legitimate area of research in physics over the past two decades. It is intimately bound with two other important trends.

On the one hand, researchers in psi are now coming to a consensus that present day physics should be able to cope with and explain psi; if not now, then in the very near future. On the other hand, psi and consciousness have been bound intimately together by researchers in paraphysics and parapsychology as well as scientists who are not involved in the study of psi.

A simple example of this can be found in Robert Ornstein's book *The Psychology of Consciousness*, published in 1972. Although Ornstein is not a parapsychologist, nor does he specifically study psi phenomena, his study of consciousness has led him to a short discussion of psi. (Ornstein, 188-191, 220-224) These three trends cannot be dissociated from the evolution and proliferation of alternate views of quantum reality which differs from the standard Copenhagen Interpretation as well as the recent revival of unified field theories based upon the field theoretic view of general relativity. Nor are the individual trends just cited completely independent of one another, even though they are all distinguishable as unique in their own rights.

Should all these trends come to a head at or near the same point in time, in the form of a single broad-based theory, then the Third Scientific Revolution will have blossomed. Since the field seems to be narrowing to physical theories of psi based on the quantum or hyperspace perspective, the relationship between the quantum and consciousness on the one hand and hyperspace and consciousness on the other are of vital importance. In both the hyperspatial and quantum models of psi, a consensus that psi and consciousness are intimately related has developed.

This melding of psi and consciousness is easily seen as an important scientific consequence of the new styles of the quantum mechanical theories since consciousness seems necessary to our physical reality, or at least a willful act is necessary to collapse the wave function in any experimental arrangement. The same is not true for the hyperspatial theories. In this case, the reason why studies of psi and consciousness are converging is neither so clear cut nor easily discerned.

Whatever consciousness is, or however it is to be defined, it is certainly something more than the mechanical sum of the parts (chemicals and chemical reactions) of a living machine called the human brain, just as there is something more to life itself. Therefore, it would seem to any knowledgeable person that something more than our four-dimensional space-time continuum, in which the mechanical parts of living organisms are constrained to interact and give a semblance of life is needed to explain consciousness.

A simple analogy between consciousness, and an extra dimension of the physical continuum, therefore, is quite evident. However, an analogy alone is not enough to warrant the adoption of dimensional model. Such a model cannot be adopted unless there are coherent and compelling logical reasons to do so, as in the case of a prediction from experiment that can distinguish between this model and alternative theories.

Physical evidence of the existence of a new dimension that has been discovered within some other context would also be considered compelling evidence. Since an extra dimension seems to be the stage upon which psi acts, and psi and an extra dimension of space have been known to display similar characteristics all the way back to Zöllner, then psi and consciousness would also be intimately related in an extra-dimensional physical theory. While this logical argument is not as ironclad as the quantum theory's claim to a connection with consciousness, it is not without precedent and indicates that hyperspace theories are valid contenders in the physics of consciousness.

The other trend mentioned is that of pure physics, independent of any speculations or hypotheses regarding psi or psi functioning. This trend plots the course by which physics is moving toward psi on its own accord, independent of any physical speculations regarding psi. More specifically, consciousness has become a legitimate subject of inquiry in physics in the past few decades and physicists who would have had nothing to do with psi research under other circumstances in the past few decades, but are interested in the physics of consciousness, are now speaking of psi in relation to consciousness.

For example, in his recent book on consciousness and physics, Euan Squires has begrudgingly come to the simple conclusion that the evidential existence of extrasensory communication (ESP) and psychokinesis are implied in the relation of consciousness to physics.

> A much simpler question, both to pose and to answer, is whether mental states can influence some elementary, external, physical system. Roughly speaking, can I move something by thinking about it? More precisely we can consider a physical system which is isolated from all external physical influences. This will behave in a predictable, calculable, manner (because of quantum effects, see chapter 10 we would have to average over many similar systems, but this does not affect the argument). The question then is whether it is possible to alter this behaviour by "mental activity". ... Conversely, we might even expect that a "non-physical" conscious mind should be able to affect objects external to the brain. Why should something non-physical be limited by the existence or otherwise of physical connections? If my conscious mind can have an effect on the atoms in my brain, why cannot it also have an effect on other atoms? Since we do not have any idea of the nature of the supposed link between the conscious mind and the brain (see section 5.6), we cannot answer such questions. Of course, there has to be some way in which my consciousness is more associated with my brain than with anything else; it tells me that I consist of this body. (Squires, 104)

Squires' argument is based purely on logic, but it still opens the door for the action of psi without his being forced to directly mention psi.

His chief concern is that consciousness might have "non-local influences," which is to say that consciousness can directly affect physical phenomena, at least at the level of collapsing the wave function, at a location in space-time that is well outside of the confines of the human brain. Such an action would be judged paranormal by any standards and correspond to a psi event. The same trend has been evident in the more recent theoretical work of the physicist David Bohm. Bohm's unique worldview evolved from his 1951 work on 'hidden variables.' He later concluded that a new kind of field must exist at the quantum level of reality to completely explain quantum phenomena. This field, which he called the 'quantum potential,' pervades all of space, but its influence does not diminish with distance like other fields.

By assuming the existence of this field, Bohm was able to account for experimental quantum effects as well as reinterpret quantum mechanics itself. Within his field, all material particles are non-locally connected in a manner that seems to violate the normal laws of nature. About the same time that Bohm developed this new view of nature, the psychologist Karl Pribram developed a 'holographic' theory of the human mind. Bohm further concluded that his field also had the same characteristics as a hologram. Thus, the concept of the 'holographic universe' arose as mind and matter were unified in this newly evolving worldview.

Contrary to the classical worldview, Bohm concluded that the parts are organized by the whole as represented by the quantum potential field. The existence of the quantum potential implied an unsuspected 'wholeness' in the universe. This idea solidified the concept of non-local interconnectedness of individual material particles (entanglement) in a manner that had not been predicted by either quantum mechanics or classical physics.

This view was akin to the simple relativistic view that the position of particles and thus the rate of change of position of particles depended on the other bits of matter scattered throughout the universe, except that space must have some potential of substantiality (or potential substantiality), which it did not have in the relativity theory.

The very concept of location in space disappeared at the level of the quantum potential where all points of space became equal to all other points of space. This interpretation presented yet another view of the concept of non-locality. Within this framework, it was found that distant particles in normal space-time were connected in a manner such that a change of state in one particle would immediately affect a change of state in another particle without a signal passing between them.

The immediacy of such twin or coupled changes of state would occur in direct violation of special relativity's ban on communications traveling faster than the speed of light. So, not only did this concept offer an alternative interpretation of quantum theory, but it also resolved the EPR paradox and predicted new experimental results.

In 1982, Alain Aspect and his colleagues performed a series of experiments whose results indicated that signals between macroscopically separated photons either traveled faster than the speed of light, in violation of special relativity, or acted non-locally by some other unexplained mechanism. These results implied that particulate matter is connected in some fashion or manner that is not limited by the tenets and principles of normal science.

Other experiments have also indicated that photons and other particles are non-locally connected in some unspecified manner if not in the specific manner that is indicated in Bohm's theory. While these experiments do not prove Bohm's theory, they at least render it far more plausible than would otherwise be expected. Aspects and similar experiments may also support theories where the interconnectedness of particles is an element or aspect of hyperspace. As of this time, Aspect's results do not provide a method of testing between different theories that feature interconnectedness.

The holographic worldview adopted by Bohm at first offered a metaphor for picturing and understanding the new level of order within the universe. According to Bohm, this order is not apparent at our physical level of perceiving events and phenomena but is 'enfolded' in the field. So, the order behind the probabilistic chaos presented in quantum mechanics, a physical reality that Einstein had suspected beyond the limits to knowledge set by the Heisenberg uncertainty principle, were implied by phenomena and events in our space-time.

This is the 'implicate' order of the universe. What quantum theorists saw as a wave packet of probability collapsing when an event was consciously observed to yield our physical reality is in Bohm's view an 'enfolded' possibility in the 'implicate' order which has been 'unfolded' and thus rendered 'explicate.' Through 'unfolding,' the 'implicate' becomes 'explicate,' the portion of reality that we sense as physical.

These views were presented in Bohm's book *Wholeness and the Implicate Order* in 1980. Bohm further developed the concept of a 'holomovement' to describe the flux of 'enfolding' and 'unfolding' within the 'implicate' and 'explicate' orders. The 'holomovement" gives rise to the dynamic progression in time of our physical reality rather than let his universe remain a static hologram. The observer has no independent existence in Bohm's model as in quantum theory since the observer is part of the system and always interacts continuously and constantly with all other parts of that system. There is no need to speak of consciousness interacting with matter to 'collapse the wave packet' just as there is no need for consciousness as part of the interaction of material bodies in the sense defined by quantum mechanics, so consciousness is not a necessary ingredient in Bohm's model.

However, Bohm still believed that consciousness is necessary as an awareness of the holomovement. Consciousness was thus rendered as a subtler form of matter which interacts with normal matter deep within the 'implicate' order, rather than when the 'implicate' unfolds to create the 'explicate,' as in the collapse of the wave packet in quantum mechanics. This holistic and holographic view of reality offers innumerable avenues for the explanation of psi phenomena. Since consciousness interacts with matter within the wholeness of the 'implicate' order, both prior to and after the moment of physical materialization of the present moment inherent in the 'explicate' order, every point in space-time is continuously and simultaneously connected with every other point in space-time.

Everything exists within the 'implicate' order, so psi phenomena such as ESP, PK, clairvoyance, precognition, remote viewing are reduced to conscious human interaction with consciousness in the 'implicate' order. Consciousness and matter are in direct contact in the 'implicate' order and humans need only find how to become aware of that contact or connection to utilize psi. Bohm's theory is purely physical, such that it is not a theory of psi per se, but the potential it offers for an application to psi is obvious.

The application of Bohm's theory to psi phenomena has been documented in Michael Talbot's book *The Holographic Universe* among other publications. Some physicists, such as Russell Targ, who are interested in a physical theory of psi, believe that Bohm's theory of the 'implicate' order offers the best chance for science and physics to explain psi. While Bohm did not venture into the study of psi on his own, he was known to have either directly or indirectly supported others' speculations on psi within the context of his theory of the 'implicate' order.

The suspected non-local action of consciousness is implicated or implicate (no pun intended on Bohm's unique use of this word) in Bohm's theory. Although Squires' distinction was independent of Bohm's model, Squires came to essentially the same conclusion. In all such considerations of the nonlocal interaction of consciousness and matter, science has been forced to consider the possibility that psi exists. When that possibility is considered, those phenomena and events that have been termed paranormal and thus outside the scope of science must be considered as legitimate.

The Nobel Laureate physicist, Brian Josephson, openly supports the study of psi while Henry Stapp has become embroiled in the debate over the existence of psi as a defender of the necessity of scientists to investigate reported phenomena. In both cases, their views regarding psi stem from their work on the relationship between consciousness and physics.

Josephson and the Greek physicist Fotini Pallikari Viras recently authored a paper on the "Biological utilisation of quantum nonlocality," firmly committing themselves to a physical explanation of psi. They argue that consciousness can act non-locally, while psi has the same characteristics as consciousness when consciousness acts non-locally. So, a quantum mechanical explanation of the non-local action of consciousness would amount to a physical explanation of psi. Bohm's theory of the implicate order explains just such non-local actions and Josephson has indicated that he believes Bohm's concepts could represent the actual case of physical reality. (Josephson, 1993, 2)

Elsewhere, Josephson gives more details on the type of theory that he envisages to account for both life and psi within physics. But life itself is concerned with meaning rather than form. Life emphasizes the qualitative nature of reality and meaning, both of which have been ignored by science for the most part. Life cannot easily be placed within a scientific model even though it has its own "potentialities" which must eventually be included in any physical model of reality.

But the self-consistent and completely logical multiple-description view of knowledge advocated here, an alternative to the conventional view that all knowledge may be reduced to quantum mechanical knowledge, allows life to have its own potentialities, beyond what the constraints of "good scientific method" will allow, for knowing and for acting on the basis of such knowing. Included in these categories of acting and knowing are psychic functioning. (Josephson, 1991, 7 of 8) Life itself corresponds to a quantum mechanical system or ensemble, which could be represented by a non-standard distribution function. This would allow for the action of consciousness non-locally.

The more precise information could be accessed psychically by intelligent life forms giving psi a physical basis. Unfortunately, including living organisms in a valid quantum mechanical model presents great difficulties. (Josephson and Pallikari-Viras, 7 of 8) These difficulties are probably the same as those facing von Lucadou and Kornwachs in their own theoretical work on psi. Finding the proper distribution function (or functions) to describe living organisms will be extremely difficult and finding the valid function for intelligent life exhibiting consciousness will be far more difficult. Despite these difficulties, Josephson and his colleagues believe it is possible and psi will eventually be shown related to the non-local action of consciousness.

Even with the support of physicists with Josephson's stature, the concept of psi is still poorly received in some corners. After four centuries of objective science, claims of the existence of psi still raise emotional, subjective, and often unscientific responses by its critics. These problems are not to be confused with legitimate scientific concerns over the existence of psi. Some scientists rule out the possibility of psi based wholly upon unspecified principles rather than scientific logic. This point has been well illustrated by an incident involving the physicist, Henry Stapp. He recently agreed to run a special series of experiments with Helmut Schmidt. Schmidt hoped to demonstrate to Stapp that anomalous phenomena associated with PK do occur at the quantum level of reality.

Schmidt only approached Stapp because of Stapp's work on the physics of consciousness, once again demonstrating the close relationship between the physics of consciousness and the concept of psi. Stapp's portion of the experiment yielded null results and Stapp published his findings in the *Physical Review*, a mainstream professional journal of physics. However, he proposed a generalization of Steven Weinberg's nonlinear quantum mechanics to demonstrate how "a reported violation of the predictions of orthodox quantum theory" could account for Schmidt's anomalous effects. (Stapp, I) In other words, Stapp offered a model of PK that is completely compatible with modern thinking in quantum theory and thus stepped over the bounds of physics in the mind of a few scientists.

Jonathan P. Dowling, a U.S. Army physicist, in a letter to Physics Today magazine, attacked his paper. Dowling leveled charges of "pseudoscience" and commission of "pathological science" against Stapp. He also stated that "Articles dealing with parapsychology should not be published in the PRA [Physical Review A] - period," (Dowling, 2 of 11) clearly implying that censorship was needed in mainstream physics publications.

Stapp defended his position by stating that a refusal "to look at such physical evidence on ideological grounds would [itself] be pseudoscience." Although he did not fully agree with Schmidt's interpretation of the experimental results, Stapp certainly defended both Schmidt's and his own interpretation of the physical data as scientific until science could prove one theory or another. As Stapp stated, science cannot *a priori* rule out PK or similar anomalous phenomena when plausible physical models can account them for. The fact that the 'Schmidt effect' can be accounted for in quantum mechanics if it is found necessary to do so implies that science must search for and either verify or repudiate the claimed effect.

Stapp's opinion on this matter offers conclusive evidence that physics has finally matured to the point where it can deal with psi scientifically, although not all scientists and scholars share in that level of maturity. So, while Stapp is neither a supporter of psi research nor inclined to accept the existence of psi without further verification, he certainly maintains the position that the existence of psi cannot be automatically precluded from physics. It must be remembered that although he does not believe in psi, Stapp has done extensive work in the physics of consciousness that has surely influenced his opinion on psi.

Other scientists are not as liberal minded as Stapp and the attitude expressed by Dowling is still endemic in a large portion of the scientific community, but as more physicists come to regard consciousness as a real and legitimate subject of inquiry in physics, more physicists have come to the realization that psi is a distinct possibility. This realization harbors a bellwether change in the scientific community's attitude toward psi. When a valid physical theory of psi is finally developed, or even a good contender for that position is proposed, most scientists in the community of physicists will quickly accept the existence of psi.

Even given these three trends and other recent developments in the science of psi, we still cannot answer the question "Is a consensus regarding the reality of psi evolving?" with any precision at this time. However, an alternate question could be considered in its stead: "Is the pre-paradigmatic period of psi drawing to a close with the development of a new paradigm in physics which may be more amenable to psi?" The historical evidence indicates that the answer to this second question is a qualified yes.

This evidence can be found in the association of psi effects with the consciousness/matter interaction, the flowering of contending theories of psi developed by physicists as well as other factors that characterize a preparadigmatic period in science. The pre-paradigmatic period is evident within the context of the three trends noted in this discussion of physical theories of psi, although they do not guarantee a change in paradigm.

These trends only indicate a coming change of scientific attitude toward psi research, which could and most probably will be part of a larger paradigm shift in physics itself. The full content of the new, suspected paradigm would include what is commonly referred to as a TOE, a 'theory of everything.' By its very definition, a TOE must include psi if there is evidence of a psi effect.

If nothing else can be concluded from an analysis of the development of theories of psi, it can at least be discerned that consciousness acting nonlocally is intricately related to the physics of psi, if it is not psi itself, rather than the psychological aspects of psi. So, physical theories of psi which account for the non-local action of consciousness at the quantum level are very likely candidates for a theory of psi within the coming paradigm shift while the interaction of consciousness and matter is at the very core of some TOE theories.

In the meantime, psi seems to be a field effect as opposed to an example of the discrete nature of matter as expressed by traditional quantum theory. Therefore, a field theory of psi also seems to be required in any future theory of psi as well as a TOE. Perhaps even the question of quantum versus field has reached a level indicating that it may soon become irrelevant, both in pure physics as well as paraphysics. The future paradigm in physics will be neither field nor quantum, but both and neither. They will coexist in a new synthesis leading to a new paradigm.

CHAPTER 4
TOEs, fingers, and the nose on your face
Unification and the basic food groups

The greatest long-term trend in all of science can be found in the attempted unification of the laws of nature that comprises the science of physics. In general, science progresses by several methods. Among these are included both the synthesis of pre-existing concepts and the explanation of newly discovered phenomena. While the synthesis of accepted theories and concepts forms the basis of unification, the explanation of newly discovered phenomena also follows the pattern of unification. When they are first discovered, an attempt is made to explain new phenomena by older accepted theories.

However, if they cannot be so explained then new theories are developed so that science can cope with the new phenomena. Later, the new theories are unified with the older theories, either as addendums to the older theories once they are expanded, or they are unified through the development of a still newer theory that incorporates both the old and new concepts in a more comprehensive model of nature. Examples of both synthesizing processes abound in the history of physics and science, so much so that unification seems to be a major, if not the primary task of theoretical physics. The development of thermodynamics offers an excellent example of this synthesis process. Thermodynamics was born in the 1840s when James Joule unified two independent branches of Natural Philosophy, the kinetic theory of matter, which explained heat and Newtonian mechanics.

This unification seemed inevitable since Hermann von Helmholtz and other scientists came to the same conclusions independent of Joule's groundbreaking unification. Joule's unification is highly significant in the history of science because it forced the emergence of physics as an academic discipline separate from Natural Philosophy. On the other hand, the development of the electromagnetic theory, which occurred over more than a century of physical research, involved the discovery of new phenomena that could not be explained within the Newtonian paradigm. Many scientists were involved with the development of the concept of electromagnetism and made significant contributions to the theory. But the two most important figures were Michael Faraday, who laid the experimental foundations of electromagnetism between 1820 and 1850, and James Clerk Maxwell. Maxwell rendered Faraday's experimental research into a mathematical model in the 1860's after appropriate modifications and additions.

What had originally been two separate branches of scientific enquiry, electricity and magnetism, had been unified into a single and comprehensive electromagnetic theory. The history of science for that century long process is virtually littered with new discoveries of electrical phenomena. However, the story of electromagnetism did not end at that moment in history. While electromagnetic theory explained many phenomena dealing with the propagation of light waves and successfully predicted still more phenomena, problems rapidly arose with phenomena that were associated with both Newtonian mechanics and electromagnetism. Primarily, those phenomena that showed evidence of an interaction between the smallest particles of matter and electromagnetic waves defied explanation by either Newtonian mechanics or electromagnetic theory.

These phenomena included the spectral lines of elements and compounds as well as black body radiation. In still another area where these two paradigms came into contact, the necessity of the luminiferous aether for the mechanical propagation of light waves, both theories fell apart as demonstrated by the Michelson-Morley and similar experiments. It was from these and similar failures of the two theories that two new unifications evolved which altered the course of physics, the developments of quantum theory and special relativity at the turn of the last century.

In the opening years of the twentieth century, quantum theory and special relativity were successful in unifying mechanics and electromagnetism at a very high price for classical physics. Fundamental changes in physics became more and more evident as each new theory progressed beyond its original formulation. The quantum theory of 1901 developed into a system of quantum and wave mechanics by 1927 and special relativity expanded into general relativity by 1916. In these forms, each new theory came to represent essentially incompatible aspects of reality. Quantum mechanics relies upon the discrete nature of reality while general relativity portrays the continuous nature of reality as represented by the concept of the field.

After the 1920s, quantum mechanics became the dominant theory in modern physics for several decades. Einstein never fully accepted the Copenhagen Interpretation of quantum mechanics and spent the remaining decades of his life in opposition to mainstream physics while searching for a unified field theory that would unite the electromagnetic and gravitational fields within a single field model. He hoped that the quantum would literally appear as a byproduct of the mathematics modeling his unified field. In this endeavor, very few physicists came to the aid of Einstein.

In the meantime, most physicists accepted the Copenhagen Interpretation of the quantum and sought to unify physics according to their own model of reality. This line of thought culminated in such concepts as the quantum field theory (QFT) and quantum electrodynamics (QED), but these theories were never totally successful in their unification of the quantum and special relativity while gravitation theory has never been incorporated into the quantum model.

The lack of success in uniting even special relativity and quantum mechanics was well recognized by the founders of QFT.

> The ambitious program of explaining all properties of particles and all of their interactions in terms of fields has actually been successful only for three of them: the photons, electrons and positrons. This limited quantum field theory has the special name of quantum electrodynamics. It results from a union of classical electrodynamics and quantum theory, modified to be compatible with the principles of relativity. (Guillemin, 176)

As Guillemin has testified in his history of quantum theory, QED is only "compatible" with the principles of relativity. QED does not provide a framework for the unification of special relativity and the quantum. This same idea has also been confirmed by Julian Schwinger, one of the founders of quantum electrodynamics, who summed up the situation in 1956.

> It seems that we have reached the limits of the quantum theory of measurement, which asserts the possibility of instantaneous observations, without reference to specific agencies. The localization of charge with indefinite precision requires for its realization a coupling with the electromagnetic field that can attain arbitrarily large magnitudes. The resulting appearance of

divergences, and contradictions, serves to deny the basic measurement hypothesis.

We conclude that a convergent theory cannot be formulated consistently within the framework of present space-time concepts. To limit the magnitude of interactions while retaining the customary coordinate description is contradictory, since no mechanism is provided for precisely localized measurements. (Schwinger, xvii)

Schwinger clearly acknowledged in this statement that QED, the primary form of a quantum field theory, has reached a specific limit whereby it cannot be judged without reference to an outside framework of space-time. The prevalent framework of space-time, also referred to as the "customary coordinate description," at this juncture of history is that supplied by the theories of relativity, so Schwinger obviously believed that QED had not yet been unified with special relativity. Special relativity just forms a limiting condition for the mathematical model of quantum field theory that does not indicate that they have been unified in a single theory.

It would further seem that the real unification to which scientists subscribe is between quantum mechanics and GR, since both describe the motion of matter in space-time while the unification with GR would certainly include an implied unification with special relativity. New scientific advances in the 1960s and thereafter have brought GR to the forefront of physical research even as old philosophical problems which have plagued the Copenhagen Interpretation of quantum mechanics were revealing more cracks in the prevalent quantum paradigm. During the last few decades, these later developments have produced a climate of change within theoretical physics, which has resulted in a renewal of Einstein's search for a unified field theory.

However, nearly all of the modern attempts at unification follow first from quantum theory rather than beginning from field theory as represented in special relativity or general relativity. Most scientists treat relativity as something to add onto or incorporate into for unification model have been proposed with such colorful names as supergravity, Grand Unification, superstrings and finally the 'Theory of Everything' (TOE). Even accepting the possibility that a TOE might exist marks a drastic change of attitude within the scientific community. But the problem of unifying the discrete and continuous aspects of the physical world has never been resolved despite attempts to do so from both the quantum and field approaches to the quantum perspective.

Under these circumstances, more recent attempts to base a future theory of physics on Einstein's hoped nature. This dichotomy is represented indirectly within modern physics by such concepts as the wave/particle duality of both matter and light. Although the philosophical problems presented by the differences between the discrete and continuous are not universally recognized in the physics community, a few physicists have been braved enough to question the established norms of modern physics in this regard.

In his book *A Unified Grand Tour of Theoretical Physics*, Ian D. Lawrie has confirmed the uneasiness felt by physicists although he has not clearly defined the cause of his concerns beyond stating the modern physicists "do not properly understand what it is that quantum theory tells us about the nature of the physical world" even though "there are respectable scientists who write with confidence on the subject." Evidently, "the conceptual basis of the theory is still somewhat obscure." (Lawrie, 95)

Mendel Sachs is far more straightforward with his criticisms. Sachs has noted two distinct and separate strains of scientific progress within modern physics.

> The compelling point about the simultaneous occurrence of these two revolutions (relativity and the quantum) is that when their axiomatic bases are examined together, as the basis of a more general theory that could encompass explanations of phenomena that require conditions imposed by both theories of matter (such as current „high energy physics"), it is found that the widened basis, which is called „relativistic quantum field theory", is indeed logically inconsistent because there appear, under a single umbrella, assertions that logically exclude each other. (Sachs, 1988, 236-237)

Sachs is, of course, referring to the logical and mutually exclusive nature of the quantum (the discrete) and the field (the continuous). He does little to hide either this fact or his criticism of the shortcomings of present-day physics. Sachs has concluded that "neither the quantum theory nor the theory of relativity is in themselves complete as fundamental theories of matter," (Sachs, 256) since they represent incompatible fundamental concepts of the discrete and continuous aspects of nature.

These philosophical problems have physical counterparts within the mathematical model of the singularity. GR falls apart at just the point where the continuous field of gravity meets the physical boundaries of the discrete particles whose curvature creates gravity, where the curvature of the spacetime metric becomes so extreme that it becomes infinite. This singularity not only occurs at the heart of elementary particles, but also in black holes and the Big Bang, which theoretically created our universe. On the other hand, quantum mechanics deals with singularities in a different manner, although no more successfully than GR deals with them. Quantum mechanics utilizes a rather artificial method known as renormalization to deal with singularities, or divergences as they are called, and then ignores the problems created by the divergences.

In QED, each particle is associated with a field, so there are as many fields as there are different particles. This situation gives rise to an unpleasant expansion of the concept of field which some have criticized. (Popper, 194) Yet far more serious problems exist in QED. At the point where the different fields interact, one would expect to find the action and reaction of the particles as caused by forces in the classical sense of the term.

However, mathematical divergences that render the masses of elementary particles infinite and undefined exist at the precise point where the fields interact. These divergences or infinities can be renormalized to yield definite answers by applying mathematical perturbation methods. Therefore, localizing and defining a point particle in QED amounts to using an artificial mathematical method for no other physically valid reason than that the method yields finite results that can be experimentally verified.

While many scientists do not see this procedure as a problem since its predictive power makes QED one of the most successful theories ever developed in science, the artificial nature of renormalization is at the very least philosophically unsatisfying and unsettling to other scientists and scholars. The method is considered at least *ad hoc*, but otherwise a necessary evil at present. Karl Popper was very critical of this shortcoming of QED.

> Moreover, the situation is unsatisfactory even within electrodynamics, is spite of its predictive successes. For the theory, as it stands, is not a deductive system. It is, rather, something between a deductive system and a collection of

Although this criticism was leveled several decades ago, shortly after QED was first developed, it is still a valid criticism. QED is considered theoretically inconsistent because of these and other problems despite its great successes. It is only a theory of electron interactions and does not unify electromagnetism with any other of the basic forces in nature.

During the 1960s, Quantum Chromodynamics (QCD), which is a theory of nuclear interactions, followed QED. QCD unifies the Yang-Mills field, which describes the nuclear forces of binding between neutrons and protons in the atomic nucleus, with the Standard Model of quarks. QCD does not share the inconsistencies that plagued QED but does require renormalization methods like those in QED.

Renormalization methods therefore remain a general characteristic of the quantum field theories and QCD suffers from some of the same criticisms that plague QED. It is hoped by quantum theorists that newer methods and developments in quantum field theory will eventually justify or replace renormalization, rendering quantum field theory more palatable to its critics.

Yet these problems persist. A large part of the work in quantum field theory still involves finding proper renormalization procedures that yield workable solutions.

If a mathematical method of renormalization works, it is adopted as long as its results are experimentally confirmed, even if there is no physical reason for its success. This is not the best situation for physics, but physicists have accepted the method in principle and moved on to other theories. The infinite masses of elementary particles that resulted in QED and QCD before renormalization can be effectively compared to the singularity problems of GR. However, such a comparison has not been common within either physics or the philosophy of science. In spite ofDespite the lack of recognition of this 'coincidence,' physics has continued to progress toward unification.

Seemingly, the fact that two fundamentally different approaches to physical reality, the continuous field and the discrete quantum, lead to the same inconsistency would be an important clue to identifying the problem of their unification. Since this clue has gone unnoticed, unification has proceeded along other lines. Both quantum physicists and relativity physicists dance around the problem of the singularity/divergence even though this problem should represent the object of the main thrust toward unifying the two perspectives of physical reality.

Unification using a fifth dimension

The first recognized attempt to use a fifth dimension for the unification of gravitation and electromagnetic fields was made by Theodor Kaluza. Shortly after Einstein's development of GR, a few scientists expressed their dissatisfaction with the artificial nature by which electromagnetic forces were imported into the field equations. They thought that both gravitation and electromagnetism should arise from a single geometric structure, the „"unitary" or unified field, but GR did not fully represent that field. In GR, the structure of the space-time continuum is modeled by a tensor equation where R_{ij} is the contracted Christoffel tensor or the Ricci tensor, g_{ij} is the metric tensor, R is the curvature scalar, k is the gravitational constant and T_{ij} is the stress-energy (or matter) tensor.

$$R_{ij} - \frac{1}{2} g_{ij} R = -kT_{ij}$$

This equation accounts for gravitational attractions between material bodies but does not explicitly include the electromagnetic field.

Many scientists had long maintained that matter itself is electrical in nature and electrical forces create the structure space-time, a view that dates back more than a century. So, the conjecture that matter is no more than curved space-time is unsatisfying to those scientists. They assumed that electromagnetism must play as important a role in the structure of space-time as gravitation.

However, the best that could be said on this subject within the context of GR was that electromagnetism could be added to the defining equation of the space-time structure by the inclusion of an electromagnetic term, E_{ij}, such that:

$$R_{ij} - \frac{1}{2} g_{ij} R = -k[T_{ij} + E_{ij}].$$

When approached in this manner, the addition of the electromagnetic term seems rather artificial. Nor does it lead to simple solutions for charged particles in combined electromagnetic and gravitational fields although an uncharged particle can be accounted for in the combined field.

The first attempt to unify these fields was made by Hermann Weyl in 1918. He sought to keep the four-dimensional space-time continuum intact while developing a more comprehensive geometry to deal with space-time. On the other hand, Kaluza suggested that a fifth dimension could be added to the space-time continuum to account for both fields simultaneously. The earliest mention of his work came in a letter to Einstein in 1919. (Raman, 212) Einstein encouraged Kaluza to continue working on his theory and further develop his physical model, (Middleton, 2) but nothing was made public regarding Kaluza's theory until 1921 when his paper "Zum Unitätsproblem der Physik" was finally published. The published theory was simple and straightforward.

In general, when a Riemannian rank two tensor is four-dimensional it can be characterized by ten components. T_{ij} is such a tensor and represents the metric structure of our space-time continuum. The ten components of T_{ij} describe the motion of material bodies within this metric. The addition of electromagnetism to this metric field structure would require four more components than exist for the four-dimensional configuration, for a total of fourteen components. So, the four-dimensional configuration cannot include electromagnetism. However, the Riemannian structure of our space-time can still be saved, and an adequate number of independent components found by increasing the number of dimensions to five. This dimensional increase yields fifteen independent components, one more than is needed to describe the combined field.

Kaluza was the first to see this as a possible answer to the problem of developing a unified field structure. Both electromagnetism and gravitation could exist on an equal footing within the geometric structure of a four-dimensional space-time continuum embedded in a fifth dimension as specified by Kaluza. At the time of his initial development of the five-dimensional model, quantum mechanics had not yet been developed and the whole issue of the direction which quantum theory would progress was in question, so there was not any need to incorporate the quantum into the theory at that time. Nor was there any hint that there might be a fundamental problem between the quantum and relativistic views of physical reality. The only purpose of Kaluza's theory was to unify gravity and electromagnetism within a common field because he saw such unification as the primary problem in physics.

Kaluza's model yielded a geometrical representation of the generally covariant form of Maxwell's electrodynamics. By introducing the fifth dimension, he immediately raised two problems: (1) Since only fourteen variables are necessary for the combined field, the fifteenth term must be omitted or its affect nullified, and (2) Since every indication implies that our world is only four-dimensional, a five-dimensional assumption would necessitate an explanation of the absence of evidence that the fifth dimension exists.

These two problems are not characteristic of Kaluza's theory alone but form the central points of contention for any theory that utilizes a hyperdimensional structure in a physical description of the world. These same two problems also define the major lines along which Kaluza's theory has been extended by some scientists and criticized by still others. Kaluza added no physical significance (Bergmann, 1976, 254; Einstein and Bergmann, 683) to his five-dimensional hypothesis, but merely used it as a tool. In so doing, he had a great deal of latitude in overcoming both problems and was able to deal with them by assuming that the field variables, γ_{00} were independent of the fifth coordinate. They need only depend on the four coordinates of the spacetime continuum when a suitable coordinate system was chosen.

This choice also had the consequence of allowing a somewhat "atrophied" (Hoffman, 403) fifth dimension and thus a somewhat less generalized theory. Kaluza's mathematical model of the fifth dimension forced cylindricity on extensions in the fifth direction by guaranteeing that a vector in the fifth dimension, A^{μ}, satisfies the Killing equation. By further requiring that "the lines to which the A^{μ} are tangents - the 'A-lines' - have to be geodesics," (Bergmann, 1976, 258-259; Einstein and Bergmann, 686) the various A-lines in the fifth dimension were shown to be equal as well as constant. The norm of A was also constant throughout all of space and not only along the A-lines. Points in our four-dimensional spacetime were extended into the fifth dimension along the A-lines.

Under these conditions, Kaluza based his five-dimensional structure on a special coordinate system defined by the metric,

$$d\sigma^2 = \gamma_{\mu\nu}dx^{\mu}dx^{\nu}$$

where μ and ν range from zero to four and thus represent all five dimensions. This metric structure represented a five-dimensional manifold in which a four-dimensional continuum cut each of the A-lines that extended into the fifth dimension only once.

The distance along any A-line could then be used to derive a value of $\gamma_{00} = +1$, where γ_{00} is the field variable in the fifth direction, thus normalizing all other components of the field. This structure, based on the cylindrical condition and referred to as an A-cylindricity, had the dual effect of guaranteeing that there would be no physical evidence of the fifth dimension while reducing the number of variables from fifteen to fourteen, solving both problems incurred by the addition of another dimension.

There are no intuitive guidelines or experiences on which to base this model of space-time, so other guidelines must be considered to derive a model of the fifth dimension. The cylindrical condition allows the four dimensions of space-time to be independent of the fifth dimension. After a manner of speaking, the cylindrical condition thus explains why there is no physical evidence of a fifth dimension. All observables associated with physical phenomena are four-dimensional and thus independent of the fifth dimension. The cylindrical condition also limits the kind of coordinate transformations possible, allowing only those which lead to covariant field equations (Tonnelat, 1966a, 7) since the fifth coordinate must play a spatial role.

The special role played by the fifth coordinate is evident under the cut-transformation, whereby anti-symmetrical derivatives of the A-curve vanish while the anti-symmetrical derivatives of Aμ remain allowing a correlation with the magnetic field. The cylindrical condition is necessary to the successful unification of the electromagnetic and gravitational fields in Kaluza's theory. In fact, Kaluza discovered two transformations that would leave the equations invariant and thus preserve the unique character of the system. Invariance under transformation is a necessary property for both the gravitational and electromagnetic fields. The two transformations that leave the components invariant are the "four-transformation," over four dimensions, and the "cut-transformation."

The field variables on which these transformations act can be grouped into three general classifications corresponding to the results of the transformation process; the γ_{mn}, γ_{0m} and γ_{00}, where m and n vary from 1 to 4 representing our four-dimensional spacetime. The γ_{mn} correspond to the sixteen components of a matrix representing the four-dimensional spacetime continuum as in GR. They reduce to ten independent components that describe gravitation. Under the four-transformation, the γ_{mn} act as a four-tensor and under the cut-transformation they are invariant.

The γ_{0m} correspond to the eight components (or four + four) in a five-by-five matrix that represents the mixed terms of normal spacetime and the fifth dimension.

$$\begin{pmatrix} \gamma_{mn} & \gamma_{m0} \\ \gamma_{0m} & \gamma_{00} \end{pmatrix} \quad \text{or} \quad \begin{pmatrix} \text{GR} & \text{EM} \\ \text{EM} & +1 \end{pmatrix}$$

Under four-transformation, the γ_{0m} act as a four-vector, while under the cut-transformation they vary by an additive term. The variation due to the additive term allowed the introduction of the electromagnetic four-vector into the new space-time structure. The γ_{0m} were equated to the electromagnetic potentials ϕ_m since "this corresponds to the fact that the electromagnetic potentials are defined only up to additive terms which are gradients of an arbitrary function." (Einstein and Bergmann, 687)

The final term, $\gamma_{00,}$ is purely fifth dimensional. It was set equal to +1 in Kaluza's original theory. This term is invariant and constant under both transformations and thus proved to be effectively removed from the spacetime structure, as we perceive it. In this way, the dependence of the field structure on electromagnetism as well as gravitation was reflected in the metric, which defines space-time. Electromagnetism was wholly incorporated into the new field structure while all the components of the new unified field were accounted for, completing the correlation to the electromagnetic field. Kaluza's theory was simple, elegant, reproduced both the electromagnetic and gravitational fields from within a unified metric structure and yielded the geodesic equations for both charged and uncharged particles within the combined fields, but his findings were not without controversy.

Within the context of what he was trying to accomplish, the unification of electromagnetism and gravitation into a single hyper-field structure, Kaluza was moderately successful. However, within the context of a mounting tide of criticisms, the success of the quantum theory and the discovery of new forces in nature which had not been foreseen in the early 1920s, the success of Kaluza's theory was short-lived and generally overlooked for almost fifty years.

Since Kaluza placed no physical significance in his fifth dimension, using the concept only as a mathematical tool with which to derive his goal of unification, he opened his theory to excess criticism. It was severely criticized by some for going too far just by introducing the fifth dimension and by others for not going far enough and adding some physical significance to the fifth coordinate once it was used. Other criticisms were leveled concerning the variable representing the purely five-dimensional characteristics of space-time, γ_{00}. Setting this variable equal to +1 seemed either unnecessary or unwarranted. Both A-cylindricity and the correlations drawn by Kaluza between the field constants in his model and the known field constants of electromagnetism also seemed artificial in some respects.

Thus, his theory seemed rather *ad hoc*. (Graves, 257) Kaluza's theory was further criticized for not making any predictions that would allow it to be tested. It merely reproduced electromagnetism without expanding or adding to the already existing Einstein-Maxwell equations. Yet all these criticisms are indirectly concerned with the question of the reality of the fifth dimension. So, the only real criticism of his theory can be posed in the simple question: Why add a fifth dimension when all physical evidence implies a four-dimensional space-time?

After two decades of pursuing his own extension to Kaluza's theory, Einstein's final comment to this question was given in the second appendix to the fourth edition of *The Meaning of Relativity*. He stated that any hyperdimensional theory could only be considered a valid theoretical option when it could be shown why all empirical data leads to a strictly four-dimensional world. (Einstein, 1956, 166) In other words, in the absence of any observational or experimental evidence of the existence of a fifth dimension, scientists could only accept the hypothesis if there were an overwhelming reason to do so. Any scientist developing a five or higher-dimensional theory must not only contend with this problem, but also justify the basic assumption of the higher dimension by demonstrating that this hypothesis and only this hypothesis can account for observed natural phenomena.

The cylindrical condition is an extremely important component in Kaluza's theoretical model. In subsequent extensions of his theory, at least in those developed prior to the 1960s, the imposition of the cylindrical condition was an important point of criticism and constituted a major weakness in these theories. Yet the special status, or perhaps the peculiarity of the fifth component of the field, is revealed in this condition.

For this reason, the condition was interpreted by some as being too restrictive or merely an 'additional' condition which was neither necessary nor justified. (Tonnelat, 1966a, 8)

It was therefore thought possible that a condition less stringent than the cylindrical condition could be used to obtain the same results for the fourteen equations describing a combined field, while leaving the fifteenth equation intact to describe other field phenomena. So, modifying the cylindrical condition was the first method of choice for extending Kaluza's theory during the early years. This approach would render the five-dimensional theory more general in its application to the physical world.

In the projective theories, such as the theory developed by Einstein and W. Mayer in 1931, the cylindrical condition was interpreted quite naturally as a projective condition that demonstrated the purely auxiliary role of the five-dimensional space. (Tonnelat, 1966a, 7) The cylindrical condition was also thought to lead to a mere codification within a five-dimensional formalism, such that it was a mathematical convenience rather than a physical characteristic of space. In that case, it was assumed that the five-dimensional space is real or that there is a true five-dimensional geometry that can describe space-time, rather than a geometrical (mathematical only) formalism representing space-time. In this instance, the cylindrical condition could be modified or dropped altogether.

Einstein, Peter G. Bergmann, and V. Bargmann in 1941 as well as J. Podolanski in 1949 took this approach to the problem. In their theories, the extra dimension, or dimensions in the case of Podolanski's theory, was considered to be real, but of special structure. Instead of a cylindrical condition, the theory of Einstein, Bergmann and Bargmann used a fifth dimension which was closed with respect to the four dimensions of normal space-time.

Podolanski took another path and solved the problem by assuming a "laminated structure" such that all the points in each layer correspond to a given point in the four-dimensional space-time (Tonnelat, 1966b, 403) continuum. Under such conditions, these theories were able to answer those criticisms that attacked the fifth-dimension dependence on the cylindrical condition.

On the other hand, Kaluza's theory has become popular and gained a new respectability within the scientific community since the 1970s. The cylindrical condition has now become an important factor in the theory's newest incarnation rather than a point of criticism. In the latest extensions to Kaluza's model, the extent of the cylindricity is uniquely small limiting the fifth dimension to the domain of the quantum world. So, the fifth 'contracted' or 'compacted' dimension is not perceptible in the common world of the four-dimensional space-time.

It is also beyond the experimental capabilities of science at this point in history, so it is not even possible to detect the fifth component of spacetime in any manner at present. Oskar Klein first developed this extension of Kaluza's theory in 1926 and is today a part of the much grander model of physical reality known as the 'superstring' theory. In his original theory, Klein altered the cylindrical condition to accommodate the new developments quantum mechanics. In his original theory, Klein altered the cylindrical condition in order to accommodate the new developments quantum mechanics.

Klein's interpretation of the Kaluza theory

Klein has been credited with both the formalization of Kaluza's theory and several attempts to extend the theory into the domain of the quantum. His name has been so closely associated with Kaluza's theory that some scientists and authors have given Klein partial credit for the original theory and refer to it as the Kaluza-Klein theory. This synthesis is so complete that some authors have given Kaluza credit for innovations that Klein made to the theory and vice versa. Kaluza's theory seems to have lain nearly dormant until Klein's first exposition of it appeared, so it is generally thought that Klein was instrumental in popularizing the theory. This lack of recognition is even stranger given the great interest in measurable hyperspaces before the Second Scientific Revolution. It could only signal the strength and intensity of the positivistic backlash that brought about the Revolution.

This interpretation of the historical events seems even more probable since most physicists were dealing with developments in quantum theory during the early to middle 1920s and would have ignored Kaluza's modification of GR since it offered nothing new to their interests in the microscopic domain of the quantum. However, Klein's alterations and extensions of Kaluza's work into the quantum domain rendered the theory more relevant to events occurring in the rest of the world of physics.

To be sure, Kaluza's theory would have seemed inconsequential against the onslaught of quantum mechanics in the early 1920s. Most physicists would have ignored it until Klein related it to the newly forming concepts of quantum and wave mechanics whose interpretations within the larger framework of science had not yet been fully developed. Klein saw within the five-dimensional hypothesis a vehicle for introducing the quantum into the space-time continuum rather naturally, as well as a way to account for the atomicity of electric charge. He first equated the geodesic in the fifth dimension to the periodicity of the electric potential N. This implied a quantum of action, while a conjugate momentum in the fifth dimension was fixed to account for the positive and negative electrical charges.

By forming the five-dimensional Lagrangian of a particle in a combined electromagnetic and gravitational field, and then differentiating it with respect to the velocity along the fifth component, he established a relationship within the field yielding the charge-to-mass ratio of the electron. This allowed the conjugate of the fifth coordinate to appear in a manner analogous to the way that matter, and momentum were conjugates in our normal four-dimensional space-time.

The periodicity that he introduced into the fifth dimension also allowed Klein to make an association between a function in the fifth dimension and Schrödinger's wave function Ψ. Klein further derived a fundamental length of $l_o = (2k)^{1/2}(hc/e)$, where k is Einstein's gravitational constant, and h, c and e are Planck's constant, the velocity of light and the electron's charge. This configuration gave Klein's fundamental length a value of 0. x 10^{-30} centimeters. (Klein, 1926, 516) He later proposed "to relate the fifteenth quantity γ_{00} with the wave function Ψ, which characterizes matter, in order to achieve a formal unity between matter and field" (Klein in Mehra, 53) and thereby further cements the relationship between Schrödinger's wave mechanics and the five-dimensional framework.

Klein later admitted that his first theoretical attempts were not satisfactory and for the next decade he published nothing more dealing with them. (Klein in Mehra, 80) In 1939, he developed new extensions to his earlier theories of a grand unification between quantum and field theories by incorporating the newly discovered 'mesotonic' forces of Yukawa into his theory. Klein's 'mesotonic' forces are better known today as the strong nuclear force. Klein reasoned that the mathematical treatment of the Yukawa potential was analogous to his earlier mathematical treatment of the five-dimensional framework. He wrote that the "direct and general way it expresses the fundamental conservation and invariance theorems seems to make this representation a natural starting point for a general quantum theory comprising also the charged fields, which are supposed to correspond to the mesotons." (Klein, 1939, 79)

His newly extended theory included the construction of a new Lagrangian containing, in addition to the gravitational and electromagnetic components, the spinor as a tensor field component. By using a variational principle, actions between protons, neutrons, and electrons, explained by the interactions of neutrinos such as theorized in Yukawa's theory of nuclear forces, were found.

In 1947, the theory was further extended when Klein developed field equations for free mesons and derived the wave equation for nucleons. To achieve this end, he had replaced the assumption that the field quantities were independent of the fifth coordinate by an assumption that they were periodic functions of a length in the fifth direction with a period of l0. This development introduced the indeterminacy "which would exclude the use of the fifth dimension in any geometrical sense and had the practical meaning that particles of given charges have naturally coherent wave functions, as is always assumed." (Klein, 1947, 3) A new fundamental length was also introduced which was equal to the product of his older fundamental length, l_0, and a constant equal to e^2/hc.

Klein quickly became dissatisfied with his 1947 theory. He thought that his theory had "such features that it should hardly be taken literally." (Klein, 1956, 59) So he made one last attempt to include nuclear forces in his five-dimensional framework by deriving, via the same periodicity function, "a theory of more physical aspect, whereby charge invariance appears as a part of a natural generalization of gauge invariance." (Klein, 1956, 59) As a further consequence of his concept of a fundamental length, Klein calculated that a particle, approximating a quantum in a linear wave equation and with a wavelength approaching zero, would have a gravitational self-energy approaching the kinetic energy corresponding to its volume. In this manner, he hoped to do away with the remaining divergences of the electron theory.

This more stringent generalization of the theory from the quantum theory point of view implied possible states of matter with a multiple charge. When all is considered, Klein's theoretical work could best be characterized as a continuing attempt to save the basic tenets of Kaluza's space-time framework while keeping pace with the advances being made in atomic and subatomic physics. Portions of Klein's work live on in today's most advanced theoretical research where the Kaluza-Klein theory is the foundation upon which the theory of superstrings has been developed.

The super theories

During the 1960s, Steven Weinberg and Abdus Salam used weak gauge symmetry to account for the masses of W and Z particles through a spontaneous breaking of gauge symmetry. This method allowed the unification of the electromagnetic and weak forces without depending upon the same type of renormalization process that was necessary in QED to prevent infinite masses.

> These encouraging successes have led to the belief that the weak and electromagnetic forces are really two aspects of a unified electroweak force. However, ... perhaps amalgamated is a better word than unified. The crucial element in this success was the formulation of the theory in terms of gauge symmetries, and this has encouraged the theoretical examination of a variety of other gauge theories for the description of the strong and gravitational forces, and their eventual unification with the electroweak force. (Davies and Brown, 56)

With the success of this 'electroweak' theory, emphasis in the theoretical research of quantum field theories changed from inventing renormalization methods that had no physical basis but gave finite and interpretable solutions to applying the correct gauge symmetries.

Unlike renormalization in QED, which has no physical counterpart, symmetries are a common characteristic of physical bodies and systems so renormalization in the Weinberg-Salam model became physically acceptable. Capitalizing on this success, the next wave of unification theories was based upon the various symmetries inherent in nature. Grand unification theories (GUTs) and supergravity were both developments of the 1970s. GUTs were attempts to unify the electroweak theory with QCD, thus unifying electromagnetism, the weak and strong forces, by embedding the gauge symmetries of each of the individual theories within a larger all-embracing gauge group.

Unfortunately, the GUTs that were developed predicted the existence of magnetic monopoles and an extremely large but finite half-life for protons. In the ensuing years, neither of these predictions has been verified by experiment or observation, so the GUTs have been seriously hampered. The GUTs did not include the force of gravity and GR within their framework, so they were not TOEs in today's sense of the name, but the concept of supersymmetries was used during the same timeframe to include gravity within the GUT framework. This class of theories is known as supergravity.

In the older quantum field theories, gravitational forces were always mediated by particles called gravitons. The graviton has never been observed in nature. The new supergravity theories predicted that not only the graviton acted as the conveyor of gravity, but a new particle called the gravitino should also exist in that role. Like gravitons, the gravitino was very weakly interactive with matter and would therefore be very difficult to detect in nature. Despite this shortcoming, supergravity did introduce a new (or perhaps old and forgotten) concept into the search for a TOE.

The geometrical structure of space-time could be greatly simplified if the unified force of supergravity was recast within an eleven-dimensional framework. This discovery gave a new impetus to the search for hyperspatial theories of unification and physicists rediscovered the Kaluza-Klein theory in the early 1980s. When the theory of supergravity was rewritten as an eleven-dimensional Kaluza-Klein theory, all the forces of nature were reduced to nothing more than different forms or adjuncts of a single gravitational field.

In the older quantum field theories, gravitational forces were always mediated by particles called gravitons. The graviton has never been observed in nature. The new supergravity theories predicted that not only the graviton acted as the conveyor of gravity, but a new particle called the gravitino should also exist in that role. Like gravitons, the gravitino was very weakly interactive with matter and would therefore be very difficult to detect in nature. Despite this shortcoming, supergravity did introduce a new (or perhaps old and forgotten) concept into the search for a TOE.

The geometrical structure of space-time could be greatly simplified if the unified force of supergravity was recast within an eleven-dimensional framework. This discovery gave a new impetus to the search for hyperspatial theories of unification and physicists rediscovered the Kaluza-

Klein theory in the early 1980s. When the theory of supergravity was rewritten as an eleven-dimensional Kaluza-Klein theory, all the forces of nature were reduced to nothing more than different forms or adjuncts of a single gravitational field.

Since the supergravity theory is an extended Kaluza-Klein theory, it can be represented as a matrix with the first four indices (variables or dimensions) representing the space-time of GR in the upper left-hand corner, followed by electromagnetism for the fifth index and then the Yang -field and the quark-lepton view of matter.

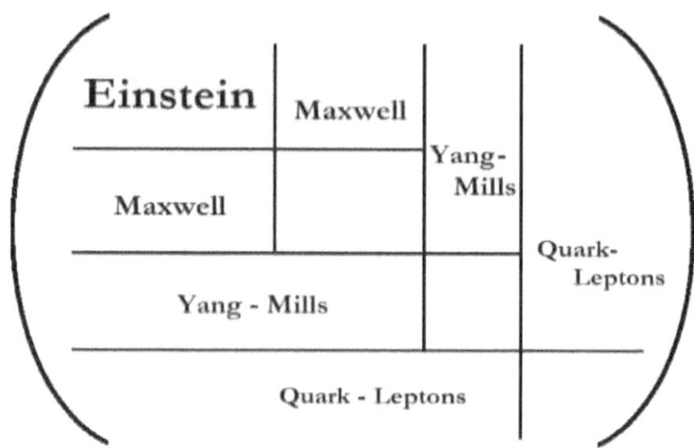

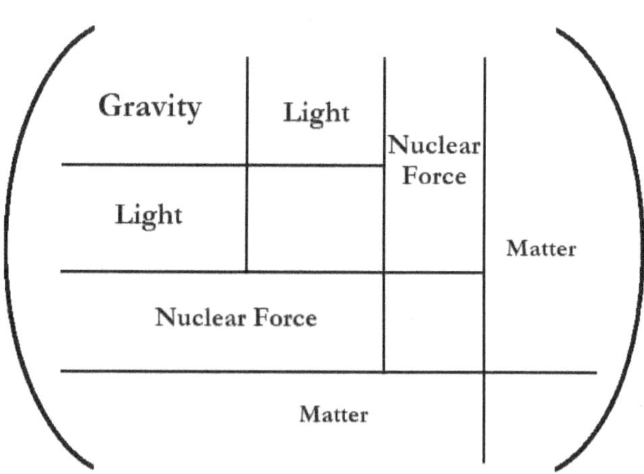

(Kaku, 146-147)

The extra dimensions had no physical meaning within the original supergravity theory. However, in the Kaluza-Klein modification they were interpreted as real physical dimensions that were rolled up in such minute proportions that they were effectively unobserved as well as unobservable in nature.

The extra dimensions were associated with various abstract gauge symmetries that were independent of the minuscule size of the extra dimensions. Unfortunately, the supergravity theories suffered from a rather crucial and perhaps fatal flaw. The weak force violates a special type of left-right mirror symmetry, referred to as violating parity. This property is called chirality and can be shown to exist only in odd-dimensioned spaces. This property therefore requires any unified field theory that includes the weak force to use a framework with an odd number of spatial dimensions plus one time dimension. Such a configuration would yield a total number of dimensions that is even. The eleven-dimensional space-time continuum of supergravity is odd, so it clearly does not fulfill this requirement.

Meanwhile, lurking in the shadows of theoretical physics was an answer to this latest predicament. Even before the advent of the supergravity theories, the concept of strings had been introduced into the physics of quantum fields. The quantized motion of a vibrating string was first used by Gabrielle Veneziano to model hadrons. Then John Schwarz and André Neveu discovered a second group of strings for modeling fermions. When QCD theory was introduced, the string model was all but abandoned by its advocates. Yet Schwarz and Joel Scherk continued to develop the string model. While strings did not seem to correspond to any of the known elementary particles found in nature, they did have properties like gravitons, which suggested that they might be ideal for a TOE.

No other quantum field theory had been able to account for gravitons and the gravitational field. Still, GUTs and supergravity overshadowed string theory and few in the scientific community paid it any attention. But string theory did profit from the prior development of these theories since they popularized and legitimized the use of hyper-dimensional models in physics and brought supersymmetries to the attention of physicists. Scientists no longer ignored theories, which assumed space-times of dimensions greater than four, and the symmetries seemed to justify their application.

Supersymmetry is deeper and more powerful than the normal symmetries of space and time. Its most endearing feature is that "it provides a geometrical framework within which fermions and bosons receive a common description." (Davies and Brown, 44) However, this supersymmetry requires the addition of five more dimensions of space to Kaluza's five-dimensional spacetime. The concept is not without problems since there is no "unequivocal confirmation in nature" of such supersymmetries. (Davies and Brown, 47) Yet the application of supersymmetry yields startling results. Simple string theory utilizes the supersymmetry to unify the four forces of nature within a single common geometrical framework and is thus called superstring theory.

In essence, the supersymmetry allows matter and radiation to be combined. Within this framework hyperdimensional strings can represent all elementary particles. Schwarz, Michael Green, and Edward Witten developed the superstring theory in the early 1980s. Superstring theory postulates a space-time continuum of either ten or twenty-six dimensions. Only these configurations give reproducible and understandable results. These extra dimensions are every bit as real as the normal three dimensions of space, but we cannot detect or otherwise perceive the higher dimensions because they are curled up or contracted to Planck length sizes, about 10^{-33} centimeters.

> These fundamental strings possess a tension that varies with the environment in which they reside, and this tension becomes large enough to shrink the loops of string to approximate points at the low energies we witness in the universe today. ... The enticing aspect of the string theories has been the unexpected discovery that the requirement of finiteness and consistency alone should prove to be so constraining. (Barrow, 31-32)

This restriction guarantees that the strings are of such minute size that they are virtually impossible to detect. Such small sizes can only be reached experimentally by energies far greater than any that science can even dream of at this time, let alone reach in high energy physics laboratories.

With the direct detection of strings so far out of the experimentalist's grasp, their existence cannot be directly confirmed. So, belief in the validity of the theory must depend on its 'beauty', simplicity, and logical structure. It is hoped that the mathematics will eventually yield predictions that will confirm the theory at much lower, and thus attainable, energies. However, even this path presents a problem since mathematics is not yet advanced enough to solve the problems associated with the superstring theory. Only approximate solutions to the superstring model exist. Many scientists consider superstrings a theory of the next century that happened to fall into this era accidentally. (Witten, 102; Kaku, 160)

Quantum field theory assumes that particles are nonextended mathematical points in space. This practice led to the infinite masses and meaningless divergent expressions that required renormalization. In this practice, quantum field theory is not alone. In both quantum mechanics and classical mechanics, in fact in all other branches of physics, particles are portrayed as points without internal structure. Science has generally shied away from questions concerning the interior portion of elementary particles. The mathematical point assumption is no longer necessary in the superstring model. Superstring theory replaces these points with one-dimensional curves called strings. One advantage of discarding points for strings is the disappearance of divergences, but there are other advantages such as the explanation of anomalies.

Speaking from his own experience, Schwarz explains that he and his colleagues were surprised at this result in superstring theory. When "the quantum corrections to gravity for string theory" were made, he and his colleagues began "to get numbers that did make sense, numbers that were given by finite expressions." (Schwarz, 75) Simply put, the theory overcomes many of the major problems inherent in earlier quantum field theories such as supergravity and GUTs. In superstring theory, fermions are particles of matter and bosons are the particles that interact between material particles as the forces of nature.

In the case of gravity, superstring theories succeed where all the quantum field theories have failed, allowing a unification of gravity with the other forces within in a single field theory. In fact, gravity is of fundamental importance to the superstring theory.

> The most remarkable feature of string theory, as we have emphasized, is that Einstein's theory of gravity is automatically contained in it. In fact, the graviton (the quantum of gravity) emerges as the smallest vibration of the closed string. While GUTs strenuously avoided any mention of Einstein's theory of gravity, the superstring theories demand that Einstein's theory be included. (Kaku, 157)

Witten's characterization of the role of gravity in superstring theory is still stronger. The theory is especially attractive to scientists because gravity is "forced upon them" by the theory itself rather than being incorporated from outside the theoretical framework. "All known consistent string theories include gravity, so while gravity is impossible in quantum field theory as we know it, it's obligatory in string theory." (Witten, 95)

The superstring framework also has intuitive appeal, which is something that quantum field theories lack. The analogy with ordinary vibrating strings is quite a strong tool for physicists. With this analogy, they have been able to develop a picture of what occurs on the sub-quantum level of physical reality. Particles are little loops of string, like 'lassos,' which oscillate about as they move through space.

As time moves forward, these loops describe "something which is rather like a tube going through space and is called a "world-sheet." That is the trajectory of a particle according to the superstring idea." (Ellis, 152) The "world-sheet"' of superstrings corresponds to the "world-line" in the spacetime framework of special relativity, which further cements the relationship between relativity theory and superstrings. While following their more classical trajectory through space and time, the string is also vibrating (non-classically) in the higher dimensions of space.

The modes of a string's vibration determine the special characteristics of the particle as well as its type and class. According to Schwarz:

> When you have a string it can oscillate and vibrate in different ways - rotate and so forth - and each of these different modes of vibration or oscillation can be thought of as describing a particular type of particle. So one can think of the electron as one mode of vibration, and a quark as another mode of vibration, and a graviton as yet another (Schwarz, 79)

Steven Weinberg agrees with this assessment and further adds that modes of vibration may even explain strange quarks and other esoteric particles in modern physics.

> The strings are vibrating in all these extra dimensions and that leads to a lot of different modes. It is the extra dimensions (or other extra physical variables) which produce the many different modes. In fact, that's one of the encouraging things about string theory. Because of that it's natural to find multiple generations of particles, not just the lowest generation with the light quarks

However, all the particles with which physicists are familiar, such as electrons, protons, and neutrons, correspond to the lowest frequency mode of vibration. Other modes, representing higher frequencies with higher energies, are not normally seen, or detected in nature. "The next mode would be hopelessly too heavy to have ever been seen." (Weinberg, 217)

And yet the common vibrational modes do not exhaust all the possible actions of a superstring that could be utilized to determine any individual particle's physical properties. For example, the electric charge of an elementary particle could arise from some type of unspecified non-vibrational action. "In fact, what we call electric charge would be some sort of collective property of the string as a whole and if the string oscillated in different ways, then it would seem to have a different electric charge." (Ellis, 154)

In this respect, the basic electric charge on particles would not be a quantity that is just experimentally determined without any apparent connection to the rest of the universe. The fundamental electric charge would have a deeper hidden meaning that is related to the motion of the string. Specific characteristics of quarks could also be attributed to these nonvibrational actions of the superstrings.

This superstring, in addition to oscillating in space, rather like a traditional violin string, also has certain internal degrees of freedom which you can't really visualize in terms of simple oscillations of space, and the actual difference between, say, an up quark and a down quark would presumably be some special yet unknown combination of these internal properties and these oscillations in space. (Ellis, 153) It is believed by its followers that there is very little in physical reality that cannot be attributed to either the various motions of superstrings or their existence.

Even the mathematical points that represent empty space may well turn out to be no more than superstrings.

In relativity theory, space and time only exist relative to the bodies with mass that constitute the universe. In the general expression of this idea, matter either curves spacetime or matter reduces to the curvature of spacetime. So, if curvature is the true reality and is associated with material particles and thus superstrings, it is not a far stretch of the imagination to speculate that relative points of space are themselves superstrings.

Each individual point of space could be no more than an object of higher dimensionality that is curled up in a little ball or loop. The group of all such loops would constitute a "stringy space-time" which is an "approximation" of a far "richer structure" of superstrings. (Green, 131) All physical reality seems to be covered within this notion. Superstrings have become the material particles, the forces between material particles as well as the relative point positions between particles, leaving little else to exist in the physical world. Thus, the superstring theories are TOEs, quite literally, if they can fulfill their promise. Superstring theories are at present the best contenders for a TOE, if indeed such a theory is possible. At least physicists are now thinking in terms of a theory that covers everything in nature. So, they believe a TOE is a distinct possibility in the very near future.

It would be prudent, then, to ask just what one might expect of a theory that seems to cover 'everything' in its wake. Davis and Brown have considered just this question.

provide an explanation for the geometry and topology of space-time, such as the number of perceived dimensions, and offer a convincing account of how the universe came into existence. But this is not all. A TOE should also unify physics. (Davies and Brown, 5)

Their criteria are simple. A TOE must explain (1) matter, (2) the forces affecting matter, and finally (3) the spacetime framework of matter as well as unify the quantum and relativity theories. Although they have placed the unification of physics last in their own wish list, it should be placed first.

The foremost task in theoretical physics should be to unify the quantum and relativity, the discrete and continuous aspects of physical reality and nature. Whether or not a theory of 'everything' is even possible under these or any criteria is still an arguable point as John Barrow has pointed out. (Barrow, 230-231, 282) But unification in physics is an essential task independent of a theory that describes all in nature. It is also questionable whether superstrings can fulfill this notion of 'everything.' The problem of defining and understanding space and time is quite formidable. Faraday ran into a similar problem a century and a half earlier when he tried to conceive the 'continuous' electric and magnetic fields between particles. He sidetracked the problem of continuity by talking about the 'contiguous' points of charge in space that carry the electromagnetic field.

Historians and philosophers of science still argue about the meaning of Faraday's use of the word 'contiguous.' There are also modern analogies to this problem. A few decades ago, quantum theorists also speculated about the discrete nature of both time and space, but no new science ever came from this speculation. The modern superstring theories carry the same stigma. At best, they can only speculate about the actual points of space as 'curled up dimensions.' However, this view is no solution to the discrete/continuous debate. It merely forestalls the debate to a later point in time and a much smaller unit of discreteness.

Beyond the already stated problems with superstrings, two more problems are inherent in these last speculations. (1) If space is no more than little 'loops' of curled up higher dimensions, then what are those little 'loops' moving within when they follow trajectories through time? And this leads to the next problem. (2) What then is time? A theory of everything should make some definitive statement about the nature of time, but superstring theory does not seem to do so.

These are only some of the fundamental problems with the theory of superstrings and they are not the only problems. There remain the obvious difficulties that are overlooked or shunted aside by stating that superstrings are a theory of the future. The makers of superstring theories can safely defer confirmation of their theory to late in the next century if not later, which seems to forestall any attempt at falsification of the theory. And finally, there is one last problem that has rarely been mentioned within the context of superstrings. During the past two decades, another important trend has developed in physics. This trend is to define or discover the relationship between consciousness and physics as well as solve the mind/body dichotomy.

Coincidentally, these questions have been raised within the same historical time frame as the change in scientific attitude toward the acceptance of the possibility of a TOE. This coincidence would seem to indicate that the development of a TOE and the discovery of a role in physics for mind and consciousness are connected, but superstring theorists have not adequately addressed these questions. So far, superstring theory is a purely physical theory without room for consciousness so it cannot be, at this time, used to make any statement concerning the mind/matter paradox.

These criticisms serve to emphasize the fact that there is surely more room in modern physics for contending theories of unification. There are certainly other options for unification without the premature declaration that superstrings represent the last word in physics and the ultimate TOE.

Universality and the single field theory

a. Philosophical arguments for an extra dimension

Supergravity, GUTs and superstring theories have popularized, sanitized and legitimized the concept of higher physical dimensions of space. However, the concept should stand on its own merits without reference to these modern interpretations. The question of why a fifth dimension (and/or higher dimensions) should be adopted in science is of far greater significance than any one theory. The concept should be considered independent of these or any single physical theory. Indeed, it is a question that has plagued science for several centuries, but which gained more immediacy only after the popularization of the new non-Euclidean geometries during the middle of the nineteenth century.

In answer to this question, there are several simple logical arguments. Although they offer no absolute proof of the existence of a higher dimension of space, they do lend some credence to the possibility. There is a very simple and straightforward argument in favor of higher dimensions that dispenses with all but the simplest of mathematical forms. It is a very well-known fact that a mathematical point has no dimensions. A line is one-dimensional, a surface is two-dimensional and a solid is three-dimensional. However, these geometrical figures are mathematical abstractions. A real point is not dimensionless just as a real physical line is not one-dimensional.

A real physical line must have a thickness, so it is at least a two-dimensional object. A two-dimensional surface must have a thickness in a third dimension to be assured of physical existence. But these are still abstractions. Physical reality has at least one more dimension than the corresponding mathematical model of reality. Real physical objects are represented mathematically by three-dimensional geometries. Therefore, by extrapolation, a real physical object must have another dimension to be physical. Real physical bodies must be four-dimensional; they must have four spatial dimensions.

This argument strikes at the heart of the problem for quantum field theories and other forms of physical theories. Superstring theory has the advantage over previous theories because it essentially adds one more dimension to the physical (and dimensionless) points those past theories assumed. This procedure is not without precedent in the history of physics, as most people should know from their fundamental physics course.

A little more than a century ago, statistical methods were employed in the kinetic theory of matter (and thus thermodynamics) to successfully derive the Ideal Gas Law. The derivation assumed that gas molecules were perfectly elastic and dimensionless point particles and were thus unaffected by short-range forces. These molecules had no volume. These properties allowed the derivation of the correct and well-known gas law. However, when these criteria were abandoned and the dimensionless molecules were given a small but finite volume which allowed the interaction of forces at close range, the kinetic theory of matter was not only able to explain the transition of gas to liquid and liquid to solid, but also describe the structure of matter within each of the three different phases.

In similar manner, the one-dimensional string of superstring theory has replaced the dimensionless physical points of mechanical theories thereby explaining gravity as represented by the space-time curvature. Superstring theories need that extra dimension just as a three-dimensional object (as perceived) necessitates a four-dimensional extension (which is not normally perceived). But even the superstring theory is an admitted approximation.

Our fundamental concept of space was developed prior to Newton with the work of Francesco Patrizi and Pierre Gassendi. Newton synthesized the views of Patrizi, Gassendi and others into a new concept of absolute and relative spaces to be used in the support of his mechanical system. Patrizi's was the first concept of space that bore a resemblance to Newton's and our own intuitive concept of space. He considered two cases for space. First, that space was a container and second that space extended beyond the boundaries of the material universe. He also countered the Aristotelian argument that space was not a 'thing.' If space was 'no-thing,' it could not be considered a legitimate subject for mathematics or science.

Since empty space was 'nothing' rather than 'no-thing' it could not be subdivided mathematically for analysis. Space was irreducible in the physical sense, differing from mathematical space and therefore impervious to physical abstraction. Of course, that may or may not be true today. Modern science has developed the relativistic viewpoint that accepts space as a 'nothing,' even though it has properties. Modern science has reduced space to the relationship of positions between real physical objects.

Given today's concept of symmetries, the 'nothing' called space has still more properties than twenty years ago. But in the older view, even an infinite number of dimensionless points could not constitute an extended space. Quantum mechanics overcame this problem by adopting probabilities and uncertainties. A probability distribution describing a particle can be continuous across an extended volume of mathematical points of space because it has no physical reality.

In wave mechanics, the wave function is continuously extended across all of space and has been interpreted in quantum mechanics as corresponding to the probability function. The physical wave function allows the mathematical probability distribution its physical presence. But the collapse of the wave function must occur at a real extended position in space, so the collapse of the wave function marks deterioration from a mathematical continuity to a discrete physical reality. Dimensions have been created out of non-dimensions to replace nonextended mathematical points with extended three-dimensional physical points. This process is highly questionable, so the probability distribution is deemed physically real instead of mathematically real. This would imply that the probability interpretation of the wave function is erroneous.

The mathematical continuity of the probability distribution merely overlaps the physical continuity of the wave function; they are not the same thing. Nor is there a concept of space in quantum mechanics that rivals the space-time of relativity. The probability distribution merely mimics the continuity of space, which presents a major paradox for quantum field theory. Since the probability distribution only corresponds to the wave function, without being the wave function, a new interpretation of wave mechanics is demanded. The wave function has a far richer physical structure than just a probability distribution, as suspected by Einstein, Erwin Schrödinger, and Louis DeBroglie. It must have a real physical interpretation since it 'collapses' to a real physical point in space. Others have noted the discontinuity problem of the 'collapse of the wave packet' as a serious problem for quantum mechanics, but never within this context.

This demand coincides with the need for another dimension of space. The present interpretation of both quantum and wave mechanics depends on mathematical points of space whereas superstring theory has demonstrated that a complete description of the forces of nature requires an extra dimension that the dimensionless point cannot contribute. A real string is not a one-dimensional object even if its mathematical model is one-dimensional. To use an old analogy, these are all cases of scientists and mathematicians mistaking the finger pointing at the moon for the moon itself.

These issues are related to the second philosophical argument for a higher dimensioned physical reality. It is modeled on Gödel's theorem. Kurt Gödel demonstrated that a mathematical system could not be proven true or untrue from within that system; or rather its logical consistency could not be decided. A mathematical system depends on stated fundamental axioms, so any proof of consistency (truth) within that system also depends on the same fundamental axioms.

If the system is 'proven' to be non-consistent, then the axioms are false, but then so is the 'proof' that was used to demonstrate the nonconsistency. A mathematical system is therefore denied falsifiability in the Popperian sense when the issue of consistency is determined from within the system. All arguments within the system are based upon the same axioms or rules that they are trying to disprove, so the consistency (truth) of the system can only be judged from outside of the system. The consistency proof of a mathematical system must rely on a larger and more general set of axioms than those used to establish the system that it is testing.

In a similar manner, a physical theory can never be proven true in our physical world. It is a philosophical impossibility to prove physical truth. That is why we have theories that can change and grow instead of absolute physical laws that never change. Yet a physical theory can be proven valid (although not true) by falsification. A theory must be falsifiable and verifiable to determine its validity. Our physical world is a logically consistent system based upon the physical equivalents of axioms. The equivalents are the conservation laws, symmetries, other principles, and various rules of nature that exist under different names. Our universe may be quite complex, but it is nonetheless a system that makes it possible to mimic physical reality with a mathematical model.

As a logical system, physical reality is also subject to Gödel's theorem even though it would be too difficult a task to find all the physical axioms upon which the system is built. It is the primary function of science to discover these physical axioms and develop a model of reality from them. To discover all the 'axioms' of this natural system is the first step toward developing a TOE.

As a self-consistent system (we must take it on faith that the universe is self-consistent or science fails), physical reality must follow Gödel's theorem. So, physical reality must rely on something outside of itself merely to exist. Applications of Gödel's theorem in fields other than mathematics are not all that uncommon, including in the field of physics and the search for a TOE.

The reason why mathematics is so successful in describing the way the world works is because the world is at root mathematical. Any limitations to mathematical reasoning, like those uncovered by Gödel, are thus not merely limitations on

Our physical reality is part of a larger logical system whose existence verifies the existence of our four-dimensional space-time continuum. At least one higher dimension is indicated by the previous arguments as providing this larger required physical system. A higher-dimensional world, or rather a higher-dimensional spatial extension of our world, is necessary for the existence of our four-dimensional physical reality.

In a rather strange sense, the criteria of a TOE even imply this. According to Davies and Brown:

> The ultimate TOE would, ideally, need no recourse to experiment at all! Everything would be defined by everything else. Only a single undetermined parameter would remain, to define the scale of units with which the elements of the theory are quantified. This alone would be fixed empirically. (In the ultimate case, experiment merely serves to define a measurement convention. It does not determine any parameter in the theory). (Davies and Brown, 7)

In an off-handed way, they have confirmed that Gödel's theorem applies to our physical reality. The ultimate TOE that they envision would not be falsifiable so there would be no way to validate the theory.

Even then, it couldn't explain everything. There would have to be one independent variable (or axiom in mathematical terminology) to justify existence, just as there would have to be an independent axiom within a larger more comprehensive system to prove the consistency of a less complete mathematical system. However, the TOE that Davies and Brown foretell is not possible. They require a single independent variable, as verified by experiment, to define the "scale of units."

But how could this be? What could it define the scale of units relative to since there is neither nothing outside nor nothing more than the 'everything' covered by the theory? What could it define the scale of units relative to, since there is neither nothing outside nor nothing more than the 'everything' covered by the theory? Their parameter could only gauge or scale measurements relative and internal to the system and would therefore be dependent upon the system, not independent of the system as assumed. What is needed for the system that is covered by their TOE is a larger more comprehensive system, another higher embedding dimension than their theory assumes.

At a more mundane level, the best scientific and practical reason for adopting a fifth or higher dimensions is inherent in the superstring theories, although not unique to them. Higher dimensions offer more degrees of freedom for explaining the more paradoxical physical properties of four-dimensional space-time. But does this mean that hyper-dimensional theories are more expedient for the theorist, or does it mean that there are real directions of space other than length, breadth and width that are perpendicular to all three at the same time?

If extra dimensions are to have any physical meaning, they must be real orthogonal extensions of our four-dimensional space-time. Otherwise, the extra-dimensions reduce to mathematical gimmicks that only explain (or explain away) unwanted or misunderstood physical properties and parameters. Whatever the case may be, Einstein's requirement remains intact as the best guide for the adoption of hyperspatial theories. No such theory can be taken seriously unless it can explain why we cannot 'experience' the higher dimension or dimensions.

While these reasons for adopting a higher-dimensional space are straight forward, they do not exactly demand such a hypothesis. There is, however, one problem in modern physics that has been grossly overlooked yet has a direct effect on the physicists" model of reality within this context. It demands a higher dimension for physical space-time. Scientists have long ignored the gravitational forces within the domain of single atoms, which means that they have also ignored space-time curvature within the confines of the atom. Gravitational forces within the atom are so small compared to the strong forces within the nucleus and the electromagnetic forces outside the nucleus, that they have been considered inconsequential. Yet the mass of electrons, protons and neutrons is an essential element of any calculated quantities within the atom and mass is related to space curvature according to GR.

Theories of the nucleus, which have never been totally successful, have always depended upon quantum explanations even though the nucleus can be represented by singularities in the space-time continuum of relativity theory. When 'everything' is considered neither the gravitational forces within the atom nor the curvature associated with the nucleus could be ignored if quantum field theory is ever to offer a 'complete' description of nature. Any theory that claims to represent a unification of the quantum and relativity, whether a TOE or not, must address this apparent paradox.

Both the questions and solutions regarding this paradox are centered on the concept of the singularity in GR. There is a strange parallel between the problems raised by singularities within the gravitational field and the convergent infinities at the point location of particles in quantum mechanics. Since both theories fall apart under the same extreme conditions, within the interior of elementary particles, one would suspect that the interior of material particles could offer a point of connection between the two theories as well as a point of unification for the concepts of continuity and the discrete nature of matter.

The singularity of GR is a discrete disruption of the smooth flow of continuity described by the field. So, understanding what happens to the space-time continuum within the boundary of elementary particles offers the best hope of unification in physics. On the other hand, a look at how physics treats other singularities in the gravitational field offers the best hope of solving the problem. In particular, the same mathematical singularities that are used to model very massive bodies such as black holes are also used for elementary particles. Yet these are distinctly different cases. What are the physical differences between the singularities of particles and those representing very massive bodies?

Mathematically, there may be no qualitative differences, but physically there must be a difference. Massive bodies are, at most, a collection of elementary particles crowded in proximity with their surfaces in contact whereas it can be assumed that the interior of elementary particles is continuously curving. The difference is difficult to understand but can be portrayed graphically.

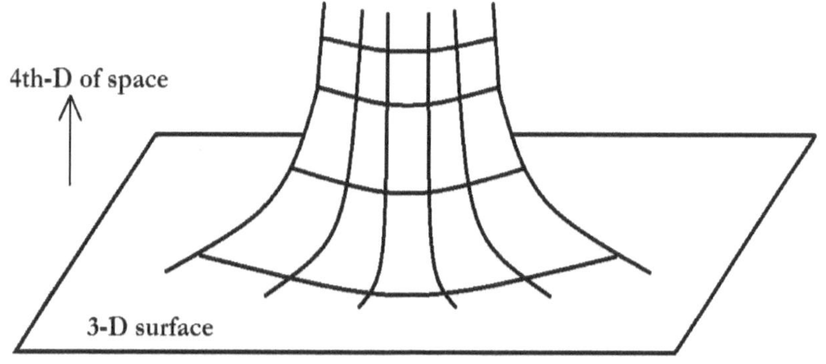

A SIMPLE SINGULARITY IN SPACE-TIME

**AN ELEMENTARY PARTICLE IS A VERY
SMALL SINGUALRITY IN SPACE-TIME**

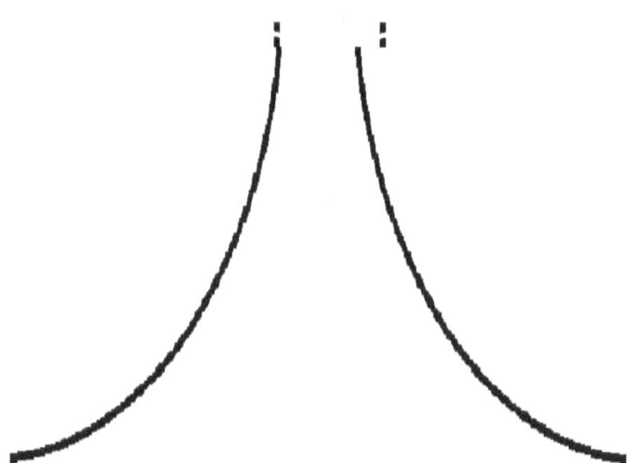

**A BLACK HOLE IS JUST A MUCH LARGER
SINGULARITY IN SPACE-TIME**

**BUT A BLACK HOLE IS ALSO A LARGE GROUP OF
TIGHTLY COMPACTED PARTICLE SINGULARITIES
WHICH DOESN'T REALLY SEEM TO FIT REALITY**

There is a grave discrepancy between these two views of a black hole, or for that matter any large accumulation of matter which creates a singularity in GR.

How can the individual particles add together to give a mathematically singular entity? To put it simply, gravity and curvature must act non-linearly with respect to the higher dimension of space, yet these diagrams depict a "real physical curvature" in a higher dimension that acts linearly in three-dimensional space but does not act non-linearly in the fourth dimension. Within the context of mathematics, curvature is a property of a space or manifold, regardless of whether it is an embedded space or not. The term 'curvature' has a specific and understandable meaning as an intrinsic property of space just as it does of an extrinsic property. However, within the above context, a 'real physical curvature' refers to a 'physicalist' concept of curvature that 'requires' a higher embedding space or manifold.

Under these circumstances, the accepted mathematical model of GR favors neither a higher embedding dimension, which implies an extrinsic space curvature, nor an intrinsic space curvature, which requires no higher dimension. The curvature of physical space-time has traditionally been treated as an intrinsic property of the four-dimensional manifold even though either the theory or the mathematical model does not require that treatment. Since the original development of GR, scientists have merely assumed that curvature is either an intrinsic property of the space-time continuum or a purely mathematical property describing space-time rather than a feature of physical space-time itself requiring a higher embedding dimension.

In the opinion of John C. Graves, GR does not stipulate the necessity of the intrinsic case of a four-dimensional curvature over the extrinsic case evident by manifolds of higher dimensions. He has tried to explain this point of contention.

> Mathematically it is probably simpler to deal with flat spaces of more than four dimensions than with non-Euclidean four-spaces. And if the notion of higher dimensions with regard to physical space seems incomprehensible, one might say the same about curved space that cannot be embedded   Finally, there is no obvious explanation within GR of why space-time should have four dimensions, no more and no less - the formalism itself makes no reference to dimensionality. (Graves, 192)

Even though an extrinsic curvature of the space-time continuum would necessitate at least a real five-dimensional manifold in which the lfour-dimensional space-time continuum is embedded while an extrinsic curvature does not. On the other hand, it could also be argued that GR accounts for an intrinsic curvature alone.

In other words, the Riemannian curvature tensor could be considered an inherently intrinsic object of four-dimensional space-time. According to the mathematician J.J. Stoker,

> In fact, throughout this chapter, as its title indicates, the inner or intrinsic geometry of surfaces will be studied extrinsically, and for two reasons: It is both interesting and important for its own sake to characterize intrinsic properties extrinsically. In fact, throughout this chapter, as its title indicates, the inner or intrinsic geometry of surfaces will be studied extrinsically, and for two reasons: It is both interesting and important for its own sake to characterize intrinsic properties extrinsically.

Such a procedure, as was remarked earlier, can be carried out in a fashion that furnishes clues for the generalization to Riemannian geometry in a manifold of any dimension without the necessity of embedding in a Euclidean space.

> A purely intrinsic treatment of Riemannian geometry is desirable for obvious reasons, and also for a reason rooted in physics: it would seem rather strange to treat Einstein's general theory of relativity, which is basically the Riemannian geometry of a certain four-space, by first embedding it in a higher dimensional Euclidean space, since Einstein's object was to investigate the character of the actual space in which we live - and to find that it is not Euclidean. (Stoker, 153)

Stoker's purpose ws not to develop a new 'physics of space-time, but to investigate an interesting mathematical problem or model. However, he does relate his studies to physics in the form of Einstein's GR and in so doing points out that using a higher-dimensional geometry to model physical spacetime would indeed seem "rather strange." Many physicists share this opinion, but it is not a requirement of the mathematics used to model space-time in GR.

On the other hand, Charles Misner, Kip Thorne, and John A. Wheeler have pointed out several methods of deriving "Einstein's field equation" as found in GR. Among these methods are included two cases based upon higher-dimensional embedding manifolds. One case proceeds from considering the "physics on a space like slice or hyperspace of simultaneity" using electromagnetism as the model. The second method proceeds from considering an infinitely dimensioned super space composed of points, each of which describes "a complete three geometry ... with all of its bumps and curvatures." (Misner, Thorne and Wheeler, 419-425) In each of these cases, the curved space-time of GR displays both intrinsic and extrinsic characteristics.

The normal interpretation of GR utilizes only an intrinsic curvature that cannot be easily portrayed or explained, as Graves has stated above. Embedding spacetime in higher dimensions allows a greater versatility in accounting for physical phenomena that occur within the normal four-dimensional space-time continuum. Again, according to Graves, one advantage to using this extradimensional embedding structure is that it "allows us to define new concepts which might help characterize our geometric structures in an especially revealing way." (Graves, 193) But this advantage has been lost since theoreticians have traditionally adopted the intrinsic model of curvature as adequate to describe physical reality.

Why then would embedding dimensions be either desirable or necessary? Kaluza's answer would be "so that electromagnetism could be unified with the gravitational field." On the other hand, if the curvature of the individual particles were somehow additive in the higher dimension, the structural differences between different physical types of singularities could be accounted for.

In the case where space-time is strictly four-dimensional and the curvature intrinsic, the additive effect of curvature could not be so easily explained. However, a real fifth dimension displaying a real physical curvature could easily account for the additive nature of the curvature.

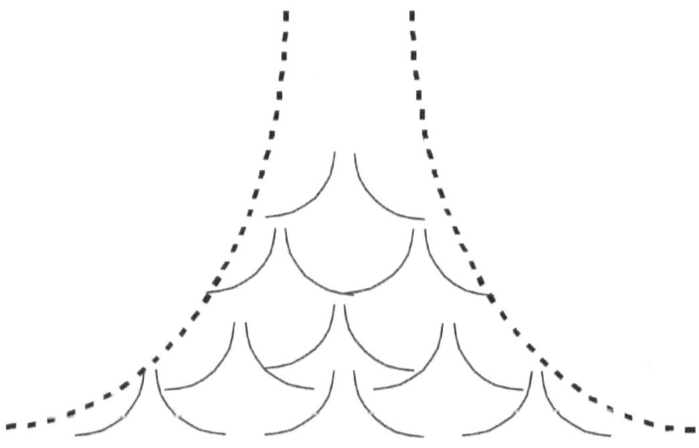

The more particles that are packed tightly together, the greater the overall curvature into the higher dimensions.

If space-time had a minute 'thickness' in the fifth direction, acting as a four-dimensional sheet, then the curvature outside of the physical boundaries of individual particles could add together in the fifth direction when the particles are near one another.

The closer material particles are packed together, up to the point where their physical surfaces come into contact, the greater the additive effect of their individual curvatures. In other words, if the fifth dimension is real and the four-dimensional space-time continuum is like a sheet, then individual particles near one another can ride up 'higher' along their neighboring particles' curves in the fourth spatial direction. The closer the proximity of particles, the higher they are pushed into the fourth spatial direction in the geometric center of the overall body.

The overall curvature of an extended material body is a consequence of its density rather than the total mass. The additive affect requires a higher embedding dimension to explain the curvature of collections of particles. The philosophical points of argument beg the question whether they have physical validity or not which is not true for the physical arguments. The philosophical points of view are scientifically valid only in so far as they agree with nature. As for providing evidence for developing or providing clues to a new theory, they can only create a little smoke from smoldering, but no fire.

The physical arguments are different. They hold more weight when developing or deciding between competing theories. Yet philosophical arguments have the power to either advance the cause of theories or destroy them. These arguments can still be used as scaffolding to support a theory, but only as ammunition for the acceptance or perpetuation of a new theory. They can help interest scientists and scholars in an idea, and they can create attitudes, climates of opinion and advertise a theory, all of which are instrumental steps in the acceptance of new theories. They can also destroy or retard the acceptance of a new theory as they did in the case of Kaluza's theory and could thus act as a detriment to science.

Undoubtedly, Kaluza had a very good idea, as modern versions of his theory seem to indicate. But his theory was ignored by all but a handful of scientists until the past two decades. His theory successfully duplicated the Einstein-Maxwell equations, a fact which cries for some answer to why it was so unpopular. It created a little heat and perhaps some sparks, but never a fire under the scientific community of his day.

The only reason that can be found for this mystery is the fact that the philosophical arguments against Kaluza's theory were overwhelming when considered against the background of the newly developed quantum theory coupled with the lack of perceiving the higher dimension upon which he based his model. The philosophical arguments provided above indicate the existence of a fifth embedding dimension to our space-time continuum and the physical arguments strengthen the case even more, justifying the adoption of Kaluza's model, with appropriate modifications.

b. The basic assumption and its consequences

Today's climate of change is more conducive to new and innovative concepts, although any new ideas must still have some scientific basis. This fact is demonstrated no more clearly and thoroughly than with the success of the multi-dimensional concept in the theory of superstrings. In the latest applications of higher dimensions, it has been assumed that Klein's interpretation of Kaluza's theory is absolute, and Kaluza's cylindrical condition describes minute cylinders of curled-up space-time. Yet, Kaluza seems to have been somewhat vague on this issue himself.

Although all our previous experience hardly provides any suggestion of such an extra world-parameter, we are certainly free to consider our space-time to be a four-dimensional part of an R_5; one then has to take into account the fact that we are

Kaluza's "cylinder condition" is only one of two options that fit the physical facts and even then, details of the cylinder's properties are not forthcoming.

At this point, Kaluza only requires that the new parameter, a measurement in the fifth direction, be small. This requirement does not necessarily guarantee the cylinder itself will be of a minute size, such that the cylinder will have a small diameter and circumference. The measurement could be small and the cylinder of a larger circumference if there were other restrictions on the measurement independent of the "cylinder condition." Klein seized upon one of the possible interpretations of Kaluza's theory. He assumed that both the measurement in the fifth direction and the circumference of the cylinder were extremely small and equal, so that the cylinder itself could account for the whole extension in the fifth direction of space-time. His assumption allowed him to both explain why the fifth direction is undetectable and relate the cylinder condition to the quantum of action.

One might think that the only possible interpretation of Kaluza's theory specifies that cylindricity is so small that the fourth dimension of space, or the fifth dimension of space-time, cannot be perceived or otherwise detected. But Kaluza's original theory only required that the A-lines, extensions of three-dimensional spatial points in the fifth direction, be of equal and constant length. He did not specify how long or short they should be.

These criteria are equally valid for space that is closed and Riemannian (a sphere) in the fifth dimension as well as cylindrical on a large macroscopic scale rather than the scale of a Planck length. Kaluza's basic theory could also be modified such that the "cylinder condition" would undergo a change of scale or cylindricity could be discarded altogether for a Riemannian curvature in the fifth direction. Einstein and

Peter Bergmann in 1938 as well as Podolanski in 1947 introduced theories of this type.

Podolanski assumed a laminate space with six dimensions while Einstein and Bergmann discarded the cylindrical condition and adopted a closed Riemannian space as the embedding structure for spacetime. Discarding the cylindrical condition with a minuscule size again invokes the question of why we cannot perceive or detect the fifth dimension. Yet the minuscule width of the cylinder, which has been discarded, could alternately be considered the 'effective width' of a four-dimensional sheet of space-time. All matter and material contact could be confined to this sheet so that the extra component of space would still be undetectable.

In this manner, our four-dimensional space-time continuum could be portrayed as a thin sheet whose spatial points are fully extended in a fifth dimension and loop around the closed figure. While the 'effective width' of the space-time sheet in the fifth direction is of minuscule extent, the loop in the fifth direction could be any length. This model assumes the reality of the fifth dimension and thus our space-time truly curves extrinsically into the extra spatial dimension. Curvature in the fourth direction could then be regarded as wrinkles in the four-dimensional sheet. This notion is not new and, in fact, predates GR by nearly a half century. The first interpretations of non-Euclidean geometry were wholly physical. In the late nineteenth century, it was generally thought that if mathematical space could be non-Euclidean, then physical space could also be non-Euclidean. Great debates over the true nature of space followed and some astronomers attempted to measure the curvature of space through parallax observations.

A half century earlier, William K. Clifford had offered the first physical theory of the type described although his primary goal was to develop a model of Maxwell's electromagnetic theory in a four-dimensional space and include gravity and other forces later. Clifford's was the first true attempt to derive a TOE in the more modern sense of the concept. In 1870, he stated that matter is no more than curved space and matter in motion is only the time varying changes in the curvature of space. A decade later, Edwin Abbott wrote a popular book about beings in a two-dimensional world that experienced contact with three-dimensional beings for the first time. The book was titled Flatland and has become a classic of the period. But Abbott was just explaining the ideas expressed in Clifford's model of space to the common people of Britain and the world.

Clifford's own model extended well beyond the few short paragraphs that he presented before the Cambridge Philosophical Society in 1870 even though this single presentation is normally all of his theory that is cited in others' work. There are strands of his theory stretching throughout his mathematical and philosophical writings. Enough is available from these other sources to reconstruct a fairly accurate model of the theory of space curvature that he was trying to develop even though he never published his theory in any one source. On the other hand, Clifford developed whole new forms of geometry based upon his biquaterions to demonstrate that dynamics in the three dimensions of space plus time reduce to kinematics in a four-dimensional space plus time which is characterized by an 'elliptical' Riemannian geometry with a positive curvature. (Beichler, 1996) His theory sounds very much like a modern TOE.

Charles Hinton also popularized and extended the intuitive model of a four-dimensional space through several books and articles over a period of two and a half decades. The model of space first proposed by Hinton was a three-dimensional sheet of ether in which atoms were embedded. The complete structure was curved within a fourth dimension. The material atoms were likened to threads passing through the sheet from outside the three dimensions of the sheet, the points of intersection representing the individual atoms. (Hinton, 1980, 16-20) In this model, Hinton could only account for some of the fundamental properties of matter. In another of his essays, he introduced "twists" as mechanical models of electrical activity. (Hinton, 1980, 36-37, 74-75)

Aspects of Hinton's four-dimensional model resemble portions of today's superstring theory. His notion of threads intersecting the three-dimensional sheet of space are like superstrings and the twists that he later introduced to explain electricity are like the nonvibrational modes of superstrings which are suspected of yielding other physical properties. However, the notion of twists was not new with Hinton. The "twist" had been developed by Clifford as the fundamental physical point of space, again in a manner like the superstring theories of our own era.

W.W. Rouse Ball developed a theory of gravitation very similar to Hinton's in the 1890s. Ball noted the similarity between his and Hinton's theories and stated that they were parallel developments, surrendering priority for the idea to Hinton. (W.W. Rouse Ball, 1891, 21) Hinton's later work overlapped the beginning of the Second Scientific Revolution and Einstein's first work on special relativity, but there is no evidence that

Hinton's and Einstein's ideas ever came together in any form. They were never associated with each other, at least in written published forms that still survive. So, it is not known if they influenced one another or that they even knew of each other's work.

Many scientists and scholars prior to World War I supported these ideas, although their supporters formed a small minority relative to the larger scientific and academic communities. Worried about the null effect of the luminiferous ether, as supported by the Michelson-Morley experiments, the astronomer Simon Newcomb theorized that our three-dimensional space consisted of two parallel sheets of three-dimensional ether separated by a small distance in a four-dimensional space. Physical events occurred between the two sheets. (Newcomb, 1888/1891, 514-515) Robert S. Ball, another astronomer, and a follower of Clifford, tried to detect the curvature of space (R.S. Ball, 1882, 519) and developed the mathematical "theory of screws" to model the higher-dimensional space. In these and the other instances, scientists assumed a real four-dimensional space characterized by curvature as opposed to the flat Euclidean space of Newtonian physics.

In these earlier models, a more intuitive and less analytical concept of geometry was applied rather than the analytical tensor calculus that Einstein later used to model GR. The tensor calculus had either not yet been developed or was in its early stages of development when these theories were proposed. In fact, these geometric models of physical space and similar ideas formed part of the impetus to develop the new analytical systems of geometry. These earlier notions of curved four-dimensional space were overwhelmed by the advances in quantum and relativity theory during the early twentieth century and all but forgotten.

The only assumption made in this new theory is the existence of a real five-dimensional extension of our spacetime continuum. All other aspects of this model come from physical theories that are already accepted by the scientific community such as GR and Kaluza's five-dimensional model. Although dimensions higher than the fifth may well exist, only the fifth dimension is necessary to explain four-dimensional phenomena at this time. Beyond the assumption of a real fourth dimension of space, a fifth dimension of space-time, the rest of the theory is derived from describing phenomena and using experimental and perceptual observations to determine the type, shape and characteristics of the model that will explain how physical events occur. This method may be criticized as an ad hoc fitting of the model to observed phenomena, but it is no more ad hoc than Newton's explanation of 'how' gravity works rather than 'why' gravity works.

The first order of business is an explanation why the fifth direction cannot be perceived or detected, fulfilling Einstein's requirement. Kaluza's model requires that the extension of every point in four-dimensional spacetime in the fifth direction be of equal and constant length, but he also suggests that the measurement along the extended lines be small. So, in this model the A-lines have a very long length (of macroscopic proportions) while the four-dimensional sheet, which is our physical world, has a small quantum sized measurement along those lines. The A-lines are orthogonal to the sheet at every point. The sheet has a 'practical' thickness in the extra fifth dimension, which is designated as the 'effective width' of the sheet even though each point extends beyond the sheet along an orthogonal line.

Four-dimensional phenomena are restricted to occur within this 'effective width' of space-time. So, the fifth dimension cannot be observed, normally perceived, or detected through normal experimental procedures. The fifth dimension has no material reality, the key word being 'material,' referring to the existence of four-dimensional matter or material bodies (such as the tesseract) but is a perfectly continuous field of varying densities. The 'effective width' of the sheet consists of the densest portion of the field that effectively constitutes the four-dimensional continuum. This continuum is a strictly potential and nothing else, although it varies in density along the fourth direction of space.

The portion of the five-dimensional potential field outside of the sheet is quite rarefied and becomes considerably less dense the farther away from the sheet along the A-lines. The A-lines curl around the fifth dimension, which is closed, and reconnect at the same point from which they originated on the other side of the sheet, and through the 'effective width.' In other words, the A-lines are continuous and unbroken in the fifth direction, just as they would be in the minute cylinders posited by the Kaluza-Klein theory.

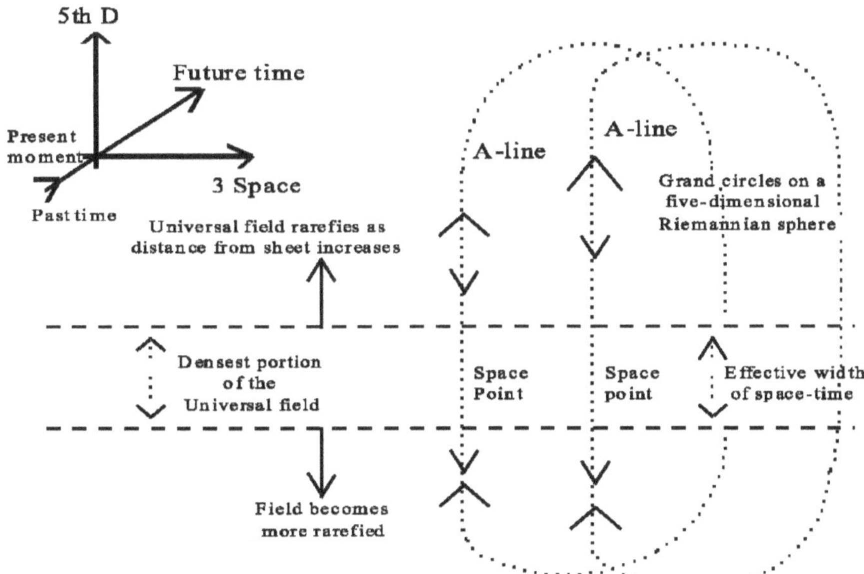

Although space-time has an effective width in the fifth dimension, each point extends into the fifth dimension and eventually curves back on itself along a grand circle on a sphere.

The fifth dimension exhibits Riemannian curvature in the large, just as the four-dimensional space-time portion of the five-dimensional world exhibits Riemannian curvature in the large as described by GR. The 'effective width' of the sheet is of the order of Klein's original fundamental length (l_o) of about 10^{-30} centimeters. The effective width is constant throughout all of space (for all practical purposes) since it is relative to all the matter in space and the quantum. The A-lines can be seen as 'propensities' in Karl Popper's sense of the word (Popper, 69-71) or perhaps a better analogy could be found in Faraday's 'lines of force' They are not physical lines, just as mathematical points have no physical reality.

The torsional portion of light waves lay along the A-lines in their fifth component extension, giving the A-lines their closest approximation to a physical reality. The wave/particle duality of light waves thus reduces to which portion of the light wave along an A-line is interacting with the rest of the world. That portion of the light wave, which is particulate, the photon, is just that part of the light wave (or A-line) which cuts across the 'effective width' of the sheet while the portion of light (along the A-line) which extends across the rest of the fifth dimension, outside of the sheet, is pure wave and exhibits wave interactions with the rest of the world. Wave/particle duality is thus a common feature of the five-dimensional space-time.

Since electromagnetic waves coincide with the A-lines across the fifth dimension, the magnetic permeability μ_o and electric permittivity ε_o of free space must be related to the five-dimensional field, but active within the four-dimensional sheet. The permittivity of free space is an interstitial connectivity constant which acts across three-space between contiguous points of space-time in the sheet. It therefore acts in a direction perpendicular to the fifth dimension and the A-lines.

The permeability is also an interstitial connectivity constant of the field, and it acts in a direction perpendicular to the A-lines, but it acts torsionally between contiguous points within the sheet. These two field constants guarantee that the maximum speed of light, all electromagnetic waves and the speed limit to matter itself in free space is constant. As field constants, these two quantities provide the connectivity between contiguous points of space and time in the direction of the normal four-dimensions. The connectivity of space-time is thus related to the speed of electromagnetic waves through empty or free space and limits the speed in that medium to $c = (\mu_o\varepsilon_o)^{-1/2}$, as Maxwell originally defined it.

The universal gravitational constant G is also a special type of connectivity constant, but it acts between material particles rather than points in space. It extends across five-space (the fourth spatial dimension) from particle to particle in a direction parallel to the average macroscopic 'flatness' of the space-time sheet. The sheet is not 'flat' (Euclidean) over macroscopic distances, but the curvature is so large that it approaches relative 'flatness' over shorter distances. G is constant and unvarying since it acts across five-space (along the fourth direction of space), whereas the permittivity and permeability act within the space-time sheet and are thus affected the local presence of matter. Permittivity and permeability vary from their values in 'free space' due to the local presence of matter, or more accurately they vary from their values in non-curved space-time due to the presence of local variations in curvature.

In Kaluza's theory, the purely five-dimensional component of the field, γ_{00}, is constant and equal to +1. Setting this field variable equal to one normalizes all other components of the field, so this is, and will remain, a normalization constant in the fifth direction in the new model. As such, it guarantees that the amount or quantity of the field (its average linear density) along the A-lines in the fifth direction is constant. The value of +1 also guarantees that a particle exists somewhere along a specified A-line when the wave function collapses such that the particle has a probability of 1 (or 100%) of existing along the A-line within the 'effective width' of the sheet. Like electromagnetic waves, the wave function of a particle lies along the A-lines. The wave function is the particulate equivalent of an electromagnetic wave, but unlike an electromagnetic wave it is characterized by extension in the directions of normal space.

This requirement automatically combines the probability distribution of a quantum mechanical point particle with the wave function in the five-dimensional model. The probability lies along the A-lines in the fourth direction of space. The significance of these requirements will become apparent as the characteristics of the model are developed further and material particles are explained in more detail. With the assumption of a real fifth dimension, the curvature of the space-time continuum described by GR becomes a real curvature as exemplified by the extrinsic nature of the four-dimensional sheet and field potential. Elementary particles are curves of the four-dimensional sheet extending into the fifth dimension.

However, when a sheet with a definite fixed thickness curves into a higher dimension, it folds on itself, doubles over or buckles forming a denser area called a cusp extending from the underside of the curve at the point of folding through the center of the doubled sheet. The exact form of the high-density pocket in the curve depends on the degree (steepness) of the curvature.

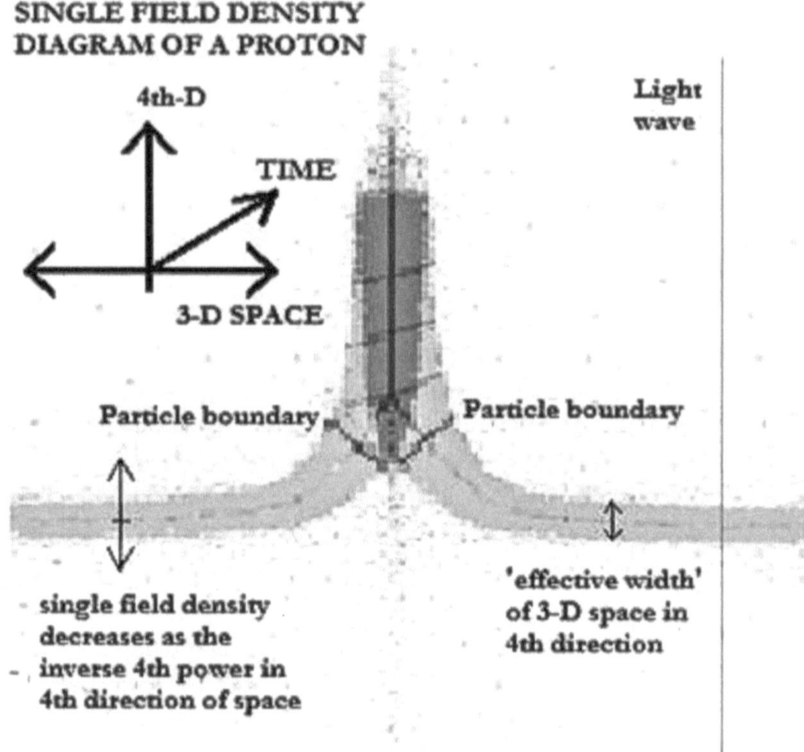

In such cases, these denser areas of the curve correspond to the four-dimensional portion of the singularities that have occurred in GR and the infinite divergences that plague the quantum theories.

The physical boundary of the particle in space-time must extend across the 'effective width' from the point of the cusp and perpendicular to the sheet. The cusp is the densest part of the field. It marks the point where the two sides of the sheet fold together and overlap each other.

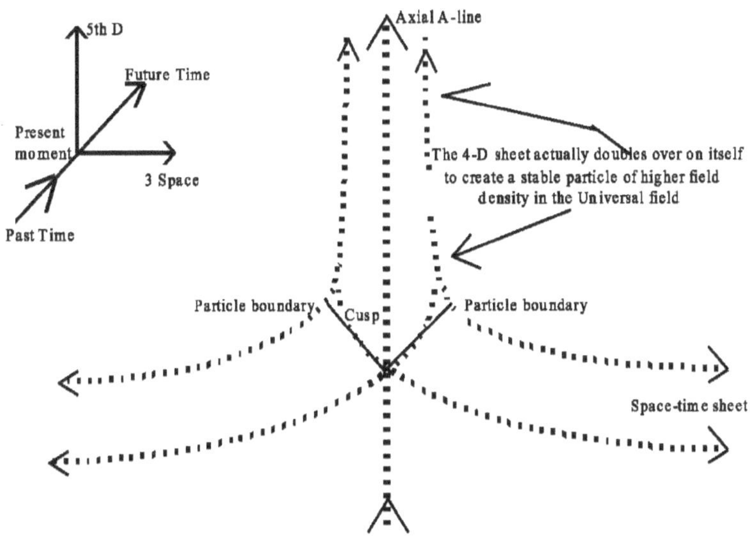

The particle boundary marks the outer edge of the singularity portion of a particle with respect to the normal four-dimensional space-time continuum. However, the singularity disappears where it extends into the fifth-dimension perpendicular to the sheet since field density decreases with increasing distance in the fifth direction from the four-dimensional sheet.

Since the singularity is essentially perpendicular to average flat portions of the space-time sheet and composed of a doubled or overlapping portion of the sheet something strange occurs. Each point of the overlapped sheet is still associated with A-lines, even the doubled over portion which is perpendicular to normal space-time. So, the A-lines associated with points along the axial A-line within the singularity radiate outward from the singularity parallel to the four-dimensional space-time sheet and extend through the fifth dimension. A-lines extending from the particle itself reach out across the fifth dimension to connect with all other material points throughout the universe.

This structure represents what is called 'quantum entanglement' of the wave functions in normal quantum theory. This means that there is a superstructure to the universe whereby all material particles are interconnected outside of normal four-dimensional space-time continuum.

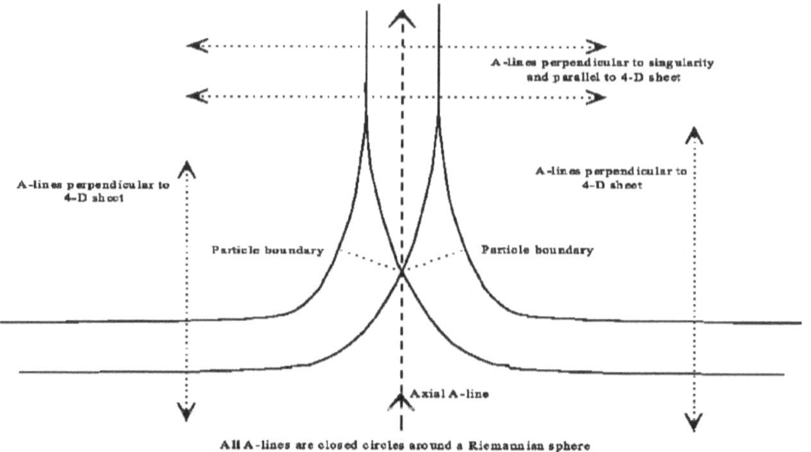

Since the A-lines have equal and constant length, the A-lines emanating outward from material particles must wrap around five space and force the spherical four-dimensional space-time to have the same overall structure in the large scale as the spherical five-dimensional spacetime, or vice versa.

The total length of the A-lines depends on the overall large-scale curvature of the four-dimensional universe and the fifth dimension is Riemannian with constant curvature of the same scale as the four-dimensional Riemannian universe. The mass of an individual particle is finite, due to the curvature up to the particle boundary, even though the curvature becomes perpendicular to the sheet as a singularity. A mathematical singularity is not the same as a physical singularity but models the physical singularity because the particle follows the A-lines that come around through the fifth-dimension Riemannian sphere and close on themselves. The particle is symmetric (it appears to be spherical) in three-space around a central or axial A-line which marks the point of the cusp in the center of the particle.

The axial A-line coincides with the densest part of the field in this case. The axial A-line is continuous around the grand circle of the Riemannian five-dimensional sphere which defines the particle across the fifth dimension. The probability distribution usually associated with the wave function of a particle corresponds directly to this axial A-line. The wave itself marks a point in space or space-time which collapses to a physical particle which is extended symmetrically in five-space about the axial A-line. The particle is a reality independent of the probability distribution and the wave function, but the particle can be represented by the wave function in quantum considerations.

The idea of a 'collapse of the wave function' has caused philosophical problems for both quantum mechanics and quantum theory in general because it creates a discontinuity of action. The discontinuity is a mathematical artifact not a physical reality because the probability corresponds to the wave function, but it is not the wave function itself. The discontinuity occurs where the abstract mathematical point designated by quantum theory devolves into a physically real point in space-time. This model distinguishes between the two. The abstract mathematical point of quantum theory corresponds to the A-line's intersection with a geometrical plane at the center of the sheet.

The wave function itself corresponds to the singularity surrounding the axial A-line of the particle which is physically real. So, the wave function can be considered the five-dimensional 'volume' of the particle. The physical boundaries of the particle mark the classical physical reality as related to and described by Newton, Faraday, Maxwell and Einstein and their respective theories. These four theories are unified with both quantum mechanics and wave mechanics in this single model of an elementary particle as signified by a five-dimensional component. The five-dimensional component of this model is the 'hidden variable' that was first suggested by Bohm.

Bohm's physical model progressed from the 'hidden variable' concept to a hypothetical underlying reality, which he called the 'quantum potential field.' The sheet and its corresponding A-lines in this model correspond to Bohm's 'quantum potential field.' But the 'wave collapse' of classical quantum mechanics also marks a progression in time, a strictly dynamical view of the universe. The above diagrams do not adequately depict this time progression, which lies outside of the flat surface of the paper.

The normal method of diagramming relativistic progression in time is via Minkowski space-time diagrams whose components are the space and time axes, the light cone marked by a past and future extending above and below the origin, and a world-line tracing the particle's history (the past, present, and future of a particle's existence). The world-line is the path of events that occur during the particle's lifetime.

The five-dimensional model of a particle can likewise be viewed in a new diagram that is no more than the space-time diagram turned on its side. When rotated through ninety degrees into the paper and turned on its side, the light cone of the past emanates out in front of the page and the light cone of the future stretches out behind it.

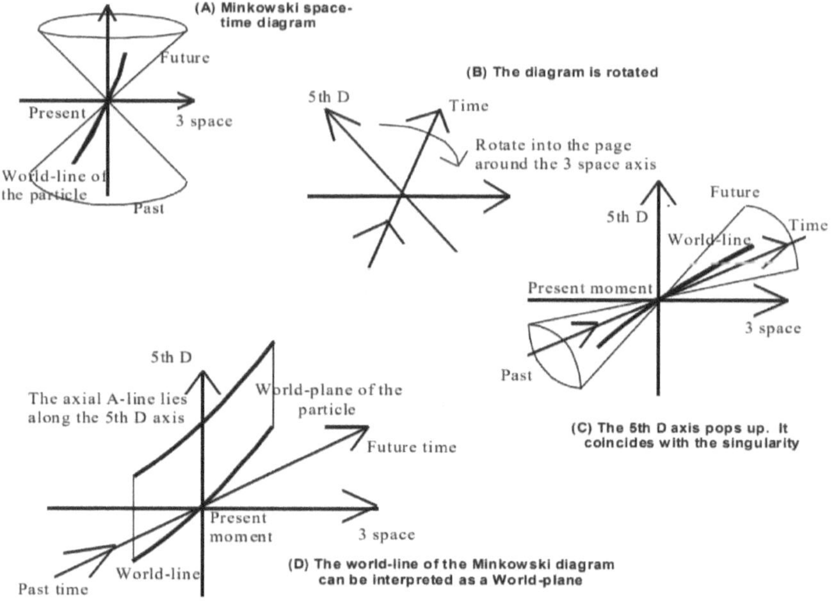

As a particle moves into the future at a constant rate of time, it follows a line along the axis perpendicular to the axial A-line and the three-space axis. The axial A-line representing the particle becomes an axial A-plane along the particle's 'world-line.' This axial A-plane is a diagrammatic representation of the history or lifetime of a point particle. But the reality of the particle is an axial sheet. In other words, the world-plane has an 'effective width' which defines the sheet of the four-dimensional space-time continuum although its width diminishes the further away from the four-dimensional sheet that one goes in the fifth direction.

The particle is restrained from moving at the speed of light or faster by μ_o and ε_o, the connectivity (interstitial) constants of space-time. So, the particle's future path is limited to an existence between the imaginary lines making up the light cone. As a particle moves into the future, it deconstructs the universal field at the present moment of any given situation, but the particle must reconstitute its curvature in the sheet at the next moment in the future. The reconstitution is made from the future portion of the sheet. In other words, the axial A-line marking the particle's fifth component has a projection into the future along the time axis, which represents the state of non-motion if the particle is at rest or, a state of constant velocity if the particle is moving. If the reconstitution does not vary from the projected state of motion of the axial A-line, there is essentially no change in the field. This is the physical representation of Newton's first law of motion.

To vary a particle from its projected path into the future, an abstract A-plane or 'world-sheet' that is not physically real until the moment passes, the particle must accelerate. This action causes a curvature change relative to the projected axial A-line or A-plane. The change of curvature from moment in time representing an acceleration corresponds to a deconstruction of the four-dimensional sheet curvature of the moment past to the newly reconstituted curvature of the moment future.

This process changes the relative field with respect to the rest of the universe as time progresses into the future. The deconstruction/reconstitution process of the spatial change of position in time is a disturbance of the field that is commonly called inertia. The whole field of the universe must react/interact (Newton's third law of motion) with a particle that changes its projected state of motion or relative projected position as time progresses into the future. The process of interaction between the whole field and the changing Aline at a moment causes inertia (Mach's Principle). The action necessary to cause the universal field's interaction (inertia) is the force (in the Newtonian sense) and is proportional to the amount of change of position, relative to time, also called acceleration. Thus, we have F = ma, Newton's second law of motion.

In other words, the universal field resists changes and the inertia of a material particle is the response of the whole field (which in a sense is the universe) to change. In this manner, the connectivity of the points in the field undergoes a change that is constrained by the permeability and permittivity constants. The universal field cannot react to the change or acceleration at a greater rate than constrained by these field constants within the sheet, so matter is limited by the constants in the form of the speed of light.

The Newtonian concept of inertia as a resistance to a change in motion is thus explained. Inertia is the resistance of space-time to changes in the deconstruction/reconstitution process. The amount of destruction/ reconstitution necessary to effect a change is directly proportional to the curvature which gives the particle mass; the greater the curvature, the greater the deconstruction/reconstitution necessary during the process of change of position. Since gravitational acceleration is just the reaction of a particle to space-time curvature near a mass associated with the curvature, there is a direct relationship between gravitational mass and inertial mass. Inertial and gravitational masses are mathematical equivalents although they represent different concepts and are thus philosophically different quantities. This explanation places Einstein's equivalence principle on a whole new footing.

The whole process of deconstruction/reconstitution, which corresponds to the classical concept of motion, also corresponds nicely to Bohm's concept of Bohm's implicate and explicate orders. Bohm's implicate is the extension of the sheet both backward and forward in time, the past and the future. Bohm's explicate is the actual portion of the axial A-plane or the 'world-sheet' which separates the past and future as the present moment. When Bohm's implicate becomes Bohm's explicate reality, the next future moment is reconstituted as the present and the moment just past by is deconstructed. The implicate consists of both the history of all the individual world-sheets that are coupled together or 'entangled' to create the universe as well as the future portion of all world-sheets, as bounded by the present which is the explicate.

At this point, the general development of the five-dimensional model of the universe is complete. It could be argued that this model is nonmathematical since neither mathematical formula nor equations have been offered, but the theory is completely mathematical. This theory and its model have added no new analytical mathematics other than the mathematics already used in Newton's theory of motion, Einstein's GR, Maxwell's electromagnetic theory, Schrödinger's wave mechanics, Kaluza's five-dimensional model and other theories discussed.

The elucidation of this model also implies a direct relationship to superstring theory. The world-line of a particle can be approximated as the trajectory of the cusp moving through time. The cusp can then be equated to a superstring which is vibrating in the fifth dimension along the axial A-line of the particle. The axial A-plane, also termed the world-plane, can be equated to a 'membrane' (brane) as described in the most recent mathematical models and extensions to superstring theory. These recently developed mathematical entities, superstrings and membranes, are only mathematical approximations to the real physical particles as portrayed in this model.

The mathematical analysis of this theory, in the form of equations and formulas, has already been expressed in these other theories. This model provides the "rich structure" for which other scientists have been searching in vain. The most surprising feature of this theory and model is the obvious reduction of all other theories to a single theory with only the assumption of a real fifth dimension and the accompanying answers to questions about how a real fifth dimension could work given our present physical 'laws and theories. Once this model is adopted, the problem will not be finding answers to explain physical phenomena. The central problem of this theory is how to pose the proper questions to ask nature within the context of this theory. The answers become self-evident once the proper questions are asked. The answers are as simple as the nose on your face.

c. Specifics of the model

The representation of electromagnetic waves/photons in the five-dimensional universal field model is both simple and straightforward. As already stated, the electromagnetic wave lies along an A–line, with the photon/particle portion being that segment of the line that cuts across the 'effective width' of the space-time sheet. When two or more waves interact across the whole length of their lines in five-dimensional space, wave phenomena are observed. On the other hand, when the portion of the wave crossing the 'effective width' of the sheet, the photon portion of the wave, interacts with matter, the electromagnetic wave acts as a particle.

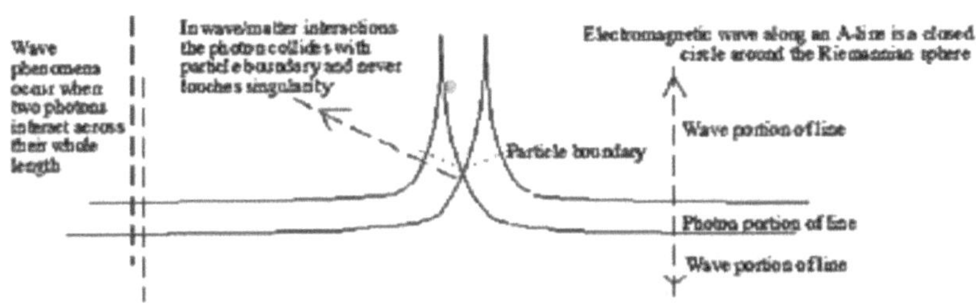

There is no true wave/particle duality.

The wave is a complete closed line on a Riemannian sphere, cutting across the sheet while it remains perpendicular to the sheet. It has only two ways of interacting, one of which is wave to wave and the other that is wave to particle. An electromagnetic wave is restricted to interact with a particle as a photon because it encounters a particle by physically touching the particle boundary within the 'effective width' of the space-time sheet before it can encounter the perpendicular surface of the singularity portion of a particle.

Another curious feature of electromagnetic waves is the ability to propagate through space either as a spherical wave surface or a photon. The reality along the A-line is both. In the above diagram, the A-line is only depicted in one of the three dimensions of normal space. Since the electromagnetic wave is not restricted to a curved surface of the sheet in space-time, although it is connected and continuous with all else in the universe across A-lines in five-space as are particles, and the connectivity/interstitial constants (ϵ_o and μ_o) are equivalent in all directions of space-time, the electromagnetic wave is not restricted to a single position in three-space but can propagate naturally in all directions.

There is a corresponding decrease of intensity at any position in three-space according to the inverse square of distance, but this is a characteristic of the structure of space itself. Electromagnetic waves can thus propagate as a wave front. However, when individual waves interact with matter, as described above, anywhere along their wave front, they do not continue to spread out in space. Instead, they collapse to photons at three-dimensional pints in the sheet. This action is equivalent to the 'collapse of the wave function' in material particle interactions.

Since the collapse takes place only in five-space in a direction perpendicular to the space-time sheet, the connectivity constants play no role in the collapse, which occurs instantaneously. Under special conditions, photons can be coupled together to act in concert, such as in a laser beam. Electromagnetic waves can interact with matter in yet another way, through pair production. If an electromagnetic wave/photon comes close enough to a particle or group of particles without boundary contact, it still follows the space curvature that extends beyond the boundary of the particulate object.

If the curvature is steep enough, the A-line, which the wave/photon follows, will come into contact with the space-time sheet on both sides of the particulate object simultaneously.

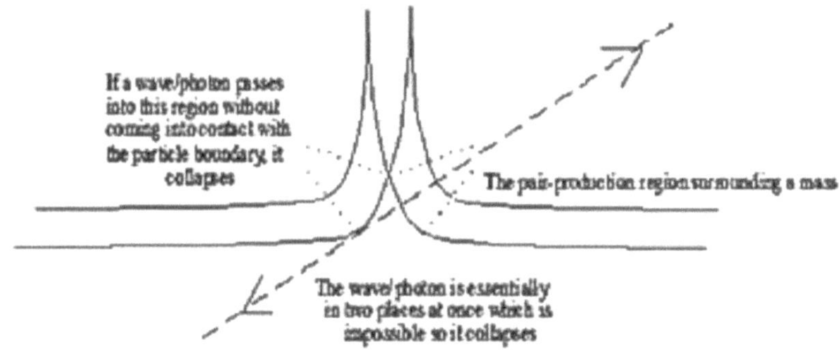

The wave/photon cannot exist in two places at once, so it collapses with its energy being equally shared by an electron and an anti-electron. Since symmetry must prevail and both charge and energy conserved, these two particles will be the same as far as normal space is concerned except that they will have opposite electrical charges.

These two particles will, however, have opposite curves in the spacetime sheet. The particle having a curve on the underside of the sheet is an anti-particle. All anti-particles are characterized by having curves on the underside of the sheet. The underside of the sheet is just that portion of the sheet that lies on the inside of the five-dimensional Riemannian sphere. Due to the conservation/symmetry rules, the particles are identical in all respects except for their opposite charge and oppositely directed curvatures extending from the space-time 'sheet.' More will be said about anti-particles below, after the fundamental particles are discussed.

The elementary particles are the proton, the electron, the neutron and the neutrino. The most common particle, as described in the diagrams above, is the proton. A proton occurs when the space-time sheet is curved so radically that it is doubled over itself or folded (as pictured above). The overlap of two portions of the sheet gives the proton incredible stability. An elementary particle is symmetric in all directions in three-space and is therefore spherical as experimentally detected during high-energy proton-proton collisions. As the sheet doubles over to create the proton, it splits open above the particle and the field density decreases with respect to distance from the space-time sheet in the fourth direction of space. So, while the particle may seem to form a mathematical singularity that is infinite, the physical singularity is not infinite.

The proton will have an 'effective height' in the fourth direction, after which the field density becomes so small that the axial A-line (from the particle's center of mass) is nearly all that is left of the singularity until the line closes around the closed embedding space Riemannian sphere on the other side of the sheet at the same position of the particle.

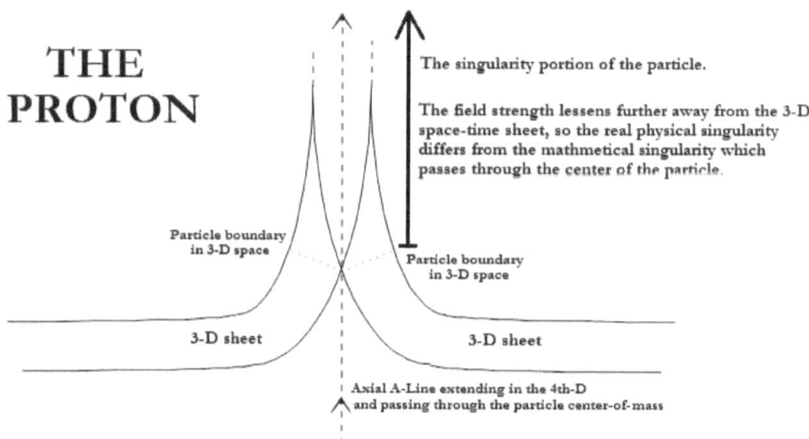

All A-Lines are close circles of equal length around a Riemannian sphere

The A-lines which make up the particle, correspond to mathematical points internal to the particle's boundary and all curve back to meet the particle on the underside of the sheet. Of all these A-lines, the axial A-line has the greatest density at any point along its closed path relative to the surrounding A-lines. In essence, the axial A-line defines the particle in five-space the greater the distance from the three-dimensional sheet.

Recent deep scattering experiments have indicated a physical three 'partedness' of hard centers within the proton, or three-sided-ness relative to the particle target of the scattering. These scattering effects would be expected in the five-dimensional model. The incoming particle would have a very high relativistic kinetic energy in order to complete the deep penetration of the proton. At such high speeds, the incoming particle would be Lorentz contracted (explained in detail later) in the direction of its motion relative to the target proton.

The high speed and resulting Lorentz contraction would cause the incoming particle to split its own dimensionality into a 1+2 configuration as opposed to its normal three-dimensional electrical configuration in normal (nearly flat) space-time when it is scattered. The incoming particle literally 'sees' and thus interacts with the partially separated constituent spatial dimensions of the proton due to its own relatively high speed. The effect is further enhanced due to the increasingly steep space-time curvature of the proton itself. This fractionalization of space into its constituent dimensions and the resulting fractionalization of the particle's electrical characteristics would seem to be a property of the proton since the proton would be moving at the same high-speed relative to the incoming particles' frame of reference.

The electron represents a different and unique geometrical and physical shape. With an electron, the space-time sheet is curved, but does not fold and double into on itself. The proton represents the point where the curvature overlaps completely, while the electron represents the point just before the overlap occurs. The electron also represents the least amount of curvature that creates a dense enough portion of the universal field to create stability.

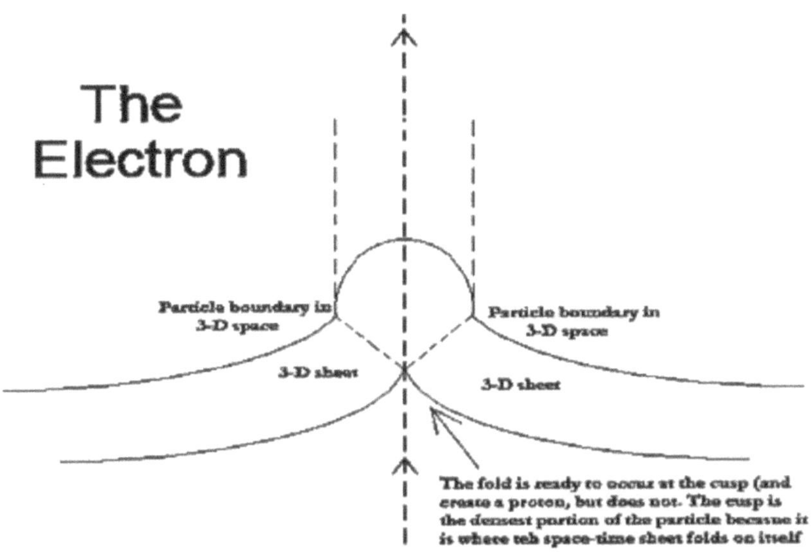

Therefore, the electron and proton are two extremes of the curved sheet. As such, they have opposite charges. There would be a tendency or oppositely directed stress across the 'effective width' of the space-time sheet at a particle's boundary with the rest of the 'sheet' for each particle at the two extremes of curvature. It would be an inward directed stress for one extreme (say the electron) and an outwardly directed stress for the other extreme, yielding the opposite charge of the other particle (for example the proton).

All electrical charges are quantized units and equal to the same value, except for sign, because the permittivity and permeability of free space are constants and the 'effective width' of the space-time sheet at the particles' boundary is constant and equal across the whole universe. Each of the three dimensions of space contributes to the charge equally, or rather the charge is dispensed though three-dimensional space equally. The overall unit of charge is a product of the 'effective width,' the three dimensions and the connectivity of contiguous points of space across the four-dimensional width of the sheet. Since the electron and proton are the two extremes of curvature, they have opposite electrical charges. At the boundary interface between particles and the space-time sheet, the connectivity can be directed 'inward', 'outward' or 'both.'

The normal connectivity of space proceeds in 'both radial directions from the particle boundary in three-space displaying electric neutrality, but the formation of a particle boundary splits the normal two-way connectivity to stress one direction or the other, three-dimensionally in or out of the particles' surface. This creates the attractive and repulsive electrical potentials observed that are observed and measured in four-dimensional space-time, which are just the spatial attempt to seek the same equilibrium of neutrality of connectivity.

The 'outward or 'inward' orientations cause a stress in the connectivity from point to point which is propagated through the interior of the space-time sheet in the form of a field, the electric field. This electric field differs from the gravitational field in a completely fundamental manner because electricity is an intrinsic property of the sheet and gravity is an extrinsic property of the sheet. This fundamental difference explains why it has been so difficult to unify electromagnetism and gravitation in a single theory. As an intrinsic or interior property of the sheet, electrical potential can be blocked or canceled by an opposing orientation in connectivity or an oppositely directed stress emanating from a particle of the opposite charge. The cut transformation of Kaluza's theory isolates and thus represents electromagnetic fields because it literally 'cuts' through or across the sheet.

There is quite a large pantheon of particles in modern physics, but only the principal particles bear mention at this time. Among these are the muon and tauon, both of which have characteristics (such as charge and magnetic moment) that are the same as the electron, but greater masses. So, these elementary particles are higher energy resonances of the electron. As such, they have greater curvatures in the fifth dimension resulting in their greater masses. However, as resonances they do not share the electron's stability. Still higher resonances will not form since higher energies are dissipated away. On the other hand, neutrons and neutrinos have their own special characteristics that render them fundamental to the five-dimensional explanation of physical reality.

Since it has a neutral charge, it would be suspected that the neutron has somehow equalized the inward and outward disparity of stresses in the continuum, and that is exactly the case. The neutron is a hybrid particle, but it is also a unique particle. It is a compound of an electron and a proton coupled together although they essentially lose their identities within the neutron. Under proper conditions, an electron and proton can combine around a single point in space to create a neutron by stacking one upon the other in the fourth direction of space. The conditions are those that will allow the axial A-lines of each individual particle to overlap or coincide geometrically. Since the two axial A-lines of the electron and proton degenerate into a single axial A-line, the neutron is a unique particle and the electron and proton lose their individual identities.

However, the neutron will share the curvature of the two and thus have the mass of both. It also inherits the connectivity of both at its boundary and thus has both inward and outward connectivity, which is electrically neutral.

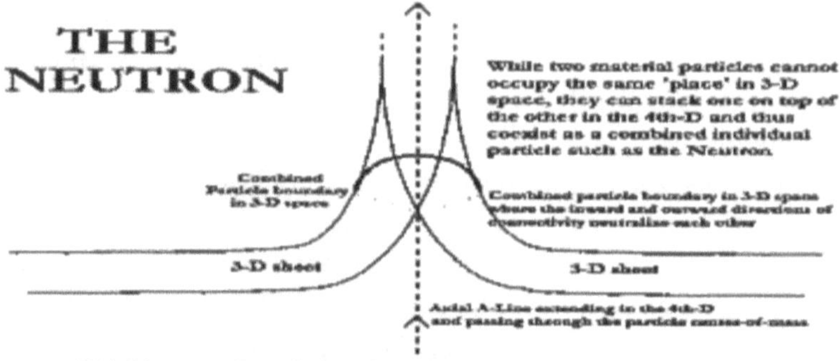

The neutron is not normally stable as a unique particle when it is not combined (stacked four-dimensionally) with protons in the nucleus. Outside the nucleus, it will decay into a proton, electron, and an antineutrino. Again, to conserve charge, the proton and electron will take opposite charges during this decay. The conditions for the decay must be those that allow the axial Aline of the neutron to split into the axial A-lines of the proton and neutron and separate the proton's cusp from the electron's cusp, the two areas of greatest field density.

The final elementary particle of any note is the neutrino. The neutrino is a by-product of decays, particle interactions and cosmic events. During these various processes and workings of nature, there may be dense portions of the sheet that are 'left over' at the conclusion of an event, *i.e.*, the decay products of a nucleus or another particle. This dense area occurs in multiples of a basic 'quantum,' or equivalent sized lumps of field density, and exists within the sheet centered symmetrically on a single A-line, just like other particles.

This 'quantum' of slightly higher density within the space-time sheet causes a small 'burble' of curvature in the sheet, which is detected as a very minute mass, but it is so small that the corresponding mass is negligible for most cases. Yet this 'burble' or bulge of curvature is enough to restrict the neutrino from moving at the speed of light, even though it has such a small mass that it can travel at very nearly the speed of light. The neutrino represents the smallest amount of local curvature of the sheet that can exist as a variance from (or independent of) the overall non-local spherical curvature due to the universe as a continuous whole.

The Neutrino

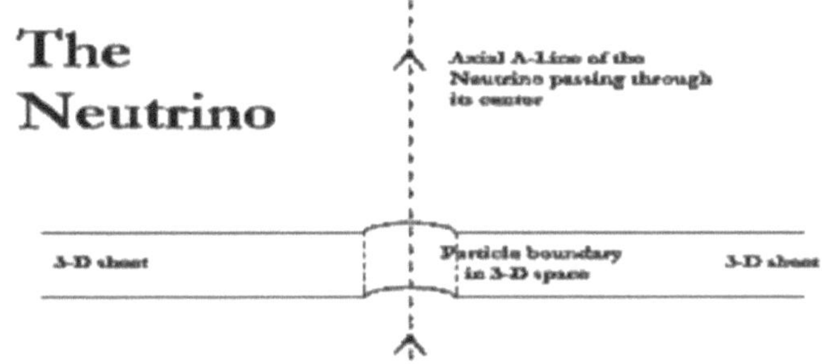

These bits of 'burble' in the 'sheet' can only occur as the result of other physical processes, but they do have the general characteristics of particles. The unique character of the neutrino results from its left-handedness. When a neutrino is created, it immediately reacts to the five-dimensional Coriolis effect of the expanding universe (explained later), much as the unique direction of a hurricane in the northern hemisphere results from a Coriolis effect, and thus the neutrino favors left handedness as opposed to a right handedness.

Like other elementary particles, neutrinos have antiparticles. Antiparticles are merely particles that are formed by a local curvature on the opposite or under side of the sheet. Pair production has been explained as the decomposition of an electromagnetic wave/photon due to its simultaneous contact with two different points in the space-time sheet. No physical entity can be in two places at the same time. It is a physical impossibility. Yet there is another phenomenon that acts in the opposite manner to pair production, the mutual annihilation of particles with their anti-particles. The only difference between particles and their anti-particles are the opposite curvatures along different sides of the sheet and opposite electrical charges. Both differences are subject to the properties of the five-dimensional model and the reality of the space-time sheet.

When a particle and its anti-particle come into physical contact, their curvatures cancel one another in much the same way that as the simple cancellation of waves known as destructive interference. Two particles cannot exist at the same place at the same time, but anti-particles can coincide for the moment before annihilation because they are positioned on opposite sides of the sheet and have opposite charges.

From the four-dimensional view they occupy the same position but from the five-dimensional view they do not. So, as soon as they come to coincide four-dimensionally, they mutually disintegrate or annihilate each other so that the two views match, as required by relativity. The basic principle of relativity requires that any event be the same from different perspectives or points of view. When this annihilation occurs, a photon or photons of equal energy are emitted. The energy conversion discovered by Einstein in his work on special relativity, $E = mc^2$, controls this process and all similar processes where energy and mass are converted. However, for simplicity and philosophical accuracy, it would be better to express the matter/energy equivalence in terms of the connectivity constants such that $E = m/\mu_0\varepsilon_0$ (instead of $E=mc^2$) reflecting the true nature of the continuum.

Connectivity is more intimately tied to the interchange of energy and mass than the speed of light 'c' is. During annihilation, the opposite charges of the particle and anti-particle neutralize to reestablish the connectivity of the dimensions of three-space at the boundaries of the annihilated particles. The connectivity of free space is in a sense 'repaired' within the sheet to its normal conditions of the neutral state. One of the curiosities of the particle/anti-particle pair is the fact anti-particles have all the same characteristics of their particles except for the charge. Why do antiparticles not display anti-gravity or other anti-properties? The anti-gravity problem is easy to understand within the five-dimensional framework.

It does not matter whether the curvature is outward above the sheet and away from the Riemannian sphere or inward below the sheet, gravity acts the same way from the four-dimensional perspective. In either case, as another particle of matter approaches, it will experience the same amount of grade or steepness of curvature and thus the same gravitational acceleration. Gravitational acceleration is the product of the geometry of the curvature and nothing else, so it does not distinguish between the two cases of outward or inward curvature. Since a particle and its anti-particle have the same mass, the degree or gradient of the curvature is equal at similar points relative to an axial A-line. Anti-gravity, therefore, does not exist.

Another interesting fact literally pops out of this model. When regarded as a particle, the photon acts as is its own anti-particle. This fact is easily explained because the photon exhibits no curvature and is therefore neutral to the outside/inside direction of the overall space-time curvature of the Riemannian sphere. No other theory can account for this fact. In fact, physicists have probably never asked the question, even though this theory suggests an answer. All particles and anti-particles fall into two basic groups, bosons and fermions. Fermions seem to include most, if not all, of the fundamental particles in nature, but both groups are significant in modern physics. Bosons are characterized by an even number of quantum spin units while fermions all have an odd number of quantum spin units.

The supersymmetry upon which superstring theory is based allows the establishment of a physical relationship between bosons and fermions that is missing in all other theories. However, there is no tangible evidence that nature is supersymmetric. Supersymmetry deals with the fact that fermions must be rotated through two complete turns of 720 degrees to return to their original quantum state, thus yielding a ±½ spin. Bosons return to their initial quantum spin state after only a 360-degree rotation and thus have a ±1 spin. These differences indicate that an unusual geometry must affect the physics of fermions and bosons.

There is a specific symmetry of Riemannian spaces, which fits this case quite well. There are two general types of Riemannian geometry, the single- and double-polar spherical curvature. The single polar Riemannian space has the property of reversing (left for right) any object that moves through the space and returns to a lower dimensioned space. Simon Newcomb was the first to discover this property (Newcomb, 1877), but counter claims of discovery have been made for Felix Klein (Monro, 1877) and F.W.F. Frankland. (Frankland, 1876) H.G. Wells even used this discovery in his science fiction tale "The Plattner story." (Wells, 1897)

Using a single polar geometry to describe bosons and fermions reduces the difference between odd and even numbered quantum spins to simple geometrical differences in the five-dimensional model. If the Riemannian five-dimensional sphere is single-polar, then a trip around the sphere from the top of the sheet to a corresponding point of the underside of the sheet (like a Möbius strip), a single circuit of the sphere, would alter an object by exchanging left for right and vice versa.

It would take a particle's axial A-line two complete circuits around a single polar Riemannian sphere to return to the 'sheet' in its original spin orientation. But the axial A-line only completes a single circuit of the Riemannian sphere in the fifth dimension, so it returns to the underside of the 'sheet' at the site of the particle in an opposite orientation. This yields the ±½ spins of fermions like the proton and electron. Thus, the five-dimensional model can account for the difference between the spins of fermions and bosons.

According to present theory, the universe was created in a process known as the 'Big Bang.' A primary singularity seems to have decayed, and our physical universe evolved into the world that we perceive today from its remnants. One of the major problems with this model is the existence of particles, the vast majority of which are matter rather than anti-matter composed of anti-particles. In other words, there is no reason why there should be a dominance of particles over anti-particles in our universe.

The two sided-ness of the space-time sheet can be offered as a simple explanation of the preponderance of matter in our universe at the expense of anti-matter. When the primary primordial singularity burst into our universe, a three-dimensional Riemannian sphere embedded in a fourth dimensional sphere began to develop and expand outward from the event. The first particles created (by a big Blow Out) after the expansion outward initial event (when Cosmic Inflation ended) would have been formed on the outward surface of the sphere only, that is, given the fact that curvature is a real extrinsic property of space-time and not just an intrinsic mathematical feature of the geometry.

If the initial Riemannian sphere was very small relative to the three-dimensional size of a proton when it was formed, the first particle could only have formed on the outward surface, not the inside of the sphere. As the sphere expanded, the impetus of change would have been from the inside to the outside of the spherical curvature of space-time, the direction of expansion, which would also have favored the development of particles rather than antiparticles and anti-matter. So, matter is dominant in our universe and gross quantities of anti-matter are a rare occurrence.

d. Quantum theory revisited

As an approximation, all particles obey Newton's laws of motion, Universal gravitation, and Maxwell's electromagnetism. However, Newton's laws of motion degenerate in the submicroscopic world of the quantum where gravitational forces are considered too weak to affect particle motion. Quantum and wave mechanics fill the void left by the inability of these classical theories to completely describe the interactions of elementary particles. Quantum mechanics utilizes a probabilistic interpretation of wave mechanics with which many scientists have felt ill at ease but have accepted because of the successful predictions of the theory.

It has already been argued that the probability distribution of quantum mechanics merely mimics the space-time continuum but cannot really account for space or time as field effects.

The probability distribution is merely a mathematical device used to mimic spatial extensions of point positions while the waves of wave mechanics represent some type of real physical quantity. There is no independent evidence that the probability distributions have physical reality. So, while the probability distributions and the waves themselves correspond nicely, they seem to be mathematical equivalents, they are not the same 'thing,' physical or otherwise. The problem becomes more than causally evident when the wave collapses to form a physically real particle that has spatial extension. The probability distribution that represented the particle before the collapse represented only a collection of mathematical points that constitute a mathematical space but have no real extension which is a property of a real space. So, the collapse of the wave function represents a discontinuity of concept, a paradox for quantum theory.

This paradox disappears, as does the wave/particle duality, in the five-dimensional universal field theory model. In fact, they are intimately related. In the five-dimensional model, a portion of the sheet literally turns or rotates in the fifth direction of the continuum by folding or bending. This gives the axial Aline a substantial reality as the densest part of the resulting particle. The single field potential density along the axial A-line decreases as the distance from the space-time sheet increases in the fifth direction. By the same token, A-lines extend in a lateral direction perpendicular to the axial Aline (parallel to the sheet).

The field density decreases as the distance from the axial A-line increases across the fifth dimension by the same proportion as distance from the sheet.

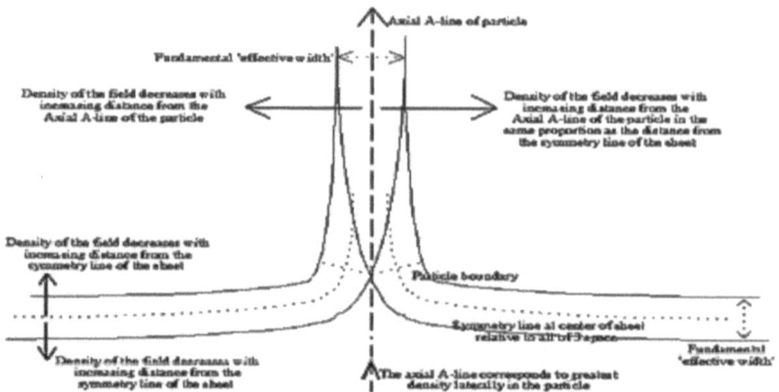

The probability distribution of quantum mechanics emanates outward from a point which coincides with the intersection of the axial A-line and the symmetry line at the center of the sheet. However, the probability distribution also coincides proportionally with the field density distribution along the axial A-line of any given particle.

The probability distribution of quantum mechanics is a three-spatial quantity which is manifest in the fifth direction as a mathematical representation of the axial A-line. On the other hand, the wave function represents the five-dimensional volume of the particle itself. In a 1928 paper, W. Wilson derived an equation that became upon substitution identical to Schrödinger's equation in wave mechanics.

Wilson used the concept of a mass 'Volume' in five-dimensional space-time, whereas the normal quantum mechanical interpretation of the wave function is a probability distribution. He continued to derive a second equation which he showed to be equivalent to Schrödinger's equation for the Hydrogen atom under a proper choice of limits. From these derivations, Wilson concluded that Schrödinger's wave function Ψ was the five-dimensional mass 'Volume' of the particle in question.

If a particle at some instant is actually within a 'volume' M_0 it will be within a volume V, which is the parallel displacement of V_0, at some time later (or earlier) instant. If its position at any

The correlation between Schrödinger's equation and those derived by Wilson was complete. But Wilson had only derived what is today known as the Klein-Gordon equation. Coincidentally, Klein and Gordon developed the equation that bears their name by considering "a zero-mass wave equation in five dimensions" in the Kaluza theory. (Appelquist, 10) Although the Klein-Gordon equation is important in wave mechanics, the Dirac equation is far more important. The Klein-Gordon equation is a second order differential equation, which describes the motion of a particle at relativistic speeds. The equation can be generalized for a particle that interacts with an external electromagnetic field. However, the equation is not fully consistent with general transformation theory.

Dirac was bothered by this lack of consistency (Schweber, 57-58) as well as the possibility of a negative probability density, so he derived a competing equation which was not only consistent with the general transformation theory, but always yields a positive probability density. The Dirac equation, as it is called, utilizes a first order derivative with respect to time instead of the second order derivative. The Dirac equation was a step forward in applying wave mechanics to physical situations. The Dirac equation has also been equated to five-dimensional space-times by other researchers. (Flint, 1966, 111; Jonsson, 1942)

The further interpretation of the probability equivalence to the fifth dimension awaits a consideration of special relativity. The dependence of a complete explanation of the quantum on special relativity marks a strange turn of events since classical quantum mechanics is independent of relativity theory and indeterministic while relativity is deterministic. However, it should have become evident by this point that there is no physical 'collapse' of the wave involved because it is a representative mathematical function alone and there is no discontinuity due to the 'collapse'. So. there should not be any further inconsistencies between quantum theory and special relativity.

That has been called the 'collapse' of the wave packet or function is only the purely mathematical transition (or mapping) of the probability to the actual wave, which merely represents two different ways of interpreting the phenomenon without explicitly demanding the physical reality of some undefined form of the 'collapse.' The emphasis therefore changes to explaining how the probability fits into the five-dimensional picture.

e. What's so special about special relativity?

Although special relativity was originally developed as a theory of electrodynamics, it is more often and more commonly regarded in its disguise as a mechanical theory explaining what happens to material particles moving at speeds closer and closer to the speed of light. Within this context, there are several formulas that describe changes of a material body's (or a single particle's) characteristics due to its relative motion.

The first of these is the Lorentz-Fitzgerald contraction. The moving object contracts by an amount dependent on the relative speed of the object to an observer. The resulting contraction of the body can be easily depicted in the five-dimensional model. As the body moves, it undergoes the deconstruction/ reconstitution process from moment to moment. But as its position in space changes by greater and greater amounts with respect to the constant progression forward in time, the deconstruction/reconstitution process is not as effective due to the connectivity constants. This is part of the same process that limits the speed of the object to the speed of light.

The reconstitution process literally stretches the curvature further into the fifth dimension as the particle approaches the speed of light.

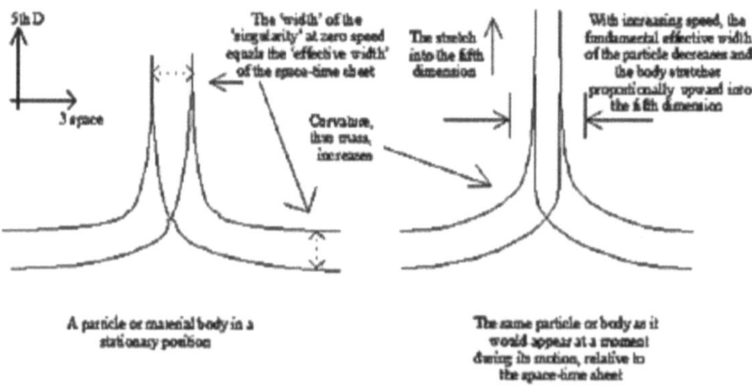

The five-dimensional volume of the particle itself must remain constant, so the particle contracts in the direction of the motion. The contraction leads to a greater or steeper curvature and hence a greater mass for the moving particle. The above diagram only depicts a single dimension in the three-space direction of motion, so that direction represents the direction of contraction. The body's boundaries remain constant in the other two directions of three-space. In essence, the particle becomes more wavelike or light-like as it approaches the speed of light.

By light-like, it is meant that the density of the field becomes spread more evenly over the entire length of the axial A-line mimicking the A-line of a light wave. However, the particle or material body cannot become truly light like since it is only contracting in the direction of the motion. Its width or measure in the other two dimensions remains the same. A light wave is unique in that it has no width in any direction in three-space, so it can spread out evenly without physical obstruction in all three directions and propagate spherically from a light source. One might say that light is Lorentz-Fitzgerald contracted to zero (or some minimum as thickness approaches zero) in all three-dimensions of normal space simultaneously.

So, while the instantaneous speed of a material particle or body can be considered the time derivative of change of position in three space as the time change approaches zero (d[x]/dt = v) such that the change of position (dx) occurs in the same direction as the contraction, the instantaneous speed of light is the change of position of the light wave in all three contracted directions simultaneously as the change in time approaches zero (d[a point in 3 space]/dt = c). Therefore, the light wave can spread out in all three dimensions of normal space simultaneously creating the spherical wavefront. The moving particle tries to mimic this behavior or quality of light at high speeds, but it cannot because its remaining two dimensions are not contracted to 'no width' at, or rather near, the speed of light. The particle may approach the speed of light, but it never reaches the speed of light because it cannot become light-like in just one dimension independent of the other two dimensions.

However, as an elementary particle moves closer to the speed of light and its fundamental "effective width" in the direction of travel contracts relative to a stationary object, its contracted width can become so small that the connectivity of space in the one direction can be detected independent of the other two directions. This means that the quantum electric charge of the particle, at the very high energy where this 'minimum effective width' is attained, would appear to split into two different charges of one-third, representing the portion of the charge in the direction of travel, and two-thirds, the portion of the charge representing the other two directions.

This splitting of the fundamental electrical charge, which can only occur at extremely high energies, has been interpreted as independent particles called quarks in modern physics. Quarks have never been detected independent of the particle that they supposedly constitute, nor will they ever be so detected because they are not independent particles but rather reflect a condition of true particles moving at very high speeds and thus carrying very high kinetic energies. Quarks have no more independence as individual particles than we can split and detect the three spatial dimensions of the space-time continuum.

As a particle's relative velocity increases from rest, the particle's center of field density extends further into the fifth dimension according to the rules of special relativity. This causes an additional stress in the spacetime continuum where it is already stressed by the existence of electrical charge. However, this added stress does not act directly between contiguous points of space extending radially outward from the surface of the charged particle in three-space, as does the electric field.

The relative change of position of a particle's center of field density in the extra direction of space acts somewhat like a torque in the extra dimension. So, like a changing torque vector it causes a stress that is perpendicular to direction of the vector, or the stress would occur in a plane surface in three-space. This is a torsional stress whose strength is directly proportional to the relative speed of the particle. This new stress, commonly called magnetism or the magnetic field, is also perpendicular to the direction of motion of the particle that caused the torque-like change in the fifth dimension. Only charged particles display this torsional stress since only charged particles cause an electrical stress in the surrounding space-time continuum.

The relatively high speed of a moving particle can explain still other quantum aspects of reality. Since the five-dimensional volume remains the same but the field density across the axial A-line changes as speed increases, the particle becomes more wavelike. So, it would be characteristic of the particle to act more like a wave. Indeed, this is exactly what it does according to its deBroglie wavelength.

A particle which has zero velocity, essentially has an infinite wavelength according to deBroglie's formula $\lambda = h/mv$. Written as a deBroglie frequency, this expression has far more meaning. Using the dispersion relation of E = hf, the deBroglie frequency can be written as f = $p^2/2m\lambda^2$. This formula works for a nonrelativistic particle. If the relative speed of an object with respect to an observer is zero, then the object has a deBroglie frequency of zero.

In the case where a particle moves at relativistic speeds, the dispersion relation becomes …

$$(hf)^2 = (m_0 c^2)^2 + (\frac{hc}{\lambda})^2 .$$

If we say that the wavelength goes to zero as the speed approaches that of light, then the frequency becomes infinite. In this formula, the second term on the right would become infinite as the wavelength goes to zero. So, it is impossible for a material object to travel at the speed of light. But as a particle or material object approaches closer and closer to the speed of light, its deBroglie wavelength would become smaller and smaller and the second term in the dispersion relationship would dominate.

At slower relative speeds, the rest mass energy dominates the frequency relationship. This whole system is compatible with SR only in so far as the frequency of a matter wave has any physical meaning. In the case of an electromagnetic wave, the frequency denotes a real 'vibration' or 'oscillation' of the combined electric and magnetic fields, but this is not true for matter waves. The frequency of a matter wave only makes sense within the context of the deconstruction/ reconstitution process, which changes with changing relative speed.

Both the object and the observer undergo the same rate of deconstruction/reconstitution as time progresses forward when their relative speed is zero. But as the object moves relative to the stationary observer, its deconstruction/ reconstitution rate differs from that of the observer by a factor of the deBroglie frequency. The maximum difference between the deconstruction/reconstitution rates would occur as the relative speed between two objects approach the speed of light. This may seem to imply the discrete nature of time as opposed to the continuous nature of time, but time remains continuous. Since progress forward in time is continuous, one would think that the deconstruction/reconstitution frequency would be infinite, or rather, the process would occur instantaneously. The fact that it can be changed by a constant calculable factor implies that the frequency is not infinite, but discrete and thus time is discrete.

However, the particle's physical boundary extends into the past and future just as it extends to the right and left or up and down. The space-time sheet has a thickness in the dimension of time, so the particle boundary at the point in time that a deconstructing particle from the moment just past overlaps the reconstituting particle in the next moment of the future, guaranteeing continuity while the frequency of the deconstruction/reconstitution process can remain finite and still be increased by a constant factor.

The constant factor of frequency increases in proportion to the amount of stretch of the particle in the fifth direction, or rather the increase of effective extension in the fifth dimension. The amount of stretch or extension is not infinite since the A-lines are closed circles on a large, but finite closed Riemannian five-dimensional sphere. There is a practical limit to the frequency change factor and a 'minimum width' that the Lorentz-Fitzgerald contraction can attain which is directly related to the distance around the Riemannian sphere and thus the overall curvature of the universe. This value is related to the Planck length, $L_p = (hG/2\pi c^3)^{1/2}$ or 1.6×10^{-33} centimeters.

These quantities are related to another important phenomenon, the tunneling effect. According to quantum theory, it is possible for a particle to escape through an energy barrier, out of a potential well, that it does not have the energy to overcome. Tunneling events have been 'observed' in the laboratory and are normally used to explain radioactive decay. In essence, the probability distribution of a particle is extended over space including areas of space beyond a particle's boundary and the potential barrier. There is always a small probability, even if it is only infinitesimally small, that the 'collapsing' wave function will leave the particle elsewhere than where it has the greatest likelihood of being, where the probability is the greatest. So, there is a very small possibility that the particle will end up 'outside' of the barrier when its wave function collapses.

Consciousness can cause the wave 'collapse,' as it is called in quantum mechanics, under the proper conditions, but human consciousness is not 'necessary' for a wave to 'collapse.' The deconstruction/reconstitution process could certainly be called the 'collapse' of the wave function since it is the same thing, but this process proceeds normally without the intervention of human consciousness due to the complex entanglement of the rest of the universe. It is a relative, not a quantum, process which occurs at the quantum level of the continuous field that constitutes the universe.

There are two, and only two, distinct types of energy in space-time, kinetic and potential. Potential energy is neither more nor less than energy of relative position while kinetic energy is the energy of relative change of position. (Clifford, 1880) In the five-dimensional model, this changes. There is only energy of relative position, the potential. Kinetic energy in the four-dimensional space-time continuum reduces to pure single field potential in the five-dimensional worldview just as four-dimensional space-time dynamics becomes five-dimensional potential field density-variations kinematics.

A particle with a constant relative velocity in space-time, when 'viewed' from the five-dimensional perspective will have an extended density along its axial A-line, relative to the density distribution along the axial A-line when the particle is stationary. This relative density (position) change is detected as a kinetic energy in the four-dimensional spacetime, but it is a potential in the five-dimensional space-time framework.

When a particle accelerates, the relative density shifts up or down the axial A-line, for increasing and decreasing speed respectively. So, at higher relative velocities the density dispersion along the A-line renders a deBroglie frequency (or wave) corresponding to the particle. This corresponds to a higher dispersion of the probability density perpendicular to the axial A-line in three-space, increasing the probability that the particle can be elsewhere than its primary expected position. In a sense, the particle has less of a presence in the four-dimensional sheet at higher relative speeds, so it is less 'anchored' to its fixed position in space-time and thus has a greater 'relative' probability of being (tunneling or reconstituting) elsewhere.

As the particle moves forward in time, undergoing deconstruction/reconstitution, there is a small possibility, due to its motion and/or interactions with other fields and other particles, that the particle can deconstruct at one point in space and subsequently reconstitute at a point in space that is not continuous with its previous position in space. Since external fields affect the particle's potential energy, in the classical sense, then these external fields and connections with other particles will also affect this process. Even when the particle is not moving, it is still subject to the effect of other fields, so this same possibility exists, perhaps to a much smaller degree and perhaps not, depending upon the relative positions of other particles and their associated fields. Anything which changes the aspect (universal field distribution) of a particle along its axial Aline will change the wave function of the particle as well as its probability distribution in three-space to which it is directly proportional.

In an atomic nucleus, an individual nucleon will have a higher relative aspect along its axial A-line due to the extreme degree of the additive effect of curvature coupled to the 'fundamental effective width' of space-time, so it will have a more wavelike character. Therefore, it will have a greater probability of being elsewhere outside of the potential barrier of the nucleus unless it is more tightly bound by its axial A-line to the other nucleons. Under the correct conditions, a nucleon can thus 'tunnel' out of a nucleus as has been 'observed' in the laboratory. A further discussion of this process must await a description of the nucleus in the five-dimensional model. The point is that tunneling is dependent upon the relative position and is thus explained by a combination of quantum theory and relativity theory.

f. Relating to special relativity

It is quite apparent that relativity and the quantum are intimately bound to one another. The Lorentz-Fitzgerald contraction is tied to the deBroglie frequency (wavelength) which is related to a particle's aspect (universal field distribution) along its axial A-line which is in turn related to (or corresponds to) the particle's probability distribution in quantum mechanics. The fact that $\int |\Psi|^2 dx = 1$ in quantum mechanics merely translates that the total amount of linear field density along the axial A-line is constant which is equivalent to saying that the volume in five-space (which equals the wave function) is constant, a new conservation law.

In the original Kaluza theory, the fifth field component γ_{00} was set equal to one. This act has been criticized as artificial or *ad hoc* without cause and merely a mathematical device to make the theory work, without physical meaning and perhaps even over restrictive. But the choice of this value for γ_{00} was quite meaningful, even if Kaluza did not realize its significance. While the probability distribution of quantum mechanics is a mathematical device used to locate a particle's position during interaction with an electromagnetic field and/or other particles, as in the case of orbital electrons, there is a definite probability that it does exist, *i.e.*, the sum of all the probabilities at each and every possible location of the particle is one. In wave mechanics this means that the square of the wave function is one, $\int |\Psi|^2 dx = 1$, or the trace of $\psi\psi^*$ is one.

The fact that the probability of finding a particle is unity guarantees the physical reality of the particle even if its momentum and position cannot be known simultaneously. In effect, setting the probability equal to one contradicts or countermands the very Heisenberg uncertainty principle that requires the probabilistic interpretation of physical reality. Neither scientists nor philosophers have ever questioned this paradox.

Given the previous explanation of the relationship between the probability distribution and the field density distribution along the axial A-line of a particle, the probability that a particle exists somewhere along the axial A-line, relative to the space-time sheet, must be one. This is the true meaning for the value of one for the fifth component γ_{00}. So, the choice of the value of one, which Kaluza used to guarantee that the other field components would normalize as well as guarantee that the fifth dimension could not be perceived, has the same meaning in quantum mechanics. It guarantees the existence of the particle at some point or place in space-time.

In any beginning class on quantum theory, a student learns about the Kronecker delta function (which is related to the Dirac delta function). The delta function represents a method of normalizing a non-behaving function to a value of one. The infinite plane waves that arise in quantum theory are not normalizable to unity in some cases because their eigenfunctions are not quadratically integrable over all of space. In such cases, the delta function is applied to allow the normalization to progress smoothly. The delta function is defined as equal to one in cases where the discrete variables are equivalent and zero where they are not equal.

So, a well-behaved function can be found and written for a function with undefined spots, when it is limited by a delta function, yielding the following equation.

$$f(y) = \int_{-\infty}^{+\infty} \delta(x, y) f(x)\, dx\,.$$

The delta function $\delta(x,y)$ is zero everywhere in space except where the x and y variables are very close in value, or rather in the limit where the value of x approaches the value of y. Physically, this is the same as saying that the resulting wave function f(y) represents a superposition of two waves that interfere constructively to give a well-behaved function with no infinities. (Merzbacher, 82) The infinities have been canceled by destructive interferences.

In the Kaluza theory, the fifth component γ_{00} is nature's equivalent to the delta function. According to Mach's principle, the mass (and thus the curvature of space-time) of an individual particle depends upon the mass of the rest of the universe. If that particle is represented by a wave function, as normally defined in quantum mechanics, the mass would be infinite. But taken relative to the rest of the universe, in the real physical situation, the wave function of the particle interferes with the collective wave function for the rest of the matter in the universe, such that the infinity at a point of mass is canceled.

The whole universe must be collectively considered so that the waves of all particles cancel in this manner. This act can only be accomplished in a fifth dimension, where the singularity occurs as a real curvature, with γ_{00} acting as a natural, and thus real, delta function. In other words, the universe acts to normalize the real particles even though mathematics cannot always normalize the quantum theory's point particle equivalent of a real particle. This interaction is called 'entanglement' in quantum theory, but it is the ultimate expression of Mach's principle and relativity in the five-dimensional field theory. It is the 'entanglement' or the connectivity of all matter in the universe at any given moment. The space-time sheet could thus be portrayed as the collective 'destructive interference' due to the sum of all matter waves in the universe. In reality, the interference could not really cease to exist and thus attains a minimum measurement called the fundamental 'effective width.'

While the delta function helps to explain the spatial components of reality, it does nothing for time. It is purely spatial and thus acts kinematically, not dynamically. Time dilation offers a far more unique perspective on the spacetime continuum that emphasizes the difference between the time and spatial dimensions. Lorentz-Fitzgerald contraction, energy, and mass (spacetime curvature) all return to normal when the velocity of the moving particle returns to a relative speed of zero. However, the dilated time frame of the particle and any other material body that moves relative to a state of rest is a permanent and lasting condition. This fact alone guarantees the uniqueness of time even as it causes grave difficulties in understanding that uniqueness and the nature of time. As a particle (or other material body) moves at higher and higher relative speeds, approaching the speed of light, the deconstruction/reconstitution rate in the time dimension decreases and the field density shifts further into the fifth direction along the axial A-line.

It could be said that the particle tries to detach itself from the sheet of normal space-time as much as possible, but it cannot detach itself, so it stretches itself further into the fifth dimension dragging its portion of the sheet with it. However, it cannot stretch the time component of the sheet upwards since the time direction has not yet deconstructed and reconstituted itself. The center of universal field density along the axial A-line would normally be located at the cusp where the spacetime sheet folds to mark a particle's physical boundary. At higher speeds, especially closer to the speed of light, the center of field density moves upward in the fifth direction along the axial A-line. Time is not experienced at the same rate as the center moves up the A-line.

Just as a light wave can spread out spherically in three-space due to its constant field density along its own axial A-line, without physical restrictions, the wave corresponding to a material particle can spread out over the time dimension so that a moment of time is extended in duration as compared to the rest frame. This process occurs in proportion to the mass increase and contraction and relative to the speed by the same ratio. However, the uniqueness of time and its special properties cause time duration to change as the reciprocal to the factor which affects all three processes, $(1 - v^2/c^2)^{1/2}$. Since deconstruction/reconstitution occurs only in the time dimension, and its rate decreases as the relative speed changes, the time dilation causes permanent changes in a material body's life that do not occur in the other dimensional variations with relative speed increases and decreases.

The inverse effect of time is evident in deBroglie's relationship between matter and wave and can thus be viewed as a product of the wave/particle duality. As demonstrated earlier, a material body has a deBroglie wavelength of $\lambda = h/p$ or h/mv in three-space while it has an equivalent frequency of $f = p^2/2m\lambda^2$ in the non-relativistic case. This reduces to $f = \frac{1}{2}mv^2/h$ in terms of classical kinetic energy. The position and effect of the speed v of the particle is inverted between the frequency (time component) and the wavelength (space component).

In Newtonian physics, these quantities represent momentum (mv) and kinetic energy ($\frac{1}{2}mv^2$). Both momentum and kinetic energy are measures of the combined quantity of motion and matter. In a manner of speaking, these two physical quantities, and only these two physical quantities, characterize 'matter in motion.' The mass m represents matter, and the speed v represents motion. So, these two quantities are ways of representing 'matter in motion' as a single unique measurable quantity. These two quantities are thus intimately related to one another by a single formulation.

$$\frac{d}{dv}\left(\frac{1}{2}mv^2\right) = mv$$

or in its integral form

$$\int mvdv = \frac{1}{2}mv^2 \,.$$

Kinetic energy is the time rate of change of momentum, or the time rate of change of matter in motion. In other words, if we sum up (or apply the integral $\int$) all the infinite number of infinitesimally small changes of simple moving matter, momentum (mv), over a varying velocity, we have a kinetic energy.

However, in a more realistic sense, the variations are in the position with constantly changing time. This process mathematically describes the reconstruction/ reconstitution process of a moving particle in classical terms. Kinetic energy is the overall effect of a moving particle when it is deconstructed/reconstituted as time progresses forward since mass is no more than space-time curvature and velocity is no more than the relative change in position of that curvature. This formulation demonstrates the inverse symmetry between the space and time dimensions.

Kinetic energy is the expression of the moment-to moment change of the state of motion of a material particle or body. If the state of motion is changing (there is an acceleration), then there is a moment-to-moment change in its aspect along the axial A-line. The aspect is the relative position of the cusp of a particle along the axial A-line. When the particle moves at constant relative speed its cusp remains at the same fixed position relative to its state of rest. The aspect remains constant. As a particle's speed increases, the cusp moves vertically along the axial A-line relative to its rest position.

The rest position of the cusp, the center of field density along the axial A-line relative to the symmetry line of the space-time sheet, is measured as the rest mass energy of the particle according to the relation $E_0 = m_0c^2$, or better still $E_0 = m_0/\mu_0v_0$.

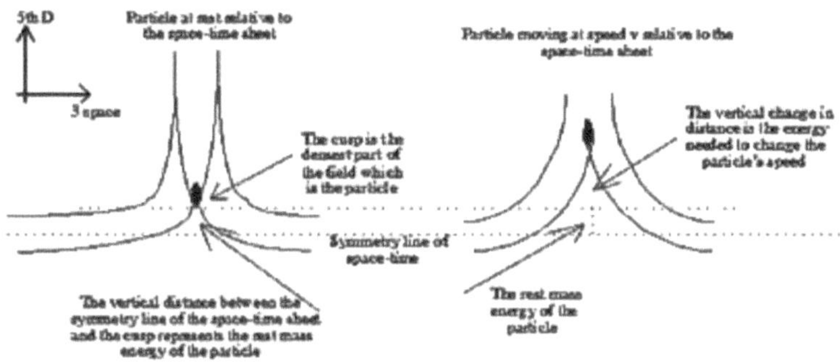

The use of the connectivity constants (permittivity and permeability) rather than the speed of light in this relationship reflects the real physical relationship between the curvature (m) the rest mass energy (relative position

n of the cusp) and the connectivity of points in the sheet. The particle's speed increases according to the amount of kinetic energy added to the particle. This kinetic energy increases the distance between the original cusp position in the fifth dimension and thus the five-dimensional potential of the particle. This change of energy has been described by Einstein's relativistic formula for relativistic kinetic energy.

$$E_{kinetic} = mc^2 + m_0c^2$$

It is easily rendered into the five-dimensional framework and interpreted accordingly as the change in relative position in the fifth dimension, or the change in aspect. E_{kin} is the relativistic kinetic energy while the total energy (or aspect in the fifth dimension) is mc^2 and m_0c^2 the rest mass energy, except that $c^2 = 1/\mu_0\epsilon_0$.

In other words, the kinetic energy of space-time becomes a measure of the position of the center of field density along the axial A-line in the fifth direction relative to its rest position. This is added to the rest mass energy, which determines the total aspect of the particle. The aspect is a measure of the potential. In the final analysis, matter is curved space-time, momentum is variations in that curvature and kinetic energy is curvature variation from the temporal perspective. The classical concept of a force translates as changing variations of curvature.

The notion that relative change in position in three-space over some measurement of time can cause a change of some type, such as a distance in another dimension (the fifth) perpendicular to space-time may seem strange, if not totally irrational, but the basic idea behind the notion is a common and well-known concept in classical physics. A torque results when a bit of matter moves, and that movement includes a change in direction. There is no torque with straight-line motion, but as motion varies from the straight line, a torque develops.

In the special case of circular motion or rotational dynamics, the change in each of the two dimensions of space is the same, so the torque is constant and perpendicular to the plane of the circular or rotational motion. In a manner that is not completely dissimilar to this classic case of torque, if any change of speed occurs during a change in time, an inter-dimensional 'torque' of sorts occurs perpendicular to both three-space and time. This 'torque' is orthogonal to the four-dimensional space-time sheet in a fifth direction, or rather in the fourth direction of space. So, an alternate descriptive view of changes along the axial A-line of a particle can be found as a 'torque' resulting from a complex motion in three-dimensional space and the normal progress of time into the future.

This torque applies to all matter in the universe, giving everything a basic spin, all the way up to galaxies and the universe itself. A very interesting question presents itself from these considerations. Did Einstein have an intuitive feeling for a model or framework such as this? Einstein worked on several variations of Kaluza's five-dimensional model when he was searching for the unified field theory. Einstein's work on those theories bears enough resemblance to this model to assume that he did have some idea of such a structure. He also assumed that the quantum would eventually pop out of the mathematics of his unified field theory.

And finally, he argued against the Copenhagen Interpretation of quantum and wave mechanics. He accepted the experimental results of quantum mechanics but had no faith in the standard interpretation of the theory. Einstein certainly had a profound, gifted and intuitive feel for nature and the physical processes that characterize our reality. His intuitive feelings told him that quantum theory was not what it seemed to most other scientists, especially Niels Bohr and Werner Heisenberg.

Einstein's primary argument against the Copenhagen Interpretation of quantum mechanics was presented in the EPR (Einstein, Podolsky and Rosen) paper of 1935. Einstein and his colleagues argued that at some point after two particles had interacted (collided), it should be possible to measure the position or momentum of one particle and the momentum or position of the other particle and then infer from these measurements the other complementary measurement of each particle.

So, it would still be possible to know the exact position and momentum of a particle simultaneously in principle, even if it was not practically possible. This would constitute a direct violation of the Heisenberg uncertainty principle, which prohibits the simultaneous knowledge of a particle's momentum and position. Einstein thought this was ridiculous. The EPR argument carried enough weight within the scientific and general academic community that it was not forgotten over the intervening years, despite the spectacular successes of the quantum theory. Eventually, EPR was reformulated and reinterpreted to cover the quantum states of spin for two interacting particles.

It was thought that particle spin represented an equivalent method of measuring the validity of the EPR argument as opposed to the Copenhagen Interpretation in the 'gedanken' experiment suggested by EPR. Experiments were designed and conducted along these new lines of thought and seem to have demonstrated the inaccuracy of the EPR argument in favor of the standard interpretation of quantum mechanics. However, experiments testing the spin of particles that interact at the quantum mechanical level can neither prove nor disprove the EPR argument. They do not really settle the dispute between Einstein's worldview and the Copenhagen

Interpretation of the world of the quantum. The fact that quantum mechanics is not 'complete' in the sense that Einstein originally intended is still a valid issue.

Einstein was a relativist, through and through. He was weaned on Mach's principle and Mach's argument against the necessity of Newton's absolute space, based on Mach's 'gedanken' experiment limiting the universe to a single particle. To Einstein, the EPR argument must have been an extension of Mach's argument. If space is relative, as Einstein assuredly thought, then two particles and only two particles would be necessary to determine the relativity of space and time. The existence of relativity would dictate that the position and momentum would both exist, simultaneously, independent of any quantum mechanical argument to the contrary, or the two particles could not act or interact relative to each other.

The Heisenberg uncertainty principle and the Copenhagen interpretation of the principle did not allow this possibility, so they were at odds with the most basic concept of relativity in Einstein's eyes. Yet the 'proven' existence of relative space and relative time was overwhelming enough for Einstein to conclude that the quantum mechanical worldview by which the Heisenberg uncertainty principle limited our knowledge of the world was 'incomplete.'

Neither spin nor 'hidden variables' were necessary to support the EPR arguments even though there was nothing in the EPR argument to expressly prohibit such interpretations. Both have become popular at one time or another in pursuit of determining the validity of Einstein's arguments. Nor is there any reason to believe that the spin of two particles determines the relativity of space, so the spin experiments do not answer the questions raised by Einstein and his colleagues. The essence of Einstein's point of view has never been fully understood nor explored, since later developments led to other tangential interpretations of EPR.

In this new five-dimensional theory the tables have been turned. The relativity of space and time is inherent in the model. Particles are coupled together across the fifth dimension and therefore must act in concert, relative to one another. The space-time sheet is not an absolute, but its overall internal structure or form, characteristics and properties are determined relative to the universe as a whole and thus conform to Mach's principle.

Even cases where matter waves, in Schrödinger's sense of the term, are concerned, the connectivity of space, from axial A-line to axial A-line independent of the space-time sheet, assures that the five-dimensional model is completely relative. It would even seem that if our scientists cannot determine the position and momentum of particles simultaneously as Einstein suspected, the universe in its wholeness knows the momentum and position of the particles, as well as all their other physical properties.

The complete entanglement of everything in the universe is complete and overwhelming and gives the universe its 'oneness.' It is not just from material body to material body, independent of time, but from discrete point-to-point simultaneously. Both the wave and the probability distribution that correspond to a particle are interacting with the rest of the universe at every moment of time.

A rather interesting historical and relevant fact should be noted at this point. At the time that the EPR argument was leveled, Einstein was in the middle of a decade and more of theoretical work on unified field theories based upon a five-dimensional model of physical reality. So, Einstein may have had an intuitive feeling for such a theory as this even if he was unable to express his ideas.

Tying up loose ends

Every theory has loose ends that need to be 'taken care of' and clarified. A TOE would have no such loose ends, or it would not be a theory of 'everything.' Physics is presently filled with many such loose ends and two of the most disconcerting operate in opposite realms of the universe. Physics has no adequate models for either the atomic nucleus or the spiral shape of galaxies. Both are considered serious problems in their respective fields. There are some models by which we can understand some facets of these phenomena, but no single model completely explains the nucleus or spiral galaxies. Although these two mysteries have defied the full explanation that is their due, the present five-dimensional model does give some insights into their solutions. There is a very important clue in the five-dimensional model of particles, as presented, which can lead to an effective (if not complete) theory of the nucleus.

The neutron is essentially a combination of an electron and proton superposed and stretched into the fifth dimension, although it is a unique particle. This structure of the neutron provides the clue to the structure of the nucleus.

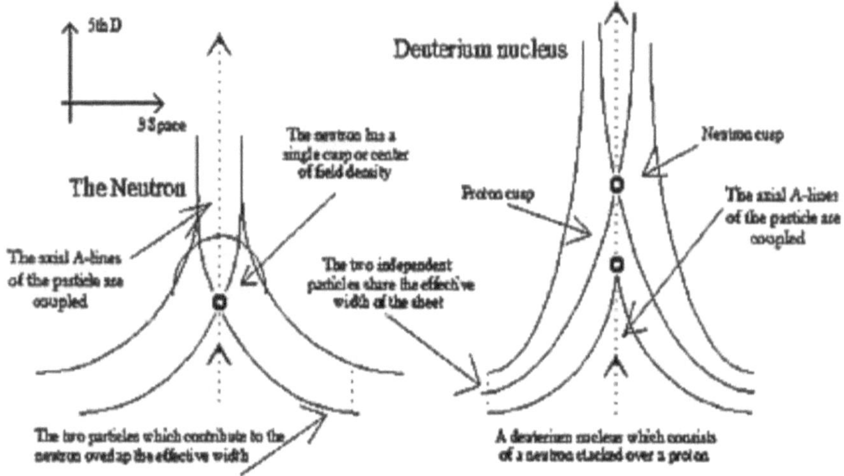

There is also an old dictum in science that says no two particles can occupy the same space at the same time. This 'rule of thumb' is essentially true in a four-dimensional space-time, but it is not necessarily true in a five-dimensional space-time as the existence of the neutron demonstrates.

The nucleus follows the example of the neutron and utilizes the fifth dimension for its structure. A nucleus is a single mass with a single outer three-dimensional spherical shape, but it consists of neutrons and protons that are stacked one upon the other in the fifth dimension or fourth spatial direction. However, unlike the electron and proton that constitute the neutron, the neutrons and protons retain their individual identities in the nucleus, while each shares a position and presence within the three-dimensional surface of the nucleus.

This model demands that the axial A-lines do not combine into a single A-line for the whole nucleus. Their A-lines may coincide or overlap, but their centers of density or cusps do not combine at a single point in the fifth dimension. As the neutrons and protons stack one upon the other in the fifth direction, their overall curvatures add because their mass is collective, but their three-space boundaries across the width of the sheet must taper into the same single fundamental 'effective width' at their collective border with the four-dimensional space-time sheet.

So, they also spread outward in three-space to increase the overall nuclear diameter while retaining their combined curvature. The nucleus thus remains spherical, rather than lumpy, as it would be if the particles would just smear or clump together in normal three-space when they came into contact. The whole structure is like stacking traffic cones one on the other for storage. There is a particular stacking order for neutrons and protons that determines the relative stability of any given nucleus. The stacking order is obvious if you start at the beginning of the periodic chart ($Z = 1$) and work upward, including all the various isotopes of the elements. The most stable nucleus is one in which the protons and neutrons alternate in the stack and are in equal numbers. The Tritium configuration of neutron-proton-neutron only has enough stability to exist; otherwise, it is unstable and will eventually decay. It has too much neutron for its proton.

The normal Helium atom with two neutrons and two protons is the most stable nuclear configuration possible and this has been designated the alpha particle which demonstrates its innate stability. As nuclei grow larger, more neutrons become necessary to stabilize the growing number of protons and thus stabilize the nuclear structure. This method works well if there are roughly an equal number of protons and neutrons, but after the number of protons climbs past twenty, the number of neutrons relative to the number of protons becomes disproportionally larger.

Calcium has a stable isotope with an equal number of neutrons and protons (twenty of each) and is the last such stable isotope as the atomic number Z increases.

For light nuclei, the neutron and proton numbers are roughly equal. However, for heavy nuclei, the Coulomb repulsion term $Z(Z-1)$ begins to grow rapidly, so extra neutrons are needed to supply additional binding energy. Thus, heavy stable nuclei all have $N > Z$. There are no stable nuclei 4 with $A = 5$ or 8. The alpha particle $^{4}_{2}H_{2}$ is a particularly stable nucleus ($B/A = 7.07$ MeV); a nucleus with $A = 5$, such as $^{5}_{2}He_{3}$ or $^{5}_{3}Li_{2}$, will quickly (10^{-21} s) disintegrate into an alpha particle and a neutron or proton, and a nucleus with $A = 8$ such as $8_{4}Be_{4}$, will quickly break apart into two alpha particles. (Krane, 240).

Several other characteristics that are displayed by atoms and their isotopes also impact the stability of the nucleus.

A nucleus can be very stable if it has a specific number of either protons or neutrons. The special numbers that seem to guarantee stability are called the magic numbers. Nuclei with 2, 8, 20, 50, 82, and 126 neutrons or protons are both more stable and more numerous. The last nucleons to complete these "shells" have higher binding energies while the energies for the first excited state of these nuclei are larger than for nearby nuclei with nearly the same numbers but not having magic numbers. There is also a tendency for nuclei with even numbers of protons to form more different stable isotopes than nuclei with an odd number of protons. This last tendency seems to indicate that proton pairs form more stable nuclei than odd numbers of protons, while the magic numbers further indicate this same tendency but stress that the secrets to the true structure of the nucleus are more complex and lie deeper within the numbers.

The existence of the magic numbers led to the development of a 'shell' model of the nucleus based rather loosely upon the concept of the electronic shells that surround the nucleus. On the other hand, C. von Weiszäcker noted that the nuclear properties connected with mass, binding energy and size are like those for a drop of liquid, so he proposed the liquid drop model of the nucleus in 1935.

This five-dimensional model shares aspects of both the fluid and the shell models of the nucleus, but it is neither. It is a fluid model due to the blending or tapering of individual nuclear particle boundaries to fit a single 'effective width' of the space-time sheet. In a sense, each particle is 'spread' throughout the whole nucleus' three-dimensional surface, like a fluid, from the point of view of the nuclear boundary in three-space. Yet the five-dimensional model is also like a 'shell' model since the neutrons and protons are stacked in a specific order in the fifth direction. There is no other possible physical structure that could unite the shell and liquid drop models of the nucleus as does this theory, thus solving one of the great mysteries of physics.

If the axial A-lines are coupled or overlap without discrepancies, then the nucleus is stable. But sometimes they do not overlap totally, and the nucleus is unstable. If the nucleus is unstable, then it can decay during the deconstruction/reconstitution process with a particle reconstituting outside of the nucleus. As indicated above, the nucleus (also called an alpha particle) is particularly stable, so it forms a primary stability that acts as a template for other nuclei.

The deuterium nucleus can also be regarded as a template. The alpha particle consists of two deuterium nuclei that are very tightly bound or coupled together in a stable configuration of axial A-lines. Three alpha particles can bind together to give a very stable and common isotope of carbon ($^{12}_{6}C$) and four are bound together in the most common form of Oxygen ($^{16}_{8}O$). However, two alpha particles yield a very rare form of Beryllium which has a half-life of only 0.07 femtoseconds or 10^{-15} seconds, so there is more to building the most stable nuclei than just adding together alphas. The most abundant form of Beryllium has a nucleus of two alpha particles with an extra neutron. Each individual alpha template comprising the Beryllium nucleus is so stable that they need the extra neutron to hold them together. This structure indicates the necessity of extra neutrons in larger nuclei to hold the nucleus together.

The magic numbers can be duplicated within a nucleus by building successive complex units of the basic alpha particle and throwing in an occasional Deuterium nucleus for stability. This nuclear model is simple, but far from comprehensive. There is a great deal more to explaining the nucleus, if for no other reason than because this number game does not include an explanation of the stability of nuclei with an odd number of protons. It cannot be considered a complete theoretical model quite yet, at least not until an explanation is given of the stability of the alpha particle and deuterium nucleus in terms of the five-dimensional theory.

Another problem is raised when the nucleus is compared to large extended objects that are composed of tightly packed particles, *i.e.*, black holes. A black hole is a singularity, as is an atomic nucleus, but the two are entirely different except for their singular nature and a few other obvious facts. This model accounts for the singularity that represents individual particles as well as the singularity that constitutes atomic nuclei. But back holes offer a still different phenomenon that must be considered in a different manner. The particles in black holes do not stack in the fifth dimension, but the particles are in boundary-to-boundary contact with their neighboring particles.

In this case, the collective gravitational attractions of an unbelievably large number of tightly packed particles have grown so strong, despite the relative weakness of gravity compared to electromagnetic forces, that the repulsive electric forces between individual particles is overwhelmed. The overall curvature of the singularity in this case results from the additive effect of the 'effective width' of the spacetime sheet. Each particle inward toward the center of the black hole sits higher in the fifth direction than the next particle outward toward the physical edge of the black hole.

There is no special coupling between the axial A-lines of individual particles as exists in the nucleus. So, the black hole has an event horizon that is a product of the individual curvatures of particles that extends well beyond the physical boundary of the black hole. An event horizon for a nucleus coincides with the outer nuclear boundary in three-space because there is five-dimensional stacking, and the axial A-lines are coupled. The concept of a black hole transplants the dynamics and subatomic structure of matter to the macroscopic realm of nature. In the macroscopic realm of space-time, the galaxy is the one of the largest, if not the largest, type of distinguishable material structures. The dynamics that lead to the basic spiral form exhibited by most galaxies is a mystery to astrophysicists and cosmologists. The fact that the spiral is so prevalent in the galactic world would seem to indicate a pattern that bears explanation.

This spiral forms because of the collective nature of particle spin. All the elementary particles in the galaxy have half-spin, which is not a motion but an orientation, so the galaxy as a single whole has the same orientation. So, the five-dimensional model can account for the predominance of the spiral shape of galaxies when the observed expansion of the universe is considered because the particle spin plus the expansion along the axial lines yields a right-handed torque relative to the five-dimensional structure of the galaxies.

If a real fifth dimension of time-space (a fourth dimension of space) is assumed, as above, then the expanding universe will represent a new motion whereby the four-dimensional spherical sheet is growing larger in the direction of the fifth direction. The notion that small material bodies moving in the four-dimensional space-time sheet produce a virtual 'torque' that extends into the fifth dimension at each point, stretching the particle singularity further in the extra dimension of space as relative speed increases, is also true for astronomical bodies as much as it is for galaxies.

In so far as the stars and collective material body systems that form a galaxy act in concert with each other, they form a loosely bound physical 'body.' This gives them a spin orientation that manifests as a collective 'torques' relative to the fifth dimension. This 'torque' is moving as the universe expands. Just as any vector whose position in space-time is changing, the change will induce as added component of rotational acceleration in the collective 'body' all material systems in the universe. As this acceleration is 'felt' by individual stars and formations in the galaxy, a pinwheel or spiral is formed. The 'body' of the galaxy cannot mechanically spin around its center as a wheel would since the gravitational forces binding the individual stars in the galaxy are far too weak, so spiral shaped galaxies result. It is just an orientation.

In other words, there is a weak precessional action on large combinations of stars which cannot make them move collectively as a solid body because they are too loosely bound to the 'body' even though they are tightly enough bound to act in concert as a galaxy. The group of stars collectively acting as a galaxy experiences a Coriolis Effect in a manner like a waterspout, a hurricane, a tornado or even the great red spot of Jupiter. This Coriolis Effect is due to the complex of different motions within the spacetime sheet and the motion of the sheet as a single whole due to the expansion of the universe. Under conditions where the balance of binding forces is not as weak or strong as those between stars in a spiral galaxy, other galactic shapes occur, such as globular clusters. This torsion acts on all bodies of matter of all sizes and proportions.

The same Coriolis Effect occurs within smaller material systems, like small eddies within the flowing waters of a river or stream. It affects the evolution of individual star systems such that planets evolve around the central heavier star. The evolution of a star's system of planets, asteroids, comets and other planetoids from vast gas clouds speckled with dust particles depends on this torsion. So, planetary systems are the rule rather than unique occurrences within the universe. The makeup of any one planetary system depends on the relative forces acting within the system as it evolves, which is in turn dependent on the relative proportions of different forms and kinds of matter and their distribution at a given moment that the evolution process begins.

This torsional effect will also add a small component of precession to individual planetary bodies since they act as single entities or cohesive bodies. Again, the course that this effect takes on any individual planet will depend on that body's internal makeup and composition. And finally, this torsional effect acts all the way down to the level of atomic nuclei and individual material particles, although the effect is much smaller in this realm.

The whole of the universe exerts this twist orientation on both nuclei and individual particles that is perceived or detected as a virtual torsion. In a nucleus, this twist or torsion acts to bind the individual axial A-lines to a slightly greater degree than they would be bound without it and is the source of the weak nuclear force. This twist is so small that evidence of it can only be detected when nuclei decay and the energy from this torsional twist is released.

CHAPTER 5

Strange facts find a theory: A new dimension for psi

That's life!

Many physicists and scholars believe that there is a great wall between the physics of centuries past and the physics of this century. The older style of physics is 'classical' while the physics of this century is 'modern.' Today, many also claim that 'their' physics is 'post-modern' which must mean that it is somehow radically different (and advanced) from normal twentieth century physics. But perhaps they are just too quick to dismiss the older styles of physics. Despite modern protestations, our science remains mechanistic in the sense that scientists' primary concern is discovering laws, rules and principles that govern 'matter in motion' at the most fundamental level of nature.

In this regard, even modern science is following strictly Newtonian ideals and pre-Newtonian ideals. There are glimmers of hope that science may be breaking out of this mold, but they exist only at the edge of science. Parapsychology and paraphysics are growing along this fringe area of science. There are also trends in science, already discussed, that although they are not breaking the mold of mechanism, they are at least stretching it to cover new areas of nature. The recent incursion of consciousness into the study of physics is the best example of this change in attitude.

Consciousness entered physics through the quantum mechanical process of 'collapsing' the wave packet. It might seem that consciousness is necessary to 'collapse' the wave packet, but such a belief cannot be sustained under very close scrutiny unless consciousness exists in every discrete point in the universe and most scientists and scholars would never accept such an assumption.

The logical conclusion to this dilemma would be to accept the alternate 'fact' that consciousness can 'collapse' the wave packet but does not permeate all of matter. This has led many to the unsatisfactory conclusion that humans are the only conscious beings in the universe, and we literally create physical reality and the universe by our own individual thoughts. This position is also absurd. The only true conclusion to this dilemma is to recognize that consciousness is sufficient to 'collapse' the wave packet, but it is not necessary for the 'collapse.' The process can proceed in the absence of consciousness.

The fact that scientists can design and conduct an experiment to 'collapse' wave packets at some level of reality, and thus manipulate nature, does not guarantee that only consciousness can 'collapse' a wave packet, and then only human consciousness. The concept of consciousness with which all scientists, and especially physicists, work is too broad and ill-defined with expectations for theoretical explanation far too broad and ambitious. The concept of consciousness with which physicists deal is not necessarily the same as that with which psychologists, doctors, psychiatrists, brain chemists or philosophers deal. The concept of consciousness upon which physicists speculate is far broader in scope than an anesthetist deals with when the anesthetist renders a patient 'unconscious' during a medical operation.

Physicists cannot, nor should not, talk about consciousness unless they can define what they mean by the term independent of the 'collapse' of the wave packet. Until this is done, the 'collapse' of the wave packet will never be understood at the same level of scientific and physicists' speculations about the process and will remain a purely mechanistic detail in a physics which is still characterized by its Newtonian overtones. Even if consciousness defies a coherent and concise definition there is enough data and information to render an approximate idea of some of its parameters. A definition of consciousness based on its observed properties would suffice to develop a theory of consciousness.

The primary and overriding fact is that consciousness seems to be associated with living, animate matter. Under these conditions, a search for consciousness must first deal with the concept of life and living matter. Associating consciousness with living matter would sidetrack any arguments about which type of complex systems are conscious: humans, dolphins, trees, dogs, cats, tulips, or paramecia. Let science fiction writers deal with that issue for the time being. A more fruitful path to follow would lead science to define animate matter rather than consciousness.

Physics is reductionist. It reduces the world that we perceive to the smallest and most fundamental bits of matter, elementary particles and the fields associated with them, and then describes how they interact. One form of that fundamental interaction would lead to animate matter and another to inanimate matter. Animate matter can be considered an extended material body that acts in a particular manner. Inanimate matter is an extended body that interacts with the rest of the universe by reacting to the basic forces of nature. But all animate matter goes beyond simply reacting to the basic forces of nature. In other words, animate matter is self-motivating while inanimate matter is motivated or moved by external forces.

In the context of physics there are two and only two classifications or types of matter: animate (or living) matter and inanimate (or non-living). Physics has never stressed the differences between the two and has therefore only considered all animate matter as acting in the same way to the external world that inanimate matter acts. A simple list of the properties of both types of matter is sufficient to illustrate the differences between them and act as a guide to developing the physics of living matter. The properties of these two types must be listed according to the reductionist manner of science and must be placed in terms conducive to the scientific study of physics.

Thus, the most common features or properties that are inherent in all forms of living matter must be enumerated. Of these various properties, self-motivation is the simplest most fundamental property that we can attribute to animate matter because procreation and negentropy are structural characteristics while ...

Inanimate matter	Animate matter
Spatial extension	All properties of matter plus Procreation
Inertia	Self-motivation (in extended bodies, not particles)
The seat of electro-magnetic and gravitational fields	Need for food or other source of energy
Mixtures and lower-level stable complexities (stars and planets)	Higher-level stable Complexities and multi-level complexities
Two particles cannot occupy the same point in space-time	Negentropy Mind? Consciousness?

Mind and consciousness may not be universal characteristics of animate matter. In other words, only self-motivation is necessary for living matter within the context of physics although it may not be sufficient to define life outside of the science of physics. Pure physics is only concerned with 'matter in motion,' not with purpose or interpretation, which are the functions of mind and consciousness.

When it comes to motion, inanimate matter always follows the field gradient. In other words, inanimate matter follows the physical laws of motion without deviation. Any deviation from the laws of motion that is displayed by inanimate matter would only demonstrate that our laws of motion are inadequate, not that inanimate matter defies the laws of nature. Physicists may take this fact on faith, but physicists do practice their craft assuming that this is true. Physics has always dealt with inanimate matter or with animate matter acting as inanimate. Physics has never distinguished between the two. Both a human body and a lump of coal will fall at the same rate in a gravitational field, but the human body has a choice to go over the cliff or not, the lump of coal does not make that choice.

Self-motivating matter has the singular ability to sense its surroundings and make a choice that allows it to either follow the field gradient or move against the field gradient. This choice represents a willful action. On the other hand, willful action requires some form of mechanism controlling device and that controlling device, whether extremely simple or quite complex, can be called a brain. The brain is just a choice-making device, but it is still a mechanical device that can be duplicated in non-living, inanimate bodies, *i.e.,* robots or simple computers. This does not mean that 'robots' are living organisms nor that computers can really 'think.' Such questions form some of the great debates of our time.

It merely means that some sort of device, at the very least a simple brain which has the conscious ability to detect and interpret simple environmental stimuli and then make a choice between *possible* modes of interacting with the environment, is a sufficient and effective device for living organisms to survive in their environment. So, while the existence of a brain or other such controlling device is necessary for living bodies, it is not unique to living bodies. It would therefore seem that something more than a brain is necessary for living matter and this would be some level of mind.

The concept of mind is different from brain because mind has an 'awareness' of the brain, the body and the environment at least at some very basic instinctual level. The brain is a physical device, however simple or complex, that is localized in the body. There is no evidence that mind is a local phenomenon, *i.e.*, it is fixed at a single position in the body, although the physical brain has nearly always been considered the seat of the mind. Mind is holistic (wholistic) rather than reductive, so it is doubtful that mind can be understood through the reductive and mechanistic logical process followed in physics. As far as we can determine, consciousness exists at a still higher level or higher order than mind, but neither is necessary for understanding simple 'matter in motion.'

All that is necessary at this level of physics is the fact that animate matter is self-motivating and inanimate matter is not. After all, physics reduces the world to 'matter in motion,' so the differentiation between self-motivating and not self-motivating is all that simple physics requires to begin the process of understanding living matter. Historically, scholars of all persuasions have long suspected that there is far more to life than could be reduced to the mechanisms described by physics. Scholars and scientists alike have invented several different quantities and qualities to describe this difference. Concepts of a 'life force,' 'vis viva' and 'elan vital' have cropped up in science several times in the past three centuries.

Today's equivalent of these concepts can be found in such terms as 'wholism' and 'complexity,' but the basic ideas that they describe are still the same. Even in these cases, the inventing of new quantities that cannot be measured are of little help in understanding the physics of motion for inanimate bodies of matter. So again, we have come back to the concept of self-motivation as the only method by which physics can account for the differences between animate and inanimate objects. Once self-motivation has been accepted as an adequate and effective basis for the physics of life, a new question emerges: How does self-motivation work in living organisms within the five-dimensional model? A living organism is a large and very complex material system.

Despite the complexity of the organism, it can still be considered a unbelievably large number of mutually interacting particles. The physical integrity of the organism is established by the specialized chemical interactions within the organism. These chemical interactions can be further reduced to the interplay of electromagnetic fields and material particles. Within this context, they represent the mechanical basis of the living organism. But in the five-dimensional model, the various particles are more than the sum of their normal physicochemical interactions.

The individual particles are connected to each other, as they are to all particles in the universe, via A-lines perpendicular to their own individual axial A-lines. These extra-normal connections (outside of the space-time sheet) cannot be equated to the special quality that makes life different from inanimate matter, but they are related to that special quality which is unique to all life. Inanimate matter would have similar connections between particles, but the complexity of chemical interactions between the various chemical substructures of living organisms, in conjunction with these extradimensional A-lines, allow a special coupling between inanimate particles within living organisms which is directly related to that which is life in any animate body. This coupling is not present between the particles that compose inanimate bodies of matter.

Living organisms are negentropic. There is a tendency of matter to form simpler complex bodies or systems that is called entropy. This tendency is explicit in the second law of thermodynamics and other physical principles. On the other hand, living organisms form more complex structures and substructures rather than devolving into less complex structures. So, they evolve in a direction opposite to entropy and are thus negentropic. Evolution is a natural process that is associated with living bodies primarily because they are negentropic. The use of the concept of evolution with a negentropic system denotes an element of time that is perfectly logical since entropy is considered the 'arrow of time' in normal physics.

Complexity is related to negentropy, and negentropy is related to living organisms. So, through some mechanism that is not yet completely known to science, living organisms reorganize matter in more complex systems and structures to grow and thrive. This negentropic effect characterizes all living bodies (or animate matter), but it also affects the Alines of the individual particles which make up a living body in a specific manner which allows the A-lines to couple together independent of their connections with other material particles in the universe.

There are ongoing chemical processes in living bodies that are unmatched in inanimate bodies. These chemical processes include the exchange of energy between particles. Normally, particles of all types of matter have various energies of motion associated with them, such as the kinetic energy of orbiting electrons in atoms and molecules. But animate organisms must also have special energies associated with the special chemical reactions and interactions that are unique to life.

These special chemical reactions follow specific repetitive patterns throughout the living body and could be interpreted as kinetic energy interchanges between the particles undergoing the specified chemical reactions. So, there are specific patterns of energy exchange between the material particles from which a living body is composed. The more complex the living organisms are, the more complex the chemical patterns or signatures of the organisms, and the more complex the patterns of kinetic energy interchange between particles.

Living organisms are characterized by specific energy signatures, which are related to the patterns of chemical interactions in their bodies. In a physical sense, these patterns or signatures could be considered resonances between the five-dimensional extensions of the material particles within living organisms and the wave functions of the material particles would form a special coherence or coherent state.

As an individual particle's speed increases or decreases during the interchange of kinetic energy of the chemical process, its extension in the fifth dimension increases or decreases according to the special theory of relativity. The particle's center of field density shifts further into the fifth dimension when its speed increases and back toward the space-time sheet as its relative speed decreases. For the case of a living organism in which specific and well-defined chemical patterns are established and repeated throughout the organism, specific patterns of five-dimensional extensions are also established. The concerted shifting of centers of field density along the axial A-lines of large numbers of particles cause specific orientations in the A-lines connecting particles perpendicular to the axial A-lines.

Since the same chemical affects numerous particles in living bodies processes, these extra-dimensional A-lines couple together in specific ways. These A-line couplings are unique to living organisms and give living organisms an extra-dimensional signature that is missing in inanimate matter and non-living bodies. The more complex the living organism, the more complex the patterns of A-line coupling, and the more evolved the living organism.

These couplings are intimately related to life and the living process, although they neither define nor constitute the quality of matter called life alone. These couplings and their related energy signatures do constitute what is normally called Chi, Qi or Ki in Eastern science and philosophical systems.

The physical framework of psi

In general, the extra-dimensional framework has offered an excellent breeding ground for theories of psi. From Henry More to J.K.F. Zöllner and the present, an extra dimension of space or space-time has been an opportune happenstance for explaining the unknown and mysterious convolutions of nature, whether real or imagined. It was obvious from the beginning that a higher dimension (or dimensions) of space could not be perceived, so it (they) offered a wonderful habitat and home for God, spirits, the soul, ghosts, other spiritual entities, and later an effective ploy for explaining various psychic phenomena. But the fact that a higher space is not normally perceivable was no more than an 'excuse' to adopt that hypothesis to explain paranormal occurrences.

The hypothesis of an extra dimension can be adopted without violating any known physical laws and clashing with other scientific models. There is nothing in nature that guarantees only three dimensions of space. Yet the fact that nature is ambivalent to the number of spatial dimensions constitutes neither a scientific nor a philosophical reason to adopt such hypotheses. After the Second Scientific Revolution, that changed. As Einstein's concept of a four-dimensional space-time continuum became a popular icon in general society outside of physics, the tradition of using a four-dimensional framework to explain some aspects of psi was born.

In 1928, Dunne put to pen his ideas that the fourth dimension of time, as expressed in relativity theory, offered a convenient method of explaining precognition through dreams. Several years later, G.C. Barnard criticized Dunne's and similar notions for misconceiving time by assuming motion in time, an absurd circular argument. (Barnard, 185-186) For his own part, Barnard had a more sophisticated approach to space-time as well as an elaborate plan for psi.

The evidence of Physics, both from the standpoint of Relativity and from that of Wave-mechanics, shows that the phenomena studied by modern physicists cannot be explained satisfactorily with less than four dimensions; whether more are needed has hardly been settled yet. It also shows that the time dimension is

Barnard's opinions came a decade after the development of Kaluza's theory during the decade that Einstein and a few others worked on extensions of Kaluza's theory, so the need for a hyperspace framework for physics has "hardly been settled yet."

But one and all, these scientists and scholars have opted for an explanation of psi by altering or twisting both common and uncommon notions of physics to their own philosophical ends. No one has ever offered a comprehensive view of reality from which psi could emerge from the physics of its own accord. Nor has anyone attempted to develop a theory of psi from physics in such a way that psi would be a natural element of our world, dictated by nature itself. If psi is a real effect, and there is more than enough evidence to support that contention, then its explanation can only come from physics, its existence demanded by nature, rather than finding a trick of mathematics to account for it.

Psi must be incorporated as a natural element in any comprehensive theory of physics, whether it is a TOE or not and the five-dimensional model is the only alternative that can support such a comprehensive theory. Recent experimental research indicates that consciousness can act non-locally and thus seems to provide a natural setting for psi. Until a comprehensive theory of physics incorporates psi as a fundamental element of our physical existence, science will remain unconvinced of the reality of psi.

Harry Settanni's views on this matter are purely philosophical and analytical. He has concluded that spacetime is real and forms an underlying reality for our particulate and material world.

[Hilary] Putnam holds that space-time is real, namely that it has an existence independent of either human cognition or of the relationships between particles of matter contained in it. This view appears reasonable since contemporary physics has demonstrated that what is called "matter" is simply a curvature in a pre-existent space-time field. In this interpretation, matter is simply a perturbation, in what otherwise is an essentially homogenous field: the field of spacetime. As both Relativity and

Although it is highly debatable whether space-time is absolute as Putnam and Settanni conclude, Settanni's ideas still contain some validity. Space-time can be real and relative simultaneously, a possibility that Settanni has not considered. Settanni's philosophical views of physical reality are further restricted to only one interpretation of quantum theory. As an independent reality, space-time proliferates and extends "itself in many dimensions." (Settanni, 69)

Settanni's limited reality is a "many-moded pluriverse," based on the 'many worlds interpretation' of quantum mechanics. In the singularities of black and white holes, he claims that we "encounter the 'seed' from which all spatial and temporal reality spring." (Settanni, 77) This "seed" is the psychic reality from which all physical reality springs. Settanni's notion is vague and not a real theory or model, just a philosophical argument, but it is at least worth mentioning because Settanni has come to this conclusion by accepting a fundamental change in his overall worldview. He recommends that the scientific community accept the reality of psychic phenomena and thus make the final break from our heritage of Newtonian mechanism.

Science is still largely Newtonian in its worldview despite a century of relativity and quantum. This lingering Newtonian perspective has blinded the scientific and academic community to the possibility of psychic phenomena. Yet the "empirical evidence for existence of these phenomena is overwhelming; it is simply a matter of changing our worldview." (Settanni, 83) Although his physics is misguided, Settanni has at least recognized the necessity of accepting the evidence of psi and incorporating it into his own perception of physical reality which has led him to the conclusion that higher dimensions of space are necessary to explain psychic phenomena.

Other modern views are no more successful than the earlier theories. Rauscher's 1977 and Puthoff, Targ and May's 1979 theories have gone without further clarification and amplification. They adopted a mathematical model that added another four dimensions of space that mirrored our own four-dimensional space-time. In essence, their four extra dimensions are just the four individual extensions of the four dimensions of our normal space-time into the single fifth dimension. Schmeidler also adopted an extra-dimensional model in 1972. She thought that perhaps a folded topology of space-time could account for psi phenomena.

Unfortunately, a folded space-time would be a gross departure from the Riemannian curvature that characterizes the topology of our normal space-time according to GR, so her theory is untenable. These theories, as those before them and after, were neither comprehensive nor complete. They were thus very short-lived. However, these theories have collectively demonstrated that there is a great deal of worth in extradimensional concepts. So, a five-dimensional model should have a great deal to offer.

No other single medium or framework can accommodate all the different aspects of psi as fully and effectively as a simple five-dimensional framework. The fifth dimension is ideal for modeling the various characteristics of psi within a physical context and many characteristics of psi have been discovered. For example, psi does not decrease in strength with distance, as do other effects such as the fundamental electromagnetic and gravitational forces. Intervening physical bodies or fields, implying that it does not travel through space in the same manner as other signals, do not block psi.

Psi communication is instantaneous for all intents and purposes. Psi action does not exhibit a transfer of energy when interacting with matter, which would also seem to restrict psi to acting outside of our four-dimensional space-time continuum. Psi seems to act holistically, universally, constantly subliminally, and broadly in the background of our sensed and experienced physical reality which would suggest that it is a field effect or otherwise associated with some type of field. And finally, psi is not time dependent, such that signals proceed both to and from the future and the past. This last property is necessary to explain precognition and retrocognition.

It would further imply that the past as well as the future already exists in some manner in a far richer structure of space-time than is currently suspected. Each of these properties, whether considered individually or collectively, indicates that the normal properties of our four-dimensional space-time continuum are somehow bypassed, abrogated or ignored in the case of psi. The five-dimensional model can accommodate all of these properties with relative ease. If a signal travels outside of the space-time sheet, then its strength will not diminish as the inverse square of the distance traveled because distance is a property of the sheet. Either physical bodies or four-dimensional fields would not block the signal because they are located within the sheet.

Energy transfer would also need to occur within the sheet or otherwise be associated with events in the sheet and the signal would not be restricted to limits on speed since it does not travel within or through the sheet. The five-dimensional view of reality is holistic by its very nature, with all particles of matter connected to each other by lateral A-lines. And time is just another dimension from the five-dimensional perspective so precognition and retrocognition would offer no special problems that could not be overcome.

The five-dimensional framework accommodates the different configurations of psi with relative ease. Each particle of matter has an axial Aline extending into the fifth dimension and each gross body has a connected bundle (so to speak) of axial A-lines which distinguishes the body from other bodies of matter in the universe.

All axial A-lines have lateral A-lines perpendicular to them extending out parallel to the sheet.

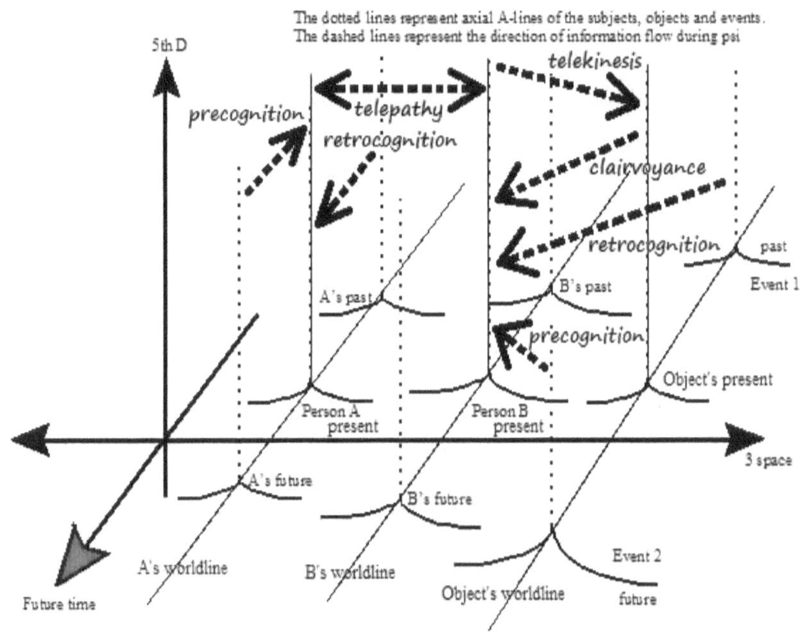

These lateral A-lines extend to other particles throughout the universe and connect all the material particles in a perfectly non-substantive manner.

When information in some as yet unknown form travels along these lateral A-lines from an object or from a person in the past, retrocognition can occur. When information travels from a future event affecting a person or some other future event to that person in the present, precognition can occur.

Telepathy follows lateral A-lines between different people within the same time frame, the present. Clairvoyance results from information traveling from objects to the clairvoyant reader in the present and telekinesis occurs when information travels from a person to an object's wave function in the present, which initiates a non-local wave collapse' for the particle or object. The information flow follows the lateral A-lines. So, lateral A-lines, as featured in the five-dimensional model, act as paths for psi. Psi is no more than an extra-dimensional form of communication between material bodies, whether animate and inanimate.

Psi is also explained by, or related to, probabilities. It acts randomly. Its random action is related to the nature of time. Although time is like a distance from the perspective of the fifth dimension, the physical reality of a moment is inextricably tied to the present moment. The present moment is just a deconstructed and reconstituted past moment waiting for deconstruction and reconstitution into the future moment as viewed by the present moment. But the 'view' from the fifth dimension would not include any notice of the deconstruction/reconstitution process of time because time would be continuous from the five-dimensional perspective just as the other three dimensions of space are continuous.

From the five-dimensional perspective, all of space and time would appear as a continuous pattern of ripples and variations in the space-time sheet. All the seeming paradoxes of physics and nature arise from the dualism inherent in our physical brain's attachment to the deconstructed/reconstituted moment and our thought's ability to surf the fifth-dimension perspective of reality. This fact is evident in our common concepts of time duration and passage of time. The difference of view between the two is literally the difference between the mind's perception of time and the experience of its own physically changing time.

The randomness demonstrated by quantum mechanics is just a product of the difference between the five-dimensional perspective of thought and the moment of the brain that is deconstructed and reconstituted, so the randomness of psi is due to the difference between physical brain and mind. This worldview is very near to that of the implicate order proposed by David Bohm. In one sense, the past and future always exist which is a good approximation of Bohm's implicate order; while we experience the physical moment-to-moment as continuous time which corresponds to his explicate order. In other words, the universe knows its future, its past and its present simultaneously because the universe is its past, present, and future simultaneously independent of mind and consciousness.

The universe is a vast interlaced connection of wave functions tied together through the universal field and its presence in the spacetime sheet. The sheet, or rather particles in the sheet, are rendered as material substance within the sheet by the simultaneous 'collapse' of wave functions throughout the universe each single moment as time passes. The simultaneous 'collapse,' which creates both particles and thus the space-time sheet, is just the reconstitution of the last moment that has just passed in time.

The wave functions themselves and their material existence in the sheet are connected within the far richer structure of their extensions extraneous to the sheet in the fifth dimension where they relate and are directly connected to all other points in space and time simultaneously. The mind interacts with these extraneous connections. When the mind becomes aware of them, we are momentarily cognizant of the far richer structure of the universe.

The fifth dimension offers the only physical model that corresponds to and can thus be used to explain GESP. In many cases, the differences between different forms of psi are not undeniability evident so the term General ESP has been adopted. However, the reason why the different forms of psi cannot be distinguished from one another has never been established. For example, a clairvoyant who knows the contents of an envelope before it is opened can either be psychometrically sensing the contents of the envelope, reading the mind of the person who put the object in the envelope, precognizing the removal of the object from the envelope or retrocognizing the object being placed in the envelope.

It is very difficult to experimentally determine which of these forms of psi action is functioning in any single specific case. An explanation of this problem is evident from the five-dimensional perspective, because all forms of psi signaling reduce to extra-dimensional communication along lateral Alines. A clairvoyant could use any of these forms of psi functioning and thus any of these sources for information about the object in the envelope, and possibly even use a combination of them. The sources of the information represent the only differences between the different forms of psi action and the clairvoyant is in contact via lateral A-lines with all the sources simultaneously.

However, if the properties of hyperspace truly reflect those of normal space-time, then there must be a corresponding principle of least action for the extradimensional extension of space-time, which would limit the expected act of clairvoyance to either reading the mind of the person who placed the object in the envelope or psychometrically detecting the object in the envelope. The experimenter could then use blind envelopes or other methods to limit knowledge of the contents of the envelope so that a true clairvoyant response would be necessary to determine the contents of the envelope. GESP is just a reflection of the similarities between the various forms of psi action within the five-dimensional model, whose only real difference is the source of the information.

A nearly complete model of psi functioning has begun to emerge from within the five-dimensional model of space-time. Psi is an integral part of physical reality and the universe. It is required by the physics of five dimensions since all bodies in the universe, both animate and inanimate, are connected extra-dimensionally by lateral A-lines. (This aspect of psi will become more evident in the next section) When a living being utilizes or cognizes this connection, a psi event occurs. Occasionally, a person cognizes either an event or the thought of another person spontaneously, lending an air of randomness to psi events. But normally, living beings are interacting with their surroundings and thus utilizing psi all the time, just not cognizing the interaction.

Psi is, in essence, an extradimensional subliminal effect that only randomly enters the human conscious mind for purely psychological reasons. Psi is also a property of all living beings, although it cannot be cognized by those living beings that have no ability for cognition. Because psi occurs 'extra' dimensionally it is 'para' normal, normal being action/reactions that occur within the space-time sheet alone.

"Eureka!" he exclaimed, with a psi of relief

The five-dimensional view of psi functioning is still missing a 'mechanism' or 'pseudo-mechanism' of action, so this theory is only partially complete. It is simply not enough to place psi within a five-dimensional framework with minimal explanation. Others have done so, and their theories have become no more than historical anecdotes and fodder for philosophers. This theory would suffer the same fate were it to end at this point.

However, a 'mechanism' for psi action is evident within the purely physical theory of a real fifth dimension. The word 'mechanism' is used within this context with some trepidation for lack of a better and more accurate term. As used here, the term is not equivalent to the Newtonian concept of a mechanism, nor does it imply anything more than an extremely vague similarity with the Newtonian worldview. The signal that passes between physical bodies and constitutes what we call psi follows the paths established by lateral A-lines in general. But psi is not a signal following the lateral A-lines in the manner that a moving electrically charged particle will follow the lines of force of a magnet. Psi is instead due to a flux or shift in the pattern of lateral A-lines extending between physical bodies.

A particle's own center of field density shifts along its axial Aline as it moves according to its rate of motion or speed. This shift sets up a changing pattern in the lateral A-lines that spread out from the particle throughout the fifth dimension. This variation in pattern is not subject to speed limits such as the speed of light. Nor is it subject to diminishing strength with increasing distance from the source or blockage by other particles or fields. This shifting pattern represents a flux in the universal field that is no more than the effect of the moving particle or object on the field, which can be detected by other particles. However, the field changes would be so minute for a single particle moving at non-relativistic speeds that the overall field variations would go undetected.

On the other hand, living bodies have ongoing chemical reactions that sustain life in the body. The chemical reactions establish a pattern of changes in the centers of field density along the axial A-lines of individual particles in the body which couple together. The coupled axial A-lines lend an air of permanence to the processes beyond the chemical signatures of the physical living bodies. These variations are carried into the fifth dimension as patterns of change in the lateral A-lines 'around' the body in the fifth dimension.

The more complex the living organism, the more complex the patterns and the more radical the field variations communicated in this manner by the body to the surrounding universal field. These changing field patterns, represented by the lateral A-lines, are transmitted throughout the universal field; or rather the universal field reacts in response to the changing individual field patterns. Any other particle or group of particles can react to these field changes, or the lateral A-line patterns, which emanate from the initiating organism. So, to say either psi is carried by the lateral A-lines from one body to another or the signal follows the path set by lateral A-lines is just an approximation for graphical rendering of the real alterations and interactions within the universal field.

The A-lines have no more physical reality than 'line of force' or 'field lines' in electromagnetic theory. If the body that is 'receiving' the 'signal' and thus senses the shifting field patterns is a living organism, it will have some form of a brain. Some brains may be too small and simple to 'interpret' the very subtle field variations. However, the more complex the living organism becomes, the more complex the brain and the more complex the pattern of field variations that can be detected by the organism becomes.

Also, a more complex living organism with a more complex brain can detect far more subtle changes in the universal field. When the mind is associated with brain, the more complex the brain and mind, the more the mind is susceptible to universal field alterations from other organisms and bodies of inanimate material. There is a clear analogy between this process and the purely electro-mechanical process of radio transmission and reception. Two elements are needed for radio transmission or reception, an antenna and a tuner. An amplifier/signal interpreter could also be included, but they are not a theoretical necessity. A receiver can pick out the waves of a unique frequency or wavelength in either of two manners.

All wavelengths of waves permeate the environment, but the receiver needs to distinguish between them, or the reception is chaotic and static. If the receiver uses an antenna, the antenna will more easily and effectively pick out those waves whose wavelengths are equal to the physical length of the antenna (the Marconi method). Otherwise, the antenna picks up all wavelengths, so a simple tuning circuit is required to pick out a specific wavelength or frequency from all the different waves that are detected by the antenna (the Lodge method).

A radio transmitter uses its tuner in the same manner, transmitting only the frequency selected. A particle or material body's extension into the fifth-dimension (extra spatial dimension above three-dimensions) acts like an antenna. The greater the extension or aspect of the axial A-line into the fifth dimension, the greater the body's ability to send and receive field variation patterns will be. But the antenna/body needs a tuner to pick out the proper signal.

Increasing its energy or rather its relative speed can extend the antenna. The body can travel at a higher relative speed to increase its energy and thus extend its fifth-dimensional antenna, but moving the body itself is not very effective because the body would need to attain relativistic speeds to cause a noticeable extension in the fifth dimension. Also, moving the body does not increase the complexity of the energy pattern as would placing the body in a state of stress, excitement, or other internal energetic state.

Individual particles within the body's molecular and chemical makeup have varying speeds and thus a broad spectrum of kinetic energies; so individual particles in a body are miniature antennas. Changing the internal energy state of a living organism is more effective than changing the energy state of the body because it creates differences or more radical gradients of energy within the living organism. These differences translate into a more complex psi pattern, which is a more easily recognized psi pattern by other living organisms. The more complex the energy pattern in the host/transmitter, the more likely it is to interrupt the normal energy patterns in the subject/receiver.

In the case of a living body or animate matter in general, the ongoing chemical processes represent a constant interchange of energy between molecules, atoms and particles that are equivalent to antennas rising and falling along their five-dimensional components in the five-dimensional representation. These antennas act as both 'transmitters' and 'receivers.' The more complex the living organism, the better it can act as a transmitter/receiver in this regard. More complex organisms are better receiver/transmitters of psi. Their more complex internal energy patterns represent better antennas for psi signaling.

Within this context, a small resonance can be set up between two living organisms; or rather the energy patterns in the organisms can promote concerted action to a small degree. The greater the physical similarity between living organisms, the greater the similarity between their five-dimensional 'antennas' will be and the greater the probability of a resonant state being established between them. These similarities are not external similarities of shape and size, but genetic similarities that affect the whole organism. The resonant state would amount to the mutual interaction of psi between the two living organisms.

Human to human extra-dimensional contact (psi) is more probable and effective than human to paramecium contact, although the latter is not impossible. Within the human species, such contact would have a greater probability between identical twins, than between a mother and her children and then between family members, before they would occur between all other humans. There is evidence, both anecdotal and experimental, that twins display a higher-than-normal degree of psychic or psi-related experiences than other people followed next by mothers and their children, and finally other family members and close friends.

At the same time, a more effective tuner can enable a receiver/transmitter to focus on one specific wavelength or frequency as in the case of a radio. The tuner in humans and living organisms is the physical brain. Nearly all other theories of psi have either assumed or concluded that the brain is the organ of psi reception/transmission, which is wrong. The whole-body acts as the receiver/transmitter antenna and the brain acts as the tuner for psi reception. However, this is only a sideline effect in the brain whose more immediate concern is with the normal functioning of the body or living organism within its local environmental surroundings. The physical brain normally coordinates or controls interaction between a living organism and its immediate environmental surroundings.

The brain and mind discriminate between different forms of environmental stimuli that are detected by the specialized parts of the body, such as the eye, ear and skin. It is no different for the case of psi reception except that the body acts as the antenna that detects the extradimensional stimuli presented by field variations. There is no specialized organ within complex living organisms for psi reception or transmission. Each cell, molecule, atom, and elementary particle in a living body is exposed to the rest of the universe in the five-dimensional perspective. There is no skin, shell, or protective organ to cover and block the internal parts of the living body from its five-dimensional environment except for the coupling of energy patterns and their complexity.

The total complex of these patterns distinguishes the living body from other bodies of matter and can thus be regarded as the 'living force' that scholars have sought for several centuries. The physical brain discriminates between environmental stimuli and directs the organism to react to those stimuli in an appropriate manner. Those stimuli include the extra-dimensional influences of other material bodies, both animate and inanimate, which are not among the normal stimuli in the immediate physical environment. In other words, the physical brain is the fine tuner for picking out the psi signals (universal field variations as represented by lateral A-line variation patterns) from other physical bodies.

The brain picks out the signals by analyzing its own body energy patterns and noting changes. It then interacts with the universal field, or rather reacts to the field in its immediate physical surroundings to compensate for the variations that it has detected. The reaction to the field is transmitted by the body as a single whole through physicochemical variations and the corresponding changes in an organism's energy signature in the fifth dimension.

The higher-level brain of a more complex organism can even detect the source of the field change (read the lateral A-line variations via its own resonances with an object) and respond accordingly. The higher level or more complex brain can be trained, through experience, to discriminate between different energy resonance patterns, so humans can be trained to use their natural psi detection abilities or just learn through experience. Humans represent a far more complex living system than many other forms of life, so they have better antennas that can pick up a greater range of resonances and universal field variations. But our greater complexity requires greater control of the living body, so human brains are also more complex than the brains or physical control organs of other animals and organisms. Resonances occur spontaneously and randomly from specifically energetic events in nature.

The spontaneous nature of psi events is a function of the transmitter more than the receiver although the energy state of the receiver can enhance a spontaneous psi event. The more energetic, repetitive, and longer the duration of a physical event, the easier it is to detect by its psi signature in the universal single potential field. The more energetic, the stronger the patterns of lateral A-lines. The more repetitive the physical process or the longer it lasts in time, the easier it is to identify and the greater the probability that the psi event will be cognized by the conscious brain mind. The more complex the brain is, the easier it is to pick out the psi signature of an event and translate it to conscious thought. A better antenna and a better-prepared tuner can even pick out the subtlest 'signal.

There are several methods used by individuals or groups of people that can enhance psi reception and signal recognition. These methods were discovered through experience rather than theoretically based prediction and experiment; however, they still reflect the nature of psi as explained above. Shamans place themselves in an excited frenzy, sometimes with the help of natural drugs to reach a higher internal state of energy and variations of energy patterns. They are extending their antennas deeper into the universal field. Then the shamans collapse into a quiet period and reach into their minds to interpret their 'visions.' The quiet contemplation, whether drug induced or not, represents a fine adjustment of the psi tuners and amplifiers so the shamans can better receive as well as cognize the psi signals.

Drugs can possibly enhance the psi experience by either creating internal excited states that increase antenna reception or increase the brain's ability to tune into the psi or both. However, drugs are just as and even more likely to destroy or scramble natural patterns of energy in the living body and/or disrupt the brain's ability to tune into and/or interpret the signals. A drug that enhances the antenna portion of psi reception may scramble the tuner and vice versa. Using and enhancing the body and brain's natural capabilities through enlightenment of some sort is a wider and more effective choice to increase psi abilities.

While an excited state of body is better for reception, a quieter more contemplative state of mind is better for both receptions, by finer adjustment of the tuner, and interpretation of the signal. This technique is used in the laboratory during remote viewing and other types of experiments. A person can be placed in a quiet room, either in the dark or under a red light, to minimize extraneous sensory input. The idea is to deprive the person of sensory distractions so the person can better concentrate on psi reception and interpretation. However, the subject is just using the natural antenna and quieting the mind so that the brain can more easily tune into the psi 'signals' that the body (antenna) is normally receiving.

The brain must interpret these natural but paranormal resonances, the changes in its own internal energy patterns, to cognize psi. Mediums and other psychics also concentrate on controlling the brain's natural capacity to tune into psi signals rather than raise their antennas by altering their internal energy patterns. The same is true for a living organism acting as a transmitter. The greater the internal energy within a transmitting organism, the more complex the 'signal' and thus the easier the 'signal' is to detect. Agitated and stressful situations in transmitting subjects increase the likelihood of a successful psi reception because the 'emitted' patterns are more complex and thus easier to cognize, recognize and interpret.

There is evidence that more stressful states in humans cause a greater occurrence of psi. Natural disasters are more readily precognized because of the distress that they cause in their victims and people with knowledge of the disaster. Recent experiments at the University of Nevada in Las Vegas demonstrate that subjects precognize viewing pictures which show stressful events or emotionally laden images more readily than they precognize pictures showing pleasant or neutral images. These and similar findings support the contention that negative emotions and stress create better conditions for transmitting psi which is easily explained by the five-dimensional model.

The brain stores patterns or templates of the body's energy levels or perhaps groups of templates for the body's energy patterns under both normal and abnormal conditions. These templates need be no more than memories of past experiences. This is how the brain recognizes emotional and physical responses to stimuli within the body. Under normal conditions within a receiver's body an event or state in another body would subtly alter the receiver's internal energy patterns that would be noted or sensed by the brain. A received psi 'signal' might even interfere or cause destructive interference with the receiver's normal energy patterns, rendering the psi 'signal' more strongly cognized by the brain.

There is experimental evidence that reaction to psi events occur in the body, not the brain. For example,

Two subjects were separated by a wall. One was continually monitored by the plethysmograph while the other, the agent, was given a group of cards upon which several names were written. Some of these names were emotionally significant to the subject, some emotionally meaningful to the agent only, some taken randomly from the phone book, and others were blank cards. The agent then tried to telepathically send the

The lesson gained from this, and similar experiments were that psi affects the subconscious mind even when the psi signal is not consciously cognized.

The underlying and unspoken assumption of this conclusion is that the brain, consciously, unconsciously, or even subconsciously, received the signal and initiated the vasoconstriction as a response. Moreover, it is also just as likely if not even more likely, that the vasoconstriction was part of the detection/reception process of the whole body but was not strong or abnormal enough to cause the brain to cognize the telepathic signal. This is an example of the chicken/egg paradox. Which came first?

Since there is no evidence that the brain is the direct transmitter or receiver of psi, then regarding the body as the antenna of a receiver or transmitter is just as likely a possibility on purely philosophical grounds. However, since there is now a theory to explain how the body as a whole (including the brain) can react to the psi signal directly and only then does the brain interpret or cognize the signal, then it must be assumed that the reported vasoconstriction is a direct result of a resonance between the energy patterns of the agent and recipient. Both the mind and the body as a single unseparated whole interpreted and responded to the psi stimulus through a resonant vasoconstriction pattern independent of the individual physical brain.

The human brain compares energy pattern variations in the body to the template or stored memory and notes the differences. The notorious difficulty in interpreting psi 'signals' results from the inherent difficulty distinguishing these rather specialized energy changes in the body and brain from non-paranormal sensations of the external world and changes in the body. The brain would have to have some such system to react to abnormal energy patterns and disruptions in energy patterns due to diseases and other abnormal conditions.

Such a wholistic view of health and energy variations within the body forms the basis of eastern medical practices. The greater control that the brain (mind) exercises over the body and bodily functions that has been aptly demonstrated by yoga masters merely confirms that there is a more specialized interplay between the body and brain than has ever been suspected. Yoga masters report a higher incidence of psi events as well as greater psi capabilities. It would certainly seem that an enhanced control of the brain over the body increases psi capabilities, which would be suspected from the view of the five-dimensional theory.

However, there is a great deal of other evidence that the internal energetic state of the body directly affects psi reception/transmission. William G. Braud has listed seven "conditions" or "symptoms" which optimize psi performance. Together, they form the model that Braud has physical termed the "psi-conducive syndrome." They are …

> (1) physical relaxation, (2) reduced "arousal or activation, (3) reduction in sensory input and processing, (4) increased awareness of internal processes, feelings, and images, (5) stressing the "receptive mode/right hemispheric hemisphere of the brain," (6) an altered view of the world, and (7) psi (or what must be accomplished through psi) must be (at least) momentarily important. (Stanford, 831)

The first four of Braud's "conditions" are antenna related and the last three are tuner related in the five-dimensional physical model of psi.

This idea can also be put in other terms: The first four conditions are paraphysical and the last three are parapsychological. Braud's purpose was not to derive differences between the physical and mental aspects of psi, even though he unwittingly accomplished that end. The very fact that he was working within the parapsychological paradigm without regard to physics and physical processes makes his model even more plausible. Braud's model directly confirms the evidence that the internal energetic state affects the psi process and thus indirectly lends more credence to a five-dimensional interpretation of psi.

The various forms of psi

Of the various manners by which psi manifests in the physical world, telepathy or thought-transference is the most common. It is so common that it would seemingly be the easiest to accomplish. There is probably a great deal more telepathy than would commonly be admitted, simply because the recipient of the telepathic 'signal' would be unable to distinguish between that 'signal' and his or her own independent thoughts in most cases.

The greatest difficulty with developing a theory of telepathy or thought transference is the vague definition of thought itself and how thought arises in the brain and body. Science does not yet know what thought is. But whatever it is, it is associated with the physical brain and electrical impulses in the brain. As such, thought causes the most radical variations of any organ in energy patterns associated with the body as a whole and thus variations in the universal field. These variations are picked up by the bodies of others but would normally go unrecognized and not cognized. If a thought is particularly stressful to its thinker, a more intense and complex energy pattern associated with the thought and subsequent shifts in the thinker's overall pattern signature would increase the probability of reception and cognition by the recipient. Still, thought transference is for the most part a spontaneous transfer of idea.

The thought is there in the universe under the guise of a pattern in the lateral A-lines emanating from the thinker/transmitter, but it would not make any sense to just anyone even if it had sufficient intensity. It would only be cognized by someone who had physical similarities (a family member), someone who had a matching template within their own memory set (such as close friends), or someone who would otherwise have some vested interest in the thought.

Clairvoyance is the knowledge of objects or events that the subject has no way of having learned through normal means. A clairvoyant would have knowledge of, or seek to gain knowledge of, a physical object or event in the present, for example, the contents of another person's wallet or an envelope. The contents of any such enclosure would have a specific energy signature with which the clairvoyant could easily be familiar, so the clairvoyant would 'feel' the identity of the object or just 'know' it.

The identity of the object or the information that it represents would 'feel' correct. 'Feeling,' in this context, is an allusion, illusion, and reference to a bodily sensation not a thought. In this case, the brain and mind are just referring to the body/antenna as the receptor of the 'signal.' The same is true for an event that the clairvoyant might sense.

Another form of clairvoyance has become popular in the past two decades. It is called 'remote viewing.' Remote viewing is literally mentally or mindfully 'seeing' or otherwise sensing some object or event at a distant location. Remote viewers usually report vague feelings about the object such as hot, cold, and moist to build an impression of the object or place upon which they are focusing their attention.

They have not really traveled, aethereally or otherwise, to the object or location where they could see it with their eyes. They also report geometrical shapes and other sensations that they associate with their target, until the view is completed. In a sense, they zero in on their target by a slow winnowing process of cognizing vague feelings and intuitions associated with the object of location. This process lends itself well to an explanation based upon the five-dimensional hypothesis. Their sensations, feelings and vague intuitions directly imply things familiar to them that evoke physical responses in the body. Their brains are comparing simple universally known patterns in the absence of specific memory patterns for the object or location.

The mind is basically orienting their thoughts together in the spacetime sheet by following a map of physical feelings and sensations. They are matching their target's energy pattern with simple universal patterns since they are unfamiliar with the complex energy pattern of the targeted object or location. They are attempting to zero in on a specific object with those energy pattern characteristics that they have sensed. They initially have no familiarity with the target, so they literally build its image from scratch and small bits of information with which they are familiar using familiar 'forms' in their memories. They systematically match their own simple memory patterns to the object or location in lieu of a whole picture of the target. This is a practical utilization of the goal-oriented nature of psi.

While telepathy, clairvoyance and remote viewing represent psi processes that occur in the present moment, there are psi functions that cross the boundary of time in both directions. Precognition is merely knowledge of a future event or events. Its opposite is retrocognition, the knowledge of a past event or past information about which the subject could not have learned in any conventional manner. Precognition, of course, does not represent knowledge of 'objects or events which could not have been learned through conventional means' because it does not represent learned or experienced knowledge, the event not yet having occurred. So, no one could have learned of the future event by conventional means.

Since precognition and retrocognition deal with facts and events outside of the present moment, they have traditionally caused more problems for theories of psi than other forms of the psi process. Simply put, the future has not yet occurred, and the past is physically out of reach. These features emphasize the unique nature of time within the space-time framework. But from the five-dimensional perspective, both the past and future 'exist' simultaneously with the present although that 'existence' may not be of the same quality and material quantity as the present. Yet it does open the possibility to explain precognition and retrocognition.

The universal single potential field does not differentiate between the time and space directions, so approximately the same process of pattern variations and resonance occurs between living organisms, objects and events in different temporal locations as occur between different spatial locations. Yet there remains the fundamental problem that the future has not yet occurred.

In every real sense, the universe 'knows' or can 'calculate' course of the future the future even though the future has not yet been made absolute, concrete, or substantial, from the point of view of the present moment as have both the whole past and the present moment. The future has not yet been fixed within the pattern of variations in the space-time sheet and the overall field, but the universe with its universal consciousness could easily account for all changes that would affect the future outcome.

According to the Newtonian worldview, as expressed by Simone de Laplace two centuries ago, given the position and velocity of every particle in the universe, the future could be predicted and thus known. The universe itself possesses that same information. In fact, the universe 'is' that information, so it, in effect, knows the future. A sense of this idea also comes from classical electromagnetic theory. Electromagnetic theory reflects the symmetry of time, forward and backwards, in the direction of the waves. Solutions of the electromagnetic equations yield answers that work in either direction; however, the solutions that show motion backward in time are normally discarded.

It is generally assumed, without any proof or confirmation, that the waves moving back in time interfere destructively and cancel each other.

But suppose the electromagnetic waves that move backward in time do not destroy one another, then either the past can be changed by the waves from the present which we know is not true or the universe (in the present) has already considered the effect of the backward traveling waves and they have already changed the past and no longer affect our present. This would seem to be the logical conclusion to make 'in light of' the backward traveling portion of the electromagnetic wave.

Therefore, our own present has already adjusted to the possibility of interfering waves from our future which have traveled into their past, our present. This line of reasoning implies that the future already exists. This argument strongly implies that even the consequences of classical theories have not been completely thought through or explored, so classical theories may have far more relevance to modern physics than they have been awarded in the post quantum and post relativity era.

The view from the five-dimensional perspective is not so different from the classical views, yet the exceptions and differences are significant. According to the five-dimensional worldview, the lateral A-line patterns spread out from particles, events, and objects in the temporal direction as easily as the spatial direction. So, these patterns, projected into the future, determine the future with a very large degree of certainty, although not with absolute certainty. The degree of certainty is open to small local fluctuations. The certainty of the future is so nearly complete, in fact, that the future is a virtual certainty except for the intervention of free will.

Free Will choices are not predictable. It is not a product of the physical brain, which bends to the will and 'laws' of nature but is a product of mind and consciousness. So, free Will must consider factors that are not included in the purely mechanical view of physical reality. Free Will is not as free as philosophers and scholars may argue or even wish, since there are many factors at work in the universe of which they have no knowledge and have not considered in their philosophical debates on the subject. But there is a hard kernel of freedom of thought that is the truly free will of sentient beings and this small amount of free Will can alter the course of events in significant ways. Free Will is not unlimited. It cannot change the future in any way that would defy physical laws and fundamental rules of the universe, even if free Will is unaware of those laws.

In other words, free Will can change the ripples and patterns in the sheet of the universal field only to the extent that it does not change the sheet and the universal field themselves. So, the future is nearly certain, but not yet completely and certain until a point in time when everything that could change the future in unexpected ways has been considered. True psychics who have honed their precognitive skills and do in fact mentally 'see' or sense the future admit that what they view, or sense is not absolute, but can be changed.

These explanations nearly complete the basic psi effects that occur on the purely mental level of ESP phenomena. However, the story is different for the class of phenomena known by the collective term PK. PK is an interaction between objects or events and people, such as remote viewing and clairvoyance. PK is, however, active rather than passive as are ESP phenomena. It is not only active, but it can be both pro-active and aggressive as any case may warrant. It has often been called 'mind over matter,' because the mind controls matter and the 'motion of matter' during the PK process.

No one would deny that the mind and brain have the capacity to control the motion of matter. However, mind and brain would normally move matter according to muscular/mechanical means. Mind acts locally to move matter. In the case of PK, the mind acts non-locally to move matter or otherwise affect its condition and/or its state of motion. There is a growing body of experimental evidence in physics that seems to confirm that consciousness can act non-locally. In this one respect, modern scientific views of reality are already moving toward the paranormal view of reality, but what is being proposed as PK in the study of the paranormal goes far beyond the simple act of consciousness causing the 'collapse' of the wave packet at nonlocal and physically unconnected positions in space-time.

Telekinesis or simple psychokinesis occurs when the mind affects the state of motion of a physical object without direct or indirect muscular/motor/mechanical contact with the object. The possibility of such motion has long been a thorn in the side of the physics of psi because moving an object takes energy. Work must be done on the object to move the object and that work will show up in the object as kinetic energy.

But energy cannot be created from nothing since that act would violate the conservation of energy, one of the most fundamental of all nature's rules. The concept of' mind over matter' would seem impossible for physics because it implies a violation of the conservation of matter. Other forms of PK also violate the very basis of modern physics. However, the redefinition of matter and motion itself within the new five-dimensional framework of space-time has changed that situation.

In the five-dimensional perspective, there is no such thing as kinetic energy. All energy is manifested by a single field potential relative to field position and distribution in five-space. Nor is there such a thing as force. The four-dimensional concept of force reduces to changing curvature and other field conditions in the five-dimensional model. There is no problem with PK from this perspective. On the other hand, everything occurring in five-dimensional space-time corresponds to events and occurrences in our normal four-dimensional space-time. So, the conservation of energy, which is purely a three-dimensional perspective of our physical reality, can be extra-dimensionally altered without being broken for PK to be understood.

As time flows toward the future, every bit of matter undergoes the process of deconstruction/reconstitution as the moments pass. Motion is a sideways displacement in space as time moves forward under restrictions imposed by the connectivity constants, μ_0 and ε_0. Consciousness and mind also can 'collapse' the wave function within this framework and the wave function corresponds to the probability that a particle can exist elsewhere than it was before the present moment in time.

At the precise infinitesimally small moment of deconstruction, before the reconstitution of a particle or body, the wave can be 'collapsed' at a new position sideways in space. Since the 'collapse' occurs before the completion of the deconstruction/reconstitution process, the particle has moved without the association of a corresponding speed. Thus, there is no kinetic energy in the change of relative position in space.

The matter curvature never changes as it would during a normal motion characterized by speed and the relative position of the center of field density along the axial A-line has remained constant relative to the fifth direction. Because of the restrictions imposed by connectivity, the particle or body could not change spatial position instantaneously but would appear to move at a constant unchanging speed, unaffected by gravitational and electromagnetic forces. This pseudo-motion would be a step-wise process whose individual steps are buried within the fundamental effective width of the space-time sheet and so appears continuously.

Two other types of phenomena are directly related to telekinesis. They are the 'spoon bending' phenomena and poltergeist activity. 'Spoon bending' and related distortions of material bodies are specializations of motion, but they are not cases where the whole body moves. Parts of the body move around a point that remains stationary relative to original body state, causing physical distortion within the material body. The object moves around a point, it bends, and the description of the process is the same as telekinesis.

Microscopic analysis of spoons bent under the influence of PK has revealed strange characteristics of metal fracture that cannot be duplicated under normal circumstances. The structural fractures that are evident in 'spoon bending' events are a direct result of a point-to-point variation of motion. Since the body of the spoon did not move linearly as a whole, the structural changes at the submicroscopic level reflect the material's reaction to momentary connectivity alterations in space-time itself. Objects that move psychokinetically as a whole, linearly, without bending or twisting carry no structural evidence or ill effects of their strange journey.

Poltergeist activity, on the other hand, is a more straightforward case of telekinesis, although it is telekinesis that occurs spontaneously in the vicinity of the source of the effect. The source of the PK activity is unaware that he or she has initiated the poltergeist events. The activity is initiated by the subconscious mind of a mentally troubled individual. A relationship has been found between poltergeist activity and the source of the effect. Poltergeist activity occurs far more often than not in the presence of a young adult going through puberty. Such a person would be experiencing radical hormonal changes which would sufficiently alter the body's physicochemical balance enough to radically change the extra-dimensional energy patterns associated with the subject. These radical alterations open the door to the spontaneous PK effects.

However, other psychological or physio-psychological problems must force the subject through the door to trigger the poltergeist activity. At this point, poltergeist activity becomes the subject of parapsychology rather than paraphysics. The subconscious aspect of the poltergeist phenomenon lends credence to the existence of the mind existing separate from the brain and the influence that mind has on variations in field patterns.

The very nature of the five-dimensional model lends itself to the interpretation of yet another form of PK, paranormal or psychic healing. Since the whole-body acts as a physical antenna and thus a mitigating component between signals leaving and entering the brain (tuner/amplifier and decoder in higher level living organisms), the brain should not only be able to control the body to a larger extent than is normally accepted by science, but also influence other living bodies through the same resonance of energy patterns. In its store of energy patterns (memories of material objects) and their templates in stored in consciousness, the brain and mind surely have templates representing the most general and healthy body. Under proper conditions of mind/brain/body interaction, this template could be used, through psychokinetic action, to help heal another person.

There are presently great numbers of 'psychic healers' using various methods that may or may not be feasible and non-fraudulent. The fact that someone claims to be a psychic healer does not mean that they are such even though all beings would have that capability to some degree. The effectiveness of any one person's abilities along this line depends upon too many variables to prove or disprove their credibility in this matter.

Many experiments have been carried out in laboratory settings on lower life forms that seem to indicate that psychic healing is a real phenomenon. Yet to take a leap of faith and accept these experiments as proof of psychic healing in more complex beings is not always wise. The clear and simple fact is, psychic healing is a real effect and the 'mechanics of the practice is just a variation of the psychokinetic process.

Exploring the forbidden country

The final types of paranormal phenomena to be considered will prove to have significance far beyond their simple explanations. These are the Near-Death Experience (NDE) and the Out-of-Body Experience (OBE). The two could be considered separately since an NDE does not guarantee that an OBE occurs, and OBEs have been reported independent of NDEs. However, they share significant qualitative properties, and their explanations are intimately bound to one another, so they will be considered as two aspects of one single phenomenon.

And finally, they lead to an explanation of 'spirits' and ghost/haunting phenomena. In the above development of the concept of life, living organisms held a unique position regarding all material bodies according to the fact that complex chemical reactions created specific energy patterns that are unique to life and living organisms. The more complex the living organism is, the more compounded the complex of interrelated energy patterns that represents the whole organism as a unique body of matter. The notion that this complex of patterns was related to the mind was also expressed and mind was noted as different from the physicochemical/mechanical organ of the brain.

The mind, since it is a whole-body existence, is both an extension of the brain and a measure of the complex energy patterns of the brain. The mind is characterized by an awareness of brain, body, and the immediate environs of both as well as their intimate relationship as expressed in the energy patterns that form a 'life force' for the body as a whole. In fact, the mind is the physical brain's link with memory patterns.

Memory patterns and templates need not be stored in the physical brain. Memory has some properties of a hologram that corresponds to the five-dimensional view of space and time. Memories could be stored in 'the universe as a whole' as part of the pattern of variation in the past portion of the space-time sheet. This idea was implied in the earlier discussion of retrocognition. This statement is not meant to sound mystical, and it certainly is not mystical, but refers to the fact that 'all of time' occurs simultaneously from the five-dimensional viewpoint.

The present 'brain' is in constant contact with the 'mind' across the whole 'length' (at all temporal points) of a being's 'world-sheet.' A portion of memory is undoubtedly stored within the physical brain, but memory is also a form of low-level personal retrocognition. Normal memory is enhanced by the mind's ability to draw on events in the subject's past via lateral A-lines extending backward in time within a person's 'world-sheet.'

Therefore, there is also a 'super memory' that is a stored pattern in the past portion of a person's 'world sheet,' a permanent retrocognitive function of mind. What is commonly called memory is a combination of the resident portion of memory in the brain and body (forming lower cognitive consciousness) in combination with the 'super memory' (a person's higher consciousness) stored in the 'world sheet' of the person. Mind has subliminal access to the 'super memory' just as the physical and material brain has access to the resident memories stored in the brain.

Under these circumstances, the mind is aware of the brain and body, which requires a higher level of reality than permitted to the physical brain alone. The mind can interact with the material brain, consciously, unconsciously, and subconsciously, so the mind must include a more 'discrete' contact with patterns of energy variation and flux in the body than does the brain. Mind is thus related to the fifth dimension even though it is technically only three-dimensional itself, but mind is also associated with choice and free Will so it must include experiences, emotions, biases, prejudices and many of the intangible qualities and elements that have defied reduction to chemical reactions in the brain.

Mind is extended five-dimensionally through its connection to the five-dimensional higher consciousness (as a template) of the physical brain and body since the brain is limited to a strictly material existence within the space-time sheet. As such, mind also connects brain with the complex energy patterns that represent the quality that was once called the 'life force' of an organism. Since the mind is an intangible quality associated with the whole body, it must exert a higher level of control of the brain and the body/organism. Brain controls the body on the four-dimensional level of space-time, but mind has ultimate control over the brain since it has a five-dimensional extension of the brain while it has direct contact and some control over the organism itself through the five-dimensional aspects of the organism.

Due to the characteristics of the organism in the fifth dimension, mind is extended beyond the physical confines of the brain itself – Mind covers the whole body even though it expresses itself through the brain. The mind is the brain's component in the complex of complexities that distinguishes individual living organisms within the universal field. The mind must be aware of 'aspect' changes along the axial A-lines of the material particles that constitute a living organism. So, it can change the 'aspects' and thus initiate motion.

The mind is at least a property of most living organisms and most probably a property of all living organisms. However, all living organisms have a brain or some form of brain function as a physicochemical basis of control within the environment of their immediate world. The mind also represents an awareness of the brain's environmental control and manipulation of the immediate surrounding four-dimensional components of reality.

But since the brain is merely a physical mechanism, the mind allows the manipulation of the local environment for purpose rather than for physical survival. Many of the individual personal intangible qualities associated with living organisms, such as compassion, rage, love, and bravery are either portions of the mind or related to the mind/brain interface. These qualities only come to the surface in personal relationships and interactions with other people, living organisms and physical bodies.

The mind deals with local environment and awareness of the brain and body's intimate and local interaction. The mind has evolved from simpler forms of the brain, just as the greater complexity of living organisms has increased over time.

During the process of evolution, organisms within a group or species reaches a specific level upon which the complexity of energy patterns within each member of the species begins to form a distinguishable higher pattern that develops into a new higher-level order which has its own individuality. This 'entity' is not separate from the physical body but anchored to the spacetime sheet as a more complex extension to the brain/body system. A special complexity of patterns within the first order complexity of energy patterns associated with life distinguishes the individuality of the brain/body mechanism.

This new higher-level complexity, the mind, is an awareness of the lower-level patterns and their functions. As the brain and mind progress, learn more of their environment, the scope of mind increases. They interact with a larger extension of their environment in the physical world and through their extra-dimensional connections until a point upon which they become aware of their nonimmediate environment, the portion of the world outside of direct physical interaction through the normal senses.

The brain/body/mind complex learns a new set of patterns and thus stores an ever larger and more comprehensive set of templates for the world at large by which it eventually attains new and higher levels of complexity. When the brain/body/mind complex becomes consciously and simultaneously aware of both its immediate (local) environment and distant (non-local in space and time) physical connections with other material bodies in the environment, a new higher plateau of awareness has been attained at a higher level of complexity called consciousness.

Since consciousness is an awareness of both the local and non-local physical environment as well as the interaction of the body/brain/mind system with them as a wealth of extra-dimensional connections corresponding to the total environment, consciousness must dwell in, or at least include knowledge of, a still higher physical level of being, a six-dimensional extension of our universe. Four-dimensional space-time is the seat of the mechanical brain and body.

Life utilizes the five-dimensional extension of the brain/body system, but both an awareness of the local environment and a partial awareness of the non-local environment come at a higher-level complexity associated with mind. The mind is in the fifth dimension as an extension to the brain and body. By analogy, consciousness represents a still higher complexity than mind, so consciousness utilizes a sixth dimension to complete the whole being.

Just as an awareness of the four-dimensional spacetime environment implies an existence in the fifth dimension, conscious awareness of mind and the Mindbody-environmental system and their intimate interactions would imply the existence of a still higher dimension. However, these higher dimensions are not to be equated to mind and consciousness. The higher dimensions are a part of nature and exist independent of mind and consciousness.

These higher dimensions are physical not particularly mental. Mind and consciousness must be aware of the higher dimensions so these higher dimensions act as fields of action in which mind and consciousness work. Consciousness need not be a property of all living beings although it is assumed that nearly all, if not all, living beings have at least a low level of mind. Consciousness is a property only of those groups or species of living beings that have reached a certain plateau of complexity.

Once this plateau has been reached through the natural evolutionary (negentropic) process, it can be explored by either an individual within his/her own unique situation or by groups of individuals collectively. These explorations mark attempts by various beings to reach a state called 'enlightenment,' 'satori' and similar states of realization and awareness of consciousness. The quantity and quality of consciousness can be increased by an individual exploration within a being's own resident dimension and lower dimensions.

If individual explorations of the higher dimensions lead to an awareness of their totality and function regarding the four-dimensional organism, new sets of memory patterns or templates can be established until a further plateau is reached, and a still higher level of complexity develops into a new and higher coupling of patterns and level of consciousness. Science seeks to learn more of the physical world by reduction of the external world to its constituent parts to increase consciousness by elaborating and increasing the numbers of memory patterns while mystics seek to move directly toward the awareness of higher forms of consciousness and thus push the individual forward and upward to higher states or levels of consciousness.

When the body dies, the physicochemical interactions that are the physical life of the body end and there are no new energy patterns or variations in the energy patterns originating in the body/brain system. The negentropic being or existence of the body ceases and the physical body reverts to its common entropic state. The body begins to decay entropically, by the normal process of physical dissipation of the orderly structure of material within the body.

The body degrades to become ordinary inanimate matter. A new discontinuity has come into being from the death of the body, but a purely negentropic state has evolved (or devolved) into a purely entropic state of matter. In one sense though, life continues since the variational patterns of energy stretch back through time in the organism's 'world sheet.' The complexity of coupled patterns which were the mind before death still exist in the past of the universe, but they also exist through their interactions with the rest of the universe in the present and future after the death experienced by the body.

With death, all that has really ended is the association of the energy patterns with the physical body at the very moment of death and thereafter along the temporal direction of space-time. The connection of the physical body with mind is severed by the death of the body. Higher consciousness and its mind template are no longer connected with the four-dimensional spacetime sheet, so the body reverts to an inanimate state whose complexity of chemical reactions, no longer interacting as life, makes the body a rich source of food for the bacteria which cause decay.

In this very narrow sense, the mind still exists as a template in higher consciousness, which does survive, after death albeit in at least a rather nonattenuated and incoherent form. If mind were all that had evolved from life, body and brain, then the death of body and brain would be the end of the individual being. But there is more to a comprehensive reality than just the dissipation of simple mind into the surrounding universe after the moment of death and the initiation of decay.

From the higher-dimensional perspective of consciousness upon physical death, the brain/body system appears as a mere scaffolding about which more complex structures of existence have been erected, the mind and consciousness. Upon physical death, the scaffolding is merely dismantled, dissembled by the natural processes of the physical world. The scaffolding is no longer necessary, but the complex structures of mind and consciousness continue to exist.

If the mind alone could exist after death, it would be in a somewhat dissipated or non-attenuated form after the moment of death and separation from the chemical processes that first created the energy patterns. If the evolutionary process ended with mind alone, then physical death of the body would mean the end of mind as a coherent 'entity' with an individual existence. But consciousness exists as a higher order of complexity than just the mind and holds the remnants of mind together as a whole. Consciousness connects the dissipating extensions of mind even after physical death.

So, higher consciousness survives death as a unique and recognizable pattern of complexity that forms an individual 'entity,' if a loose definition of 'entity' is applied. As a higher order awareness of mind and the extra dimensionality of a living organism, consciousness survives death by giving mind the coherence that had been supplied the mind by the brain and body before their death, albeit in a form which has no complete counterpart in our physical world.

Because of its higher level of awareness across time (the awareness of awareness), consciousness is aware of the concept of death and the death experience well before it is time for the body to die. This awareness trickles down to the brain itself, so the brain has knowledge of death and thus a pattern or rudimentary (within its own level of understanding) template of the death state. Since knowledge of death in the brain is represented by a rudimentary pattern or template, a near death experience can be triggered by a simple set of environmental stimuli or internal symptoms that the brain associates with death even when the body is not really dying.

This turn of events epitomizes and characterizes the NDE. In the NDE, the brain fools the mind and consciousness to react as if the body and brain were dying. The brain initiates the consciousness" awareness of the death state without dying and thus the mind is momentarily forced into a death mode. In some cases, the body does die, but through human mechanical, electrical and/or chemical intervention the body is revived to a living state. In either case, an NDE can result.

The NDE is a quite common experience among a large and growing percentage of human beings. Those who have experienced and remembered the NDE report common elements to the experience despite cultural differences, but prevalent cultural templates that fade with greater distance with the physical reality of three-space cloud many of those common elements. In other words, the stronger any one person's NDE the greater the similarity of the experience with people having different cultural backgrounds will be.

However, shared properties (visions of death) do indicate a fifth dimensional perspective of reality. The white tunnel, tunnel of light, or tunnel with a bright light at the far end which is common to NDE experiencers is simply the physical brain's emergence upward along the coupled axial A-lines which constitute the five-dimensional portion of an organism's own being. The white light is a physically simulated and conceptualized light wave, the cognization of a simulated A-line perpendicular to the space-time sheet.

Another property of NDEs is the state of change (attitude and emotional) of individuals after they recover from their experience. They have a greater sense of love, compassion, and regard for all around them. From the higher-dimensional perspective of this model, as a subject moves higher into the fifth dimension along the entangled axial A-lines of its own physical body, the subject would realize or become aware of its own complex structure of connections with other living beings as well as inanimate objects by way of the lateral A-lines from the other physical bodies.

The stronger more recognizable contacts, of course, would be from other living organisms with which the subject is familiar. The subject would have a clearer resonance with family members, friends, loved ones and the people and objects most recently encountered in the physical world since they were the last things that the subject experienced before the NDE. Another property common to NDEs is the flashing by of a person's 'whole life before their eyes' in a moment's time. This result of the NDE is merely the subject's realization of those specific lateral A-lines which extend back through the subject's own 'world sheet' within the space-time continuum.

Other properties of NDEs are just as easily explained even when they seem scientifically paradoxical. Those friends, family members and loved ones who have passed on before the subject often greet the subject to the 'hereafter.' The 'spirits' who meet the NDE subject in the ND state are those individuals whose templates are psychologically the best established or strongest within the subject's own individual brain, body and/or mind. Those templates are used to pick out what is left of those individual's consciousnesses from within the extradimensional portion of the universe where they still exist in some manner. For every fifth-dimensional aspect there is a corresponding physicochemical change in the normal space-time continuum of four dimensions.

Chemical explanations and arguments against the reality of NDEs are irrelevant to the reality of the paranormal experience since there is a corresponding four-dimensional physicochemical artifact for every event or action that occurs in five-space. The fifth dimension is not something separate from the four-dimensional space-time but is an extension of space-time in another orthogonal direction, so the paranormal aspects of the NDE would trigger corresponding chemical reactions in the brain that is all that the critics are detecting. The critics of the NDEs merely use chemical explanations as excuses, ignoring the physics, for disregarding the quality and quantity of the experiences and ignoring the possibility of a larger and richer structure of the universe that would include such messy and puzzling features as those described.

OBEs have a similar explanation whether they occur concurrently with the NDE or independently. The higher-dimensional existence of individual consciousness acts in a manner like an individual extended 'entity' or a person's 'spirit.' As such, it orders the complexity of patterns that constitute the mind from above even though the mind originated from below with the bonding or coupling of energy patterns of the organisms within the four-dimensional space-time sheet.

Under special conditions, the consciousness can act as an individual 'entity' with restricted freedom within the higher dimensions. To say that it could move laterally within the higher space to a new location in four-dimensional space-time would not be accurate because motion or movement is associated with a relative location change only within the normal four-dimensional space-time continuum of our common experience. It would be far more accurate to say that consciousness and mind somehow turn off or tune out their immediate awareness and interaction with the physical body in space-time and more critically attune themselves to their inherent awareness of the rest of the universe.

This action would amount to a simulated motion through the rest of the universe. However, contact with the physical body/brain mechanism would never really be severed during the OBE. During an OBE event, a person can still cognize or become aware of their connection with the space-time continuum through the bundled A-lines or rather the dense portion of the universal field corresponding to the particles that comprise their body. The person's mind must deal with this cognition in some recognizable form, so the person senses or 'sees' a golden or silver cord attaching them to their physical body. Some OBE subjects have described such a golden connecting cord which is just a mentally derived metaphor for their continued connection to their three-dimensional material brain and body.

When the subject experiences contact with 'spirits' during an NDE, he or she selects individual consciousnesses in the state in which those consciousnesses survive within the extra-dimensional cosmos and thus the subject is visited by those personalities during the near-death state. When the consciousness acts as an individual entity during an OBE and encounters spirits, the same is true. If someone wishes to call them 'spirits,' then they are 'spirits' for lack of a better and more precise scientific term. But they could just as well be called discarnate consciousnesses. Discarnate consciousnesses are just OBEs that have become dissociated with their brain/body mechanisms due to the death of their material bodies, or even aware and non-aware higher consciousnesses of still living beings.

These 'spirits' are different from, but related to, 'ghosts' and 'apparitions.' When a person or other being dies their bodies continue in an inanimate state. The universe had grown familiar with the animate state and could therefore project probable life, along the 'world sheet,' into the future. But the situation changes upon death and the 'world sheet' essentially ends as the self-motivation aspect of the body ends and the inanimate remains decay.

The future portion of the 'world sheet' is part of the 'implicate' order, to use Bohm's terminology, and can be sensed by others. This is especially true if the person died under stressful or extremely emotional circumstances in which case the future portion of the 'world-sheet' is stronger and more easily detected by others 'para' normally. This 'implicate' impression in the future acts as a template when another person detects it paranormally. The person then taps into the previous history or 'world sheet' of the 'ghost' as if the personality of the individual had survived without a body.

This experience or situation represents a haunting. Hauntings are usually associated with places, objects and locations where people died under stressful conditions, such as murders, deaths during combat or death during some type of disaster. The more stressful or emotional conditions of the death just develop a stronger and more complex energy pattern that would be more easily detected and recognized by another person. As such, a ghost is no more than a discarnate 'super-memory' template of some tragic incident.

Some preliminary concluding remarks

So, within this five-dimensional model of the universe, paranormal phenomena are easily explained. Their explanations emerge from physics and are not dictated by the phenomena as new specialized physics create only for the moment and situation, such that new physical quantities need not be invented to precisely fit the phenomena. Psi has effectively been 'reduced' to an interaction of mind and consciousness within their respective physical realms of action.

Mind and consciousness have received new definitions based upon this new physics. They are paraphysical quantities associated with living organisms and animate matter. Consciousness is an awareness of the purpose and extra-dimensional scope of mind while mind is also a form of awareness as well as an interaction between body, brain and the patterns of life itself within a living organism. Consciousness exists at a higher level than mind alone and there appears to be still higher levels of consciousness, which can be attained by humans or other conscious and sentient beings.

This last aspect of the model corresponds nicely with the concept of Buddhist and similar forms of enlightenment. In the Buddhist tradition, a person experiences or becomes aware of psi phenomena at the lowest level of enlightenment. This would correspond to a five and/or higher-dimensional awareness of the lower dimensions of physical reality and their intimate connections. But the Buddhist doctrine and beliefs as well as some other mystic beliefs continue to teach that a person must not become entangled or stuck in this level of enlightenment or the person cannot attain higher levels of enlightenment.

Within the context of this theory and model, the Buddhist and mystical belief simply means that a person should not just explore the five-dimensional portion of the universe where psi occurs but explore the sixth dimension where consciousness acts to move on to an even higher-dimensional awareness through the development of a higher-level complexity, a new stable magnetic domain, and coupling of higher-dimensional potential field patterns which equate to matter and energy in our three-dimensional experienced world.

Within this physical model of psi there exists a simple line of demarcation between the parapsychological aspects of psi and the paraphysical aspects of psi. Paraphysics treats and explains the 'signals' between physical bodies, but it does not address the methods of amplification of 'signal' nor the interpretation of 'signal' by the material brain during cognition. The interpretation of the 'signal' is influenced by chemistry, but also by biases, emotions and cultural influences that are within the province of the parapsychologist as are the other aspects of psi that are not physical in nature.

By developing a model of psi which is physical and providing a strict line of demarcation between the physical and psychological aspects of psi, new experiments, and methods for studying psi with a greater precision can be designed and implemented. This theory and theoretical model of psi should provide a new impetus for the development of science as a single whole.

Conclusion

Solving the Universe!

"Solving the universe" is a phrase that was used by William Kingdon Clifford during the 1870s to describe his attempts to derive a theory of matter based upon curved space. Today, the phrase better fits what are called 'theories of everything' or TOEs. Even though some scientists are working on TOEs, it is questionable whether any theory can explain 'everything' in the universe. The name TOE may be no more than an example of errant human egocentrism (it is very egocentric to presume that we know the whole universe from our extremely small corner of reality), scientific bravado or possibly a false hope of a limited reality which we will soon understand, but the acronym TOE is 'cute' and 'catchy', so its usage has become popular.

The five-dimensional theory proposed in this volume, dubbed SOFT for single field theory (or Single 'operational' Field Theory if the 'o' needs elucidation), comes far closer than any other theory to the mystical concept of a TOE, but it is not truly a TOE. It could only be considered a TOE if we wish to belittle what the term 'everything' refers to, the wholeness, oneness and continuity of the universe.

Several philosophical problems arise if the notion of a TOE is even contemplated. P.C.W. Davies and Julian Brown have stated that the ultimate TOE would not be subject to falsification.

> The ultimate TOE would, ideally, need no recourse to experiment at all! Everything would be defined in terms of everything else. Only a single undetermined parameter would remain, to define the scale of units with which the elements or the theory are quantified. This alone must be fixed empirically (In this ultimate case, experiment merely serves to define a measurement convention. It does not determine any parameter in the theory). Such a theory would be based upon a single principle, a principle from which all of nature flows. This principle would presumably be a succinct mathematical expression that alone embodied all of fundamental physics. In the words of Leon Lederman, director of Fermilab - the giant particle accelerator facility near Chicago - it would be a formula that you could 'wear on your T shirt'. (Davies and Brown, 7)

So, if the ultimate TOE that scientists presently seek explains 'everything,' it can be neither verified nor falsified as the true picture of nature. This conclusion would seem logical given Gödel's theorem whereby we would have to somehow go outside of the universe, or outside of this ultimate TOE, to a more comprehensive physical reality to determine the validity of this mythical TOE.

Quite simply, a true TOE could never be verified or otherwise subject to falsification. If it explained everything, it could never predict new phenomena to test and therefore, could never be open to falsification. But even before that point is reached, the most advanced theories will become more difficult to falsify or verify. The more comprehensive and complete any theory becomes, the more difficult falsification will prove to be, and SOFT is no exception. As it stands, SOFT is too broad and flexible for direct falsification. In general, scientists do not like such broad and comprehensive theories since they leave nothing from which they can be verified. At the very least, scientists are suspicious of the claims made by those who propose such over-reaching theories.

But this theory represents the logical application of the single simple hypothesis of a real fifth dimension that is either falsifiable or verifiable by its individual elements, consequences, and applications, although not the whole. The problem with proving this theory, by demonstrating that it is open to falsification, is inherent in the very nature of the five-dimensional concept. It is just as Einstein stated, the five-dimensional hypothesis cannot be accepted unless an explanation of why we cannot perceive the fifth dimension is given.

Expanding upon this idea, the linear one-dimensional extensions of the dimensionless mathematical points in the fifth dimension which represent real positions in the four-dimensional space-time continuum require that every occurrence, event and phenomenon in the normal four dimensions have a direct five-dimensional counterpart or component. On the other hand, any five-dimensional occurrence, event, phenomenon or (subtle) influence such as psi invokes a corresponding four-dimensional action at the mechanical/ chemical/physiological level. This requirement is a new version of the chicken-egg paradox. There is no easy way to distinguish action, interaction, or influence in the fifth dimension if a corresponding four-dimensional explanation can be derived.

The fifth dimension and the four normal dimensions of space-time are not separate things. So, demonstrating the validity of the five-dimensional hypothesis is difficult since the 'cause' of any event could just as likely be found in either the normal four-dimensions of experience or the fifth-dimension structure. Scientists and philosophers will always choose the explanation built upon that which is normally perceptible over one that which is not. However difficult falsification may be, it is not impossible since SOFT is not a true TOE, or rather it is not the ultimate TOE. It is subject to justification, verification and falsification by determining physical phenomena or events that depend directly upon the geometrical characteristics of the five-dimensional structure.

Four obvious areas, which will lead to falsification, were implied in the earlier discussion of this theory. Each of these areas or aspects of SOFT represents what would have appeared as independent theories if not for the umbrella of SOFT. They are a comprehensive theory of the atomic nucleus, a theory of galactic spiral formation, a quantitative physical theory of life based upon the chemical structure of a simple living organism, and the relationship between psi and human emotional states. These individual theories were either briefly explained within the text or directly implied in the foregoing explanation of the overall theory.

The purpose of this work was not to give a complete and comprehensive exposition of individual consequences of the five-dimensional hypothesis, but to support the hypothesis of a real fifth dimension by demonstrating its logical necessity, illustrate how the five-dimensional hypothesis can be used to unify other physical theories and solve logical paradoxes between those theories, and explore how the fifth dimension can account for a vast array of physical phenomena that are otherwise without explanation.

The key feature of SOFT is its overall 'wholeness' - how all the elements come together or fall into place within a single logical framework. Individual consequences and applications of SOFT, such as these four theories, are to be worked out later. 'Solving the universe', as Clifford would have said, takes more time and effort than can be expended within the writing of a single volume. In this respect, SOFT is a theory that is also its own 'programme' for future development as well as its own metaphysics.

No theory of the atomic nucleus has ever been completely successful. In fact, different theories explain different physical features of the nucleus, but are not compatible with each other. The five-dimensional theory of the atomic nucleus suggested by SOFT depends on the specific geometrical characteristics of the five-dimensional structure, so it should verify the existence of the fifth dimension and render SOFT falsifiable.

In this case, the explanation of the nucleus reduces to a five-dimensional packing problem with the nucleus taking on the particular and unique geometrical characteristics of the extended dimension. The fifth dimension determines the physical characteristics of the nucleus in our four-dimensional space-time continuum and will thus lead to predictions regarding nuclear decay and the stability of nuclei. As an embedding dimension, it also determines the physics of our normal four-dimensional space-time. The predominant spiral form of galaxies results from a Coriolis-like Effect related to the expansion of the universe, a known phenomenon. Without the extra dimension, there would be no such Coriolis-like Effect. So, we could offer this Effect as 'proof' of the existence of the higher-embedding dimension.

Using a mathematical model based on chaos theory, computer simulations of galactic evolution given specific initial distributions of matter and other initial conditions will explain the variety of galactic shapes observed by astronomers. Not only will the spiral shapes be explained, but globular clusters and other formations will also be accounted for. When the same techniques are applied to smaller volumes of primordial gas and dust clouds, the evolution of stellar systems will be explained. Even the Titus-Bode law will finally be explained.

On yet a smaller scale, subtle planetary precessions such as Earth's 'Chandler wobble' will become a consequence of the same Coriolis-like Effect. As the calculational techniques are perfected, measurements of 'Chandler's wobble' and other observed phenomena will be used to predict the overall size and curvature of the universe. Such predictions can be verified by astronomical observations, thus rendering this specific aspect of SOFT falsifiable and scientifically verifiable. Predictions of the shape and distribution of matter in other planetary systems will also be verified as we discover more planetary systems through new observations.

The mathematics of chaotic systems will also provide the models necessary to explain the physical basis of life. A detailed study of the chemical interactions within the simplest living organisms and the application of the five-dimensional hypothesis will confirm the new theory of life. When all chemical interactions within the simple organism are identified and catalogued, the energy patterns representing each different class of chemical interactions can be modeled.

The levels or states of order that emerge from the seemingly chaotic energy interactions will yield further complexities upon which life itself will be modeled. Testing new variables and varying existing parameters within this simple model will allow scientists to predict the different life forms that can be found in nature. Further studies upon the simplest models will yield predictions regarding the nature of life and characteristics of individual organisms as well as the emergence of mind and brain as well as other functional organs within life forms.

The verification of psi will come from experiment rather than limited theoretical research and anecdotal evidence since psi is the result of action and interaction at a much higher level of complexity than simple life. SOFT implies that psi reception occurs best when emotional factors are considered. The relationship between energy changes and alterations in the five-dimensional single potential field, which influence non-local objects, is quite straightforward.

On the other hand, the physiological relationship between emotional states and energy fluctuations within an organism or body are also well documented. Therefore, an experimentally verifiable link exists between emotional states and psi reception/transmission. Knowledge of this link should lead to new methods and designs of experiments that will verify psi and confirm the link between psi and the physics of extra dimensions. Eventually, quantifiable models of psi will evolve from SOFT, but that possibility depends upon new advances within biology, physiology and understanding of living organisms.

While these four aspects will lead to complete theories in themselves, they are interrelated through SOFT. Each of these theories confirms another portion of the overall space-time structure represented by SOFT. This situation may well be unprecedented in physics. Soft cannot be proven as a whole, but each individual triumph for the theory builds confidence that the basic hypothesis of an extra dimension is both viable and necessary for the advancement of science and the human species. Nor do these four theories exhaust the possibilities represented by SOFT. They only represent the most obvious extensions of the five-dimensional hypothesis. More applications will be found and eventually psi will be used as a tool of exploration for advancing science further.

The fact that this theory can be extended in so many different and valuable ways, even while it covers so many various and seemingly unrelated phenomena, is its most intriguing feature. SOFT is so open-ended that it will act as its own metaphysics and guide for individual applications as it develops beyond the initial stages of its application. The relationship between physics, paraphysics and metaphysics thus alters the range and scope of possibilities of science itself.

If we take the sum of that which we can know, normal physics explains only the small portion of things that are known.

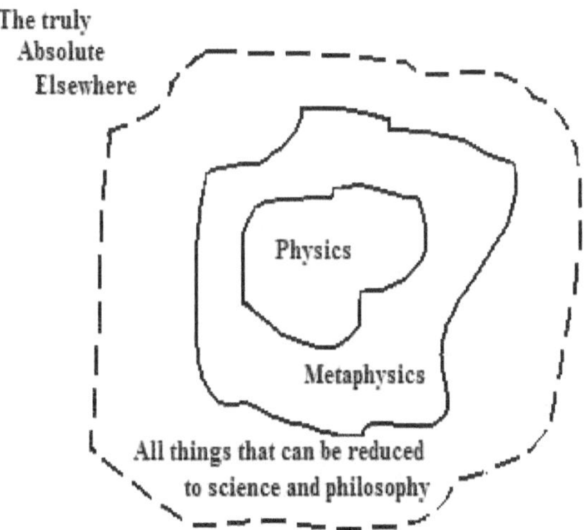

including all of those things that cannot be
reduced to science and philosophy

Metaphysics acts as a buffer between the two. Metaphysics is the academic study or the area of thought where what we know about nature and reality leads to speculation about the rest of nature and reality that are not yet known. So, metaphysics acts as a guide and a list of possibilities for future physics and science, but the establishment of paraphysics alters this basic relationship.

In one fell swoop, paraphysics intervenes between normal physics and metaphysics. It pushes the whole of physics well into the area of metaphysics and forces metaphysics to a new level of speculation and application within the whole. This new relationship results from the fact that SOFT is based upon a more fundamental concept than normal physics: Matter and 'matter in motion' which form the basis of normal physical explanation are reduced to curvature and 'variations of curvature' in paraphysics while single potential field density variations in the fifth dimension are the new physical basis of the space-time continuum in which curvature occurs.

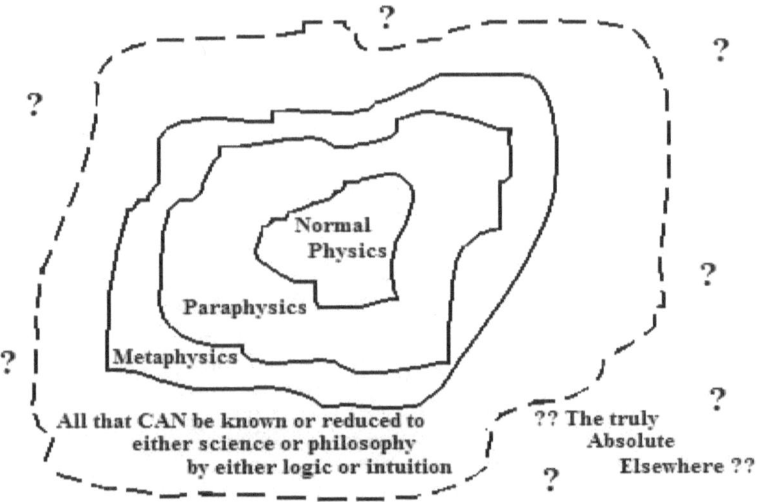

Things that exist beyond any capacity to reduce them to science or philosophy

The switch to a more fundamental concept for physics, creating the new paraphysics, renders many old problems and priorities in metaphysics into physical terms and enlarges the overall field of physics although paraphysics still remains. Physics is natural and normal, but paraphysics is natural and paranormal.

Yet not all the problems and questions in the older metaphysics are solved, a few remain for metaphysical explanation and speculation. Among those questions that remain metaphysical, free Will is probably the most tenacious and far reaching. An explanation of free Will is beyond even the concept of consciousness because free Will assumes choices based upon knowledge. The ultimate act of free Will must therefore include not only knowledge gained through psi, consciousness, mind and life, but free Will must be able to manipulate these facets of physical reality for its own benefit, to gain knowledge.

Free Will is the highest form or result of human conscious activity because true choices of free will change the whole universe at some level of reality. Free Will is the last refuge of metaphysics that cannot be reduced to paraphysics or physics. Yet free Will shall eventually be reduced to choices that are based upon the human awareness or knowledge of physical reality in our expanding concept of what is physically real.

It may be wise to differentiate between 'relative free will' (denoted by small case letters) and 'absolute free Will' (denoted by capital letters). In our everyday lives, we normally deal with free will, *i.e.*, we make choices which are not forced upon us by purely environmental (common physical) concerns: We choose how to act, react, or interact to the physical stimuli through which we sense the world around us. We assume free Will because we make choices on how to act, react and interact with some form of reality.

On the other hand, truly free Will assumes no outside influence. Yet we cannot really say that our free Will choice to eat asparagus (for example) is free of external influence. At any time, we make that, or any individual choice based upon either some event or sequence of events that occurred in our past or present environmental influences. So free Will may not always be so free. We can never isolate and identify all the influences that may affect our choices, so we only assume that some common choices are based on free will.

From the five-dimensional point of view, the situation is even more complicated. The physical influences upon which we make free Will choices are further influenced by the vast new array of physical stimuli and knowledge arriving via higher levels of conscious non-local contacts, consciousness itself and psi. Beyond even this, there is the fact that future events can influence free Will choices in the present through conscious or subconscious precognition. The past and future are equally distant and accessible from the present in the five-dimensional perspective. So free Will seems even less likely from the five-dimensional point of view. However, the fact that we can recognize the existence of free Will and contemplate the concept of free Will does imply that free Will exists. Where reason fails to confirm the existence of free will, intuition validates its existence.

We accept the concept of free Will based upon faith as well as our common experience of free will. There is no reason to doubt the existence of human free Will, which has now been raised to a much higher level of existence than ever before. In order to be a true TOE, the ultimate (proverbial big) TOE, a theory must explain or some manner account for free Will, which may be impossible. This admission does not mean that free Will is impossible; it means that an explanation of free Will may be beyond both reason and intuition. A true TOE will eventually reduce physical reality to human reason at which time 'only' intuition could move thought beyond the TOE.

If humans indeed possess free Will, then there is an extension of every human being which exists beyond anything that we can imagine at present, because true free Will assumes everything that we presently know, can know in the future, or imagine is used to make a final decision or choice. Choice presupposes a comparison of at least two possible outcomes and if one possible outcome is represented or explained via a true TOE then the other possible outcome must go beyond that TOE and the physical reality that it explains.

If that ultimate TOE includes the total of all reasonable knowledge about physical reality, then only a single final intuitive leap beyond all possible reason can give a being the second option for making a free Will choice. In other words, if everything in the universe, all physical reality, can be reduced to its most fundamental elements (represented by mathematical axioms) and described by a single concept (represented by a mathematical equation) then we would still have the choice of whether to apply this knowledge or not within the physical universe.

Reality assumes that the infinite regress of fundamental elements by which we construct our notion of the physics of reality has a logical endpoint upon which we can base that single most fundamental concept. In accordance with Gödel's theorem, confirmation of the existence of that physical reality must depend upon some still higher and more comprehensive reality (the penultimate mathematical system?). The final choice by free Will must depend on the existence of that higher reality which allows us to formulate the second option for a legitimate choice utilizing free Will. How we may wish to define that final reality would be more of a religious conviction than a scientific prediction at this moment in science and human evolution.

BIBLIOGRAPHY

Abell, George O. and Barry Singer. Eds. New York: Scribners, 1983: 56-69.

Abbott, Edwin A. [A. Square]. *Flatland: A Romance of Many Dimensions.* Introduction by William Garnett. New York: Barnes & Noble, 1963: Reprint of the 2nd edition, Oxford: Blackwell, and London: Seeley, 1884.

Abramenko, B. "On Dimensionality and Continuity of Physical Space and Time." *British Journal for the Philosophy of Science* 9 (1958): 89-109.

Abro, A d". *The Evolution of Scientific Thought from Newton to Einstein.* 2nd edition. New York: Dover, 1950; 1st edition, New York: Boni & Liveright, 1927.

Adams, George. *Physical and Etherial Spaces.* London: Rudolf Steiner Press, 1965.

Ajad. "Quantum Future Physics." Online. WWW. Accessed 4-11-1996. Available at <www.ift.uni.wroc.pl/~ajad/qf-phys.htm>

Aklom, G.M. "Some Fourth-Dimensional Curiosities." *The Fourth Dimension Simply Explained.* Manning: 134-143.

Alcock, James. "Parapsychology as a "Spiritual Science"." *Skeptic's Handbook.* Kurtz: 537-565.

Ambou, M.G. "A Unitary Theory for the Electric and Gravitational Fields." Online. WWW. Accessed 06-02-97, Available <www.ctv.es/USERS/positivo/ fields/ home.html >

Armstrong, Karen. *A History of God: The 4,000-Year Quest of Judaism, Christianity and Islam.* New York: Ballantine Books, 1993.

Archibald, Richard Claire. "Non-Euclidean Geometry." *Bulletin of the American Mathematical Society* 18 (1912): 254-258.

Archibald, Richard Claire. "Unpublished Letters of J.J. Sylvester and other new information concerning his Life and Work." *Osiris* 1 (1936): 85-154.

Ashby, Robert H. *Guidebook for the Study of Parapsychology.* New York: Samuel Weiser, 1973.

Austern, Matt. "What is the Mass of the Photon?" *psi.physics FAQ*. Online. WWW. Accessed 7-10-96. Available at <sunsite.unc.edu/lunar/mass phot.html>

Baars, Bernard J. "Can Physics Provide a Theory of Consciousness." A review of *Shadows of the Mind* by Roger Penrose. Psyche 2 (May 1995). Online. WWW. Accessed 6-28-96. Available at <psyche.cs.monash. edu.au/volume2-1/psyche95-2-08-shadows-6-baars.html>

Baez, John. "General Relativity Tutorial - Parallel Transport." Online. WWW. Accessed 6-13-96. Available at <www.math.ucr.edu/home/baez/gr /parallel. transport.html>

Baez, John. "General Relativity Tutorial - Torsion." Online. WWW. Accessed 87-96. Available at <www.math.ucr.edu/../home/baez/gr/torsion. html>

Baez, John. "A chat with Penrose." Online. WWW. Accessed 06-17-97. Available at <math.ucr.edu/home/baez/penrose.html>

Baird, Eric. "Erik's Relativity Pages." Online. WWW. Accessed 613-96. Available at <ourworld.compuserve.com/homepages/ericbaird/homepage .htm>

Baldwin, James Mark. *Dictionary of Philosophy and Psychology*. Four volumes. Gloucester, Massachusetts: Peter Smith, 1960: Reproduction of 1901 original.

Ball, Robert S. "A Sketch in the Theory of Screws," *Hermathena*, 1 (1874): 506519.

Ball, Robert S. *The Theory of Screws: A Study in the Dynamics of a Rigid Body*. Dublin: Hodges, Foster, and Co., 1876.

Ball, Robert S. "The Non-Euclidean Geometry." *Hermathena* 3 (1879): 500-541.

Ball, Robert S. "Notes on Non-Euclidean Geometry." *Reports of the British Association* 50 (1880): 476-477.

Ball, Robert S. "The Distance of Stars." Read 11 February 1881. *Proceedings of the Royal Institution of Great Britain* 9 (1879-1882): 514-519.

Ball, Robert S. "On the Elucidation of a Question in Kinematics by the Aid of Non-Euclidean Space." *Reports of the British Association, (York)* 51 (1881): 535-536.

Ball, Robert S. "Certain Problems in the Dynamics of a Rigid System Moving in Elliptic Space." Read 14 November 1881. *Transactions of the Royal Irish Academy* 28 (1880-1886): 159-184.

Ball, Robert S. "Measurement." 9th edition. *Encyclopaedia Britannica* 15 (1885). London: Adam & Charles Black, 1885: 659-668.

Ball, Robert S. "Notes on the Kinematics and Dynamics of a Rigid System in Elliptic Space." Read 9 June 1884. *Proceedings of the Royal Irish Academy* 4 (1884-1888): 252-258.

Ball, Robert S. "Note on the Character of the Linear Transformation which Corresponds to the Displacement of a Rigid System in Elliptical Space." Read 9 November 1885. *Proceedings of the Royal Irish Academy* 4 (1884-1888): 532-537.

Ball, Robert S. "Elliptic Space." *Nature* 33 (26 November 1885): 86-87.

Ball, Robert S. "A Dynamical Parable." *Nature* 36 (1 September 1887): 424429; originally published as the Presidential Address to the Physical Section of the British Association. *Reports of the British Association*, (Manchester) 57 (1887): 568-579; Reprinted in Ball. *A Treatise on the Theory of Screws*: 496509.

Ball, Robert S. "On the Theory of the Content." Read 12 December 1887. *Transactions of the Royal Irish Academy* 29 (1889): 123-182.

Ball, Robert S. "The Discussion of Quaternions." *Nature* 48 (24 August 1893): 391.

Ball, Robert S. *A Treatise on the Theory of Screws*. Cambridge: At the University Press, 1900.

Ball, Robert S. "The Geometry of Forces." A review of *Geometrie der Kräfte*. H.E. Timerding. (Leipzig: B.G. Teubner, 1908). *Nature* 81 (8 July 1909): 34.

Ball, Robert S. *Reminiscences and Letters of Sir Robert Ball*. Ed. W. Valentine Ball. London: Cassell, 1915.

Ball, William Walter Rouse. "A Hypothesis Relating to the Nature of the Ether and Gravitation." *Messenger of Mathematics* 21 (1891): 20-24.

Ball, William Walter Rouse. *Mathematical Recreations and Problems of the Past and Present Times*. London: Macmillan, 1892; 3rd edition, 1896; 6th edition, 1914.

Ball, William Walter Rouse. *A Short Account of the History of Mathematic*. New York: Dover, 1960; Unaltered reproduction of 4th edition of 1908; 1st edition published in London: Macmillan, 1888.

Band, W. "Flint's Five-Dimensional Theory of the Electron." *Philosophical Magazine* 29 (June 1940): 548-552.

Barbour, Ian G. *Issues in Science and Religion*. New York: Harper Torchbooks, 1966.

Barbour, Ian. *Religion in an Age of Science*. San Francisco: Harper, 1993.

Bargmann, Valentine. "Relativity." *Review of Modern Physics* 29 (April 1957): 169-174.

Barnard, G.C. *Psychic Phenomena*. London: Random House, 1997; Reprint of *The Supernormal*. London: Rider, 1933.

Barnett, Lincoln. *The Universe and Dr. Einstein*. New York: Bantam Books, 1968.

Barr, Frank. "The Theory of Evolutionary Process as a Unifying Paradigm." Online. WWW. Accessed 7-3-96. Available at <users.lanminds.com/~ab/ay /barr.html>

Barrow, John D. *Theories of Everything. The Quest for Ultimate Explanation*. New York: Fawcett Columbine, 1988.

Barrow, Logie. "Socialism in Eternity; The Ideology of Plebeian Spiritualists, 1853-1913." *History Workshop Journal*: 37-69.

Bateman, Harry. "The Conformal Transformations of a Space of Four Dimensions and their Applications to Geometrical Optics." *Proceedings of the London Mathematical Society* 7 (1909): 70-89.

Bateman, Harry. "The Transformation of the Electrodynamical Equations." *Proceedings of the London Mathematical Society* 8 (1910): 223-264.

Bateman, Harry. "The Physical Aspects of Time." *Memoirs and Proceedings of the Manchester Literary and Philosophical Society* 54 (1910): 1-13.

Bateman, Harry. "On General Relativity." Letter to the editor dated 10 August 1918. *Philosophical Magazine* 37 (1919): 219-223.

Beal, James B. "The Emergence of Paraphysics: Research and Applications." *Psychic Explorations*. Mitchell: 426-447.

Becker, Carl B. *Paranormal Experience and the Survival of Death*. Albany: SUNY Press, 1993.

Beichler, James E. *A Five-Dimensional Continuum Approach to a Unified Field Theory*. Unpublished master's Thesis. San Francisco State University, December 1980.

Beichler, James E. *The Riemann Program: Changes in the Concept of Absolute Space, 1854-1900*. Unpublished master's Thesis. University of Maryland, College Park, 1985.

Beichler, James E. "Ether/Or: Hyperspace Models of the Ether in America." *The Michelson Era in American Science*, 1870-1930. *American Institute of Physics Conference Proceedings*, #179. Eds. Stanley Goldberg and Roger H. Stuewer. New York: American Institute of Physics, 1988: 206-223.

Beichler, James E. "Twist 'til we tear the house down." Yggdrasil: *The Journal of Paraphysics*, 1 (Winter Solstice 1996). Online. WWW. Available at <ourworld.compuserve.com/homepages /Paraphys>

Bell, E.T. *The Development of Mathematics*. 2nd edition. New York: McGraw Hill, 1945; Reprint of 1940.

Bell, E.T. *Men of Mathematics*. New York: Simon & Schuster, 1962; Reprint of 1937 edition.

Bell, J.S. "On the Einstein Podolsky Rosen paradox." *Physics* 1 (1964): 195-200.

Bell, J.S. "On the problem of hidden variables in quantum mechanics." *Reviews in Modern Physics* 38 (July 1966): 447.

Beloff, John. "Could There Be a Physical Explanation of Psi?" Online. WWW. Accessed 09-17-97. Available at <moebius.psy.ed.ac.uk/~dualism/ papers/physical.html>; Originally published in *The Relentless Question*. McFarland, 1990.

Beloff, John. "The Geller Controversy: The Current State Play." *Research in Parapsychology* 1976. Morris, Roll, Morris: 199-209.

Beloff, John. "Mind-body interactionism in the light of the parapsychological evidence." Online. WWW. Accessed 09-17-97. Available at <moebius.psy.ed.ac.uk/~dualism/papers/index.html>

Beloff, John. *New Directions in Parapsychology*. Metuchen, New Jersey: Scarecrow Press. 1974.

Bender, Hans. Ed. *Parapsychologie Entwicklung, Ergebnisse, Probleme*. Darmstadt: Wissenschaftliche Buchgesellschaft, 1966.

Bennett, J.G., R.L. Brown and M.W. Thiring, "The Unified Field Theory in a Curvature Free Five-Dimensional Manifold." *Proceedings of the Royal Society* 198 (22 July 1949): 39-61.

Berger, Arthur and Joyce Berger. Eds. *The Encyclopedia of Parapsychology and Psychical Research*. New York: Paragon House, 1991.

Bergmann, Peter G. *An Introduction to the Theory of Relativity*. New York: Dover Press, 1979.

Bergmann, Peter G. "Unified Field Theory with Fifteen Variables." *Annals of Mathematics* 49 (January 1948): 255-264.

Bergmann, Peter G. "Topics in the General Theory of Relativity." *Summer Institute of Theoretical Physics*. Notes taken by Nicholas J. Wheeler. Waltham, Massachusetts: Brandeis University, 195: 41-42.

Bergmann, Peter G. "Physics and Geometry." *Logic, Methodology and the Philosophy of Science*. Proceedings of the 1964 International Congress. Ed. Yehoushua Bar-Hillel. Amsterdam: North Holland Publishing, 1965: 343-346.

Bergmann, Peter G. "Comments on the Scalar-Tensor Theory." *International Journal of Theoretical Physics* 1 (1968): 25-36.

Bergmann, Peter G. "Summary of the Colloque International De Royamont." *Les Théories Relativistes de La Gravitation*. Paris: Centre National de La Recherche Scientifique, 1962: 463-467.

Berkeley, Hastings. *Mysticism in Modern Mathematics*. London: Henry Frowde, Oxford University Press, 1910.

Bernstein, Herbert J. and Anthony V. Phillips. "Fiber Bundles and Quantum Theory." *Scientific American* (July 1981): 123-137.

Beveridge, W.I.B. *The Art of Scientific Investigation*. New York: Vintage Books, 1950, 1953, 1957.

Bisaha, J.P. and B.J. Dunne. "Multiple Subject and Long Distance Precognitive Remote Viewing of Geographical Locations." *Mind at Large*. Tart, Puthoff, Targ: 107-124.

Blanton, John. "The EPR Paradox and Bell's Inequality Principle." Online. WWW. 8-31-93. Accessed 5-5-96. Available at <www.desy.de/user/ projects/Physics/bells-inequality.html>

Blanton, John. "The EPR Paradox and Bell's Inequality Principle." Commentaries by Jack Sarfatti. Version 0.2. Online. WWW. 8-31-93. Accessed 6-13-96. Available at <www.hia.com/hia/pcr/ epr.html>

Bochner, Salomon. *The Role of Mathematics in the Rise of Physics*. Princeton: Princeton University Press.

Bohm, David. *Quantum Theory*. New York: Prentice-Hall, 1951.

Bohm, David. *Causality and Chance in Modern Physics*. Princeton: Van Nostrand, 1957.

Bohm, David. *Wholeness and the Implicate Order*. London: Routledge & Kegan Paul, 1980.

Bohm, David and B.J. Hiley. *The Undivided Universe*. London: Routledge, 1993.

Bolyai, Johann. "The Science Absolute of Space." *Supplement to Non-Euclidean Geometry*. Roberto Bonola. Trans. George Bruce Halsted. Intr. George Bruce Halsted.

Bonola, Roberto. *Non-Euclidean Geometry: A Critical and Historical Study of its Development*. Trans. H.S. Carslaw. Intr. Federigo Enriques. New York: Dover, 1955; Reprint of LaSalle, Illinois: Open Court, 1912; From the Italian edition of 1906; originally published in 1892.

Bork, Alfred M. "The Fourth Dimension in Nineteenth-Century Physics." *Isis* 55 (1964): 326-338.

Born, Max. *Theory of Relativity. With the collaboration of Gunther Leibfried and Walter Biem*. New York: Dover, 1965: Revised edition of Methuen, 1924.

Born, Max. "Einstein and Relativity." *Nature* 166 (4 November 1950): 751. Bostwick, Arthur E. "Four-Dimensional Space." *New Science Review* 2 (October 1895): 146-152.

Bostwick, Arthur E. "On Hyperspace." *The Monist* 18 (1908): 629-631.

Boyer, Carl B. *A History of Mathematics*. New York: Wiley and Sons, 1968.

Boyer, Carl B. "William Kingdon Clifford." *The Encyclopedia Americana. International Edition*. Danbury, Connecticut: Grolier, 1988, 7: 70.

Bragdon, Claude. "Space and Hyperspace." *The Fourth Dimension Simply Explained*. Manning: 91-99.

Bragdon, Claude. *A Primer of Higher Space: The Fourth Dimension to which is added Man the Square, a Higher Space Parable*. New York: Alfred A. Knopf, 1929; originally published in 1913 and republished in 1923.

The Brain Project. Author not listed. "On Quantum Physics and Ordinary Consciousness." Online. WWW. Accessed 06-17-97.Available at <www.merlin .com.au/brain-proj/quantum.htm>

Bridgeford, Barry. "Unified Field Theory." Online. WWW. Accessed 04-28-97. Available at <www3.sympatico.ca/barrybri/SCIENCE .htm>

Brier, Bob and Maithili Schmidt-Raghavan. "Precognition and the Paradoxes of Causality." *Philosophy*: 243-252.

Briggs, John F. and F. David Peat. *Looking Glass Universe: The Emerging Science of Wholeness*. New York: Touchstone, 1984.

Brill, B.R. "Review of Jordan's Extended Theory of Gravitation." *Evidence for Gravitational Theories*. Course XX, *Proceedings of the International School of Physics <<Enrico Fermi>>*. New York: Academic Press, 1962: 44-68.

Britannica, The Encyclopaedia. 9th edition. Edinburgh and London: Adam and Charles Black, 1875-1889.

Britannica, New Volumes of the Encyclopaedia. 10th edition. London and Edinburgh: Adam and Charles Black, 1902-1903.

Britannica, The Encyclopaedia. 11th edition. London: Adam and Charles Black, 1910-1911.

Britannica, The New Encyclopaedia. "William Kingdon Clifford." 15th edition. Chicago: University of Chicago Press, 1988: *Micropaedia*, 3: 386.

Broad, C.D. "The Notion of 'Precognition'." *Science and ESP*. Smythies: 165-196.

Brooks, Robert. "The physics inside of topological quantum field theories." Online. WWW. Accessed 09-17-97. Available at <www.iop.org /PEL/ article/cq1400513/full/>

Broughton, Richard S. *Parapsychology: The Controversial Science*. New York: Ballantine, 1991.

Brown, Julian. "Martial arts students influence the past." *New Scientist and Science Journal* (27 August 1994). Online. WWW. Accessed 7-3-96. Available at <alethea. ukc.ac.uk/CPRS/RPKP /martial>

Brown, P.R.F. "Theories of the Aether." Online. WWW. Accessed 7-22-96. Available at <magna.com.au/~prfbrown/aether.html>

Browne, Robert T. *The Mystery of Space: A Study of the Hyperspace Movement in the Light of the Evolution of New Psychic Faculties and An Inquiry into the Genesis and Essential Nature of Space*. New York: Dutton, 1919.

Brunstein, Karl A. *Beyond the Four Dimensions: Reconciling Physics, Parapsychology and UFOs*. New York: Walker, 1979.

Buchheim, Arthur. "A Memoir on Biquaternions." American *Journal of Mathematics* 7 (1884): 293-326.

Buchheim, Arthur. "On Clifford's Theory of Graphs." *Proceedings of the London Mathematical Society* 17 (12 November 1885): 80-106.

Buchheim, Arthur. "On the Theory of Screws in Elliptic Space," *Proceedings of the London Mathematical Society* 15 (10 January 1884): 83-98; 16 (13 November 1884): 15-27; 17 (10 June 1886): 240-254; 18 (11 November 1886): 88-96.

Budnik, Paul. "Measurement in quantum mechanics FAQ." Online. WWW. Accessed 7-9-96. Available at <www.mtnmath.com/faq /meas-qm.html>; The following sections are available: 1. "About this FAQ." <.../meas-qm1.html>; 2. "The measurement problem." <.../meas-qm-2.html>; 3. "Schrödinger's cat." <.../meas-qm-3.html>; 4. "The Copenhagen Interpretation." <.../meas-qm- 4.html>; 5. "Is QM a complete theory?" <.../meas-qm-5.html>; 6. "The shut up and calculate interpretation." <.../meas-qm-6.html>; 7. "Bohm's theory." <(.../meas-qm-7.html>; 8. "The Transactional Interpretation of Quantum Mechanics." <.../meas-qm-8.html>;9. "Complex probabilities." <.../meas-qm9.html>; 10. "Quantum logic." <.../meas-qm-10.html>; 11. "Consistent theories." <.../meas-qm-11.html>; 12. "Spontaneous reduction models." <.../meas-qm-12.html>; 13. "What is needed?" <.../meas-qm13.html>; 14. "Is this a real FAQ?" <.../meas-qm-14.html>.

Budnik, Paul. *Intellect and Intuition: Einstein's Revenge*. Online. WWW. 1995. Accessed 7-9-96. Available at <www.mtnmath.com /book.html, to .../ node??.html>

Burnside, William. "On the Kinematics in Non-Euclidean Space." Read 8 November 1894. *Proceedings of the London Mathematical Society* 26 (1894/1895): 33-56.

Burt, Cyril. "Psychology and Parapsychology." *Science and ESP*. Smythies: 61142.

Burtt, E.A. *The Metaphysical Foundations of Modern Science*. Garden City: Doubleday-Anchor, 1954.

Bushman, Robert. "Scientific Inquiry into OBE." Extract from the website *A Comprehensive Bibliography of the Out-of-Body Experience*. Online. 05-05-97. Accessed 06-26-97. Available at <www1.huskynet.com/ intuitive/obe/SCIENCE.HTM>

Bylok, Zdzislaw Ted. "The Field Energy Theory of Spacetime." Online. WWW. Accessed 6-26-96. Available at <www.eskimo.com/ ~mpb/spacetime .html>

Cajori, Florian. *A History of Mathematics*. New York: Macmillan, 1894.

Cajori, Florian. "Early <<Proofs>> of the impossibility of a Fourth Dimension of Space." *Archivio di Storia della Scienza* 7 (1926): 25-28.

Cajori, Florian. "Origins of Fourth-Dimensional Concepts." *American Mathematical Monthly* 33 (1926): 397-406.

Callahan, Linda. "Press Release: Tying it all together with superstrings." Online. WWW. 11-07-95. Accessed 04-07-97. Available at <www.tc.cornell. edu/er/media/1995/superstring. html>

Callaway, Joseph. "The Equations of Motion in Einstein's New Unified Field Theory." *Physical Review* 92 (15 December 1953): 1567-1570.

Campbell, Percy A. *A Non-Euclidean Theory of Matter and Electricity*. Cambridge, Massachusetts: George Henry, 1907.

Cantor, G.N. and M.J.S. Hodges. *Conceptions of the Ether*. Cambridge: Cambridge University Press, 1983.

Čapek, Milič. Ed. *The Concepts of Space and Time: Their Structure and Their Development. Boston Studies in the Philosophy of Science*, Volume 22. Dordrecht, Holland: D. Reidel, 1976: W.K. Clifford, "On the Bending of

Space:" 291-294; "On the Space-Theory of Matter:" 295-296; A. Calinon, "Geometrical Spaces:" 297-303.

Čapek, Milič. *The Philosophical Impact of Contemporary Physics.* Princeton: Von Nostrand, 1961.

Capra, Fritjof. *The Tao of Physics: An Exploration of the Parallels Between Modern Physics and Eastern Mysticism.* Berkeley: Shambhala, 1975.

Carrelli, A. "Relativity in Five-Dimensions." *Accad. Lincii. Atti.* 7 (1 April 1928): 566-567.

Carslaw, H.S. "Gauss and Non-Euclidean Geometry." *Nature* 82 (September 1910): 362.

Cartan, Élie-Joseph. *Oeuvres Complétes.* Paris: Gauthier-Villars, 1955. Cartan, Élie-Joseph. "Notice Historique sur la notion de parallelisme absolu." *Mathematische Annalen* 102 (1930): 698-706.

Cartan, Élie-Joseph and Albert Einstein. *Letters on Absolute Parallelism.* Ed. Robert Debever. Princeton: Princeton University Press, 1979.

Cartan, Élie-Joseph. "Sur Les Variétés à Connexion Affine et La Théorie de La Relativité Générale." *Ann. Ec. Norm.* Sup. 40 (1923).

Cartan, Élie-Joseph. *Leçons sur les invariants intégraux.* Paris: 1922. Cartan, Élie-Joseph. *La géométrie des espaces de Riemanns.* Fasc. 9 of Mémorial des Sciences Mathématiques. Paris: 1925.

Cartan, Élie-Joseph. *Leçons sur les géométrie des espaces Riemann.* Paris: 1928.

Carus, Paul. "Editorial Comment." *The Monist* 18 (1908): 467-471.

Carus, Paul. "Space of Four Dimensions." *The Monist* 18 (1908): 471-475.

Carus, Paul. "Editorial Comment." *The Monist* 21 (1911): 131-137.

Cassirer, Ernst. *The Philosophy of the Enlightenment.* Boston: Beacon Press, 1955.

Cassani, Walter E.R. Index to "The Wave Theory of the Field." Online. WWW. Accessed 7-8-96. Available at <www.lnet.it/ospiti/cassani/index .html>

Cayley, Arthur. "On the Non-Euclidean Geometry." *Clebsch Mathematical Annalen* 5 (1872): 630-634.

Cayley, Arthur. "Presidential Address to the British Association." *Reports of the British Association, (Southport)* 53 (1883): 3-37; Reprinted in *Collected Mathematical Papers* 11: 429-459.

Cayley, Arthur. "On the Non-Euclidean Plane Geometry." *Proceedings of the Royal Society of London* 37 (1884): 82-102: Reprinted in *Collected Mathematical Papers* 12: 220-238.

Cayley, Arthur. "Non-Euclidean Geometry." Read 27 January 1890. *Transactions of the Cambridge Philosophical Society* 15 (1894): 37-61; Reprinted in *Collected Mathematical Papers* 13: 480-504.

Cayley, Arthur. Letter to Lord Kelvin. 25 March 1889, #C63, Add. Mss. 7342. Kelvin Archives, University Library, Cambridge.

Cayley, Arthur. *The Collected Mathematical Papers*. Volumes I-VII. Arthur Cayley. Ed. Volumes VIII-XIII. A.R. Forsyth. Ed. Cambridge: At the University Press, 1889.

Cayley, Arthur. Letter to Lord Kelvin. 25 March 1889, Add.7342/C63. Kelvin Papers, University Library, Cambridge.

Chalmers, David L. "Minds, Machines, and Mathematics." Review of *Shadows of the Mind*. Roger Penrose. *Psyche* 2 (June 1995). Online. WWW. Accessed 78-96. Available at <psyche.cs.monash. edu.au/volume2-1/ psyche95-2-09shadows7-chalmers.html >

Chalmers, David J. "Facing Up to the Problem of Consciousness." Online. WWW. Accessed 06-17-97. Available at <www.artsci.wustl.edu/%Ephilos /papers /chalmers.consciousness.html>

Chalmers, David. "Online papers on consciousness." Online. WWW. Accessed 05-07-98. Available at <ling.ucsc.edu/~chalmers/mind.html>

Chambers, George Gailey. "The Meaning of the Term Fourth Dimension." *The Fourth Dimension Simply Explained*. Manning: 201-210.

Chandler, Charles H. "Transcendental Space." *Transactions of the Wisconsin Academy of Sciences, Arts and Letters*. 11 (18961897): 237-248.

Chari, C.T.K. "The Challenge of Psi: New Horizons of Scientific Research." *Journal of Parapsychology*. 38 (1974): 1-5.

Chari, C.T.K. Letter. *Journal of Parapsychology* 38 (1974): 418-420.

Chari, C.T.K. "Precognition, Probability and Quantum Mechanics." *Journal of the ASPR* 66 (1972): 193-207.

Chari, C.T.K. "Some Generalized Theories and Models of Psi: A Critical Evaluation." *Handbook of Parapsychology*. Ed. Benjamin B. Wolman. New York: Reinhold, 1977: 803-822.

Chari, C.T.K. "Comments." *Quantum and Parapsychology*. Oteri: 252-265.

Chase, Charles H. "Pseudo-Geometry." *The Monist* 18 (1908): 465-467.

Childress, D. Hatcher. *The Anti-Gravity Handbook*. Revised edition. Kempton, Illinois: Adventures Unlimited, 1993.

Chisholm, J.S.R. and A.K. Common. Eds. *Clifford Algebras and their Applications in Mathematical Physics. Proceedings of the NATO and SERC Workshop on Clifford Algebras*, Canterbury, UK, September 1985, NATO ASI Series, Series C, Mathematics and Physical Sciences, 183. Dordrecht: Reidel, 1986.

Chrystal, George. "Non-Euclidean Geometry." *Proceedings of the Royal Society of Edinburgh* 10 (1878-1879): 638-664.

Chrystal, George. "Clifford's Mathematical Papers." *Nature* 26 (6 July 1882): 217-219.

Chrystal, George. "Theory of Parallels." *Encyclopaedia Britannica*. 9th edition. 18 (1885): 254-255.

Clark, Ronald W. *Einstein: The Life and Times*. London: Hodder and Stoughton, 1973.

Clifford, William Kingdon. "On the Space-Theory of Matter." Read 21 February 1870. *Transactions of the Cambridge Philosophical Society* 2 (1866/1876): 157-158; Reprinted in *Mathematical Papers*: 21-22.

Clifford, William Kingdon. "The Existence of a Root." *The Athenaeum* 2209 (26 February 1870): 295.

Clifford, William Kingdon. "On the Aims and Instruments of Scientific Thought." Lecture delivered before members of the British Association at Brighton on 19 August 1872. *Macmillan's Magazine* 26 (October 1872): 499-512; Reprinted in *Lectures and Essays* 1: 124-157.

Clifford, William Kingdon. "Preliminary Sketch of Biquaternions." *Proceedings of the London Mathematical Society* (12 June 1873): 381-395; Reprinted in *Mathematical Papers*: 181-200.

Clifford, William Kingdon. "On Some Curves of the Fifth Class." Title only. *Reports of the British Association, (Bradford)* 43 (1873).

Clifford, William Kingdon. "On a Surface of Zero Curvature and Finite Extent." Title only. *Reports of the British Association*, (Bradford) 43 (1873).

Clifford, William Kingdon. "The Unreasonable." *Nature* 7 (13 February 1873): 282.

Clifford, William Kingdon. "The Philosophy of the Pure Sciences." Delivered as part of the afternoon lecture series at the Royal Institution on 1, 8, and 15 March in 1873. Subsequently printed in *Contemporary Review* 24 (October 1874): 712-727; Reprinted in *Lectures and Essays* 1: 254-294; "Part II. The Postulates of the Science of Space." *Contemporary Review* 25 (1875): 360-376; Reprinted in *Lectures and Essays* 1: 295-323; "Part III," *Nineteenth Century* 5 (March 1879): 513-522; Reprinted in *Lectures and Essays* 1: 324-340.

Clifford, William Kingdon. "Body and Mind." *The Fortnightly Review* 16 (July-December 1874): 715-736; Reprinted in *Lectures and Essays* 2: 31-70.

Clifford, William Kingdon. "The First and Last Catastrophe." *The Fortnightly Review* 17 (January-June 1875): 465-484; Reprinted in *Lectures and Essays* 1: 191-227.

Clifford, William Kingdon. "The Unseen Universe." *The Fortnightly Review* 17 (January-June 1875): 776-793; Reprinted in *Lectures and Essays* 1: 228-253.

Clifford, William Kingdon. "Instruments used in Measurements." Handbook to the Special Loan Collection of Scientific Instruments, 1876: 55-59; Reprinted in *Mathematical Papers*: 419-423.

Clifford, William Kingdon. "On the Classification of Geometric Algebras." *Mathematical Papers*: 397-401; an abstract of the paper was communicated to the London Mathematical Society on 10 March 1876. *Proceedings of the London Mathematical Society* 7 (1876): 135. The unfinished manuscript which was published in *Mathematical Papers* was found in Clifford's private papers.

Clifford, William Kingdon. "Instruments Illustrating Kinematics, Statics, and Dynamics." *South Kensington Handbook to the Special Loan Collection of Scientific Apparatus*. (1876); Reprinted in *Lectures and Essays* 2: 9-30; Reprinted in *Mathematical Papers*: 424-440.

Clifford, William Kingdon. "On the Theory of Screws in a Space of Constant Positive Curvature." *Mathematical Papers*: 402-405.

Clifford, William Kingdon. "On the Free Motion under no force of a Rigid System in an n-fold homaloid." *Proceedings of the London Mathematical Society* 8 (1876): 67-70; Reprinted in *Mathematical Papers*: 236-240.

Clifford, William Kingdon. "A Modern Symposium: The Influence upon Morality of a Decline in Religious Belief." *The Nineteenth Century* 1 (March-July 1877): 331-358.

Clifford, William Kingdon. "On the Canonical Form and Dissection of a Riemann's Surface." Read 14 June 1877. *Proceedings of the London Mathematical Society* 8 (1877): 292304; Reprinted in *Mathematical Papers*: 241-254.

Clifford, William Kingdon. *Elements of Dynamic: An Introduction to the Study of Motion and Rest in Solid and Fluid Bodies, Part I: Kinematics.* London: Macmillan, 1878.

Clifford, William Kingdon. "On the Nature of Things-in-Themselves." *Mind* 3 (1878): 57-67; Reprinted in *Lectures and Essays* 2: 71-88.

Clifford, William Kingdon. "Applications of Grassmann's Extensive Algebra." *American Journal of Mathematics* 1 (1878): 350-358; Reprinted in *Mathematical Papers*: 266-276.

Clifford, William Kingdon. "On the Classification of Loci." Read 8 April 1878. *Philosophical Transactions of the Royal Society* 169 (1878): 663-681; Reprinted in *Mathematical Papers*: 305-331.

Clifford, William Kingdon. *Lectures and Essays*. Eds. Leslie Stephen and Frederick Pollock. London: Macmillan, 1879.

Clifford, William Kingdon. "On Boundaries in General." *Macmillan's Magazine* (1879): 359-368; Reprinted in *Seeing and Thinking*: 127-156.

Clifford, William Kingdon. *Seeing and Thinking*. London: Macmillan Nature Series, 1879.

Clifford, William Kingdon. "Energy and Force." Eds. Frederick Pollock with J.F. Moulton. *Nature* 22 (10 June 1880): 122-124.

Clifford, William Kingdon. "Lecture Notes." From University College Library Archives, London.

Clifford, William Kingdon. *Mathematical Papers*. Ed. Robert Tucker. Pref. H.J. Stephen Smith. New York: Chelsea Publishing, 1968; Reproduction of 1882 original.

Clifford, William Kingdon. *The Common Sense of the Exact Sciences*. Ed. Karl Pearson. Newly ed. James R. Newman. Pref. Bertrand Russell. New York: Dover, 1955; A reproduction of the 1946 Knopf edition which was an unaltered reprint of the third edition of 1899; Originally published London: Macmillan, 1885; First American edition, New York, D. Appleton, 1885; Reprinted in German as *Der Sinn der Exakten Wissenschaft*. Trans. Hans Kleinpeter. Leipzig: Verlag von Johann Ambrosius Barth, 1913.

Clifford, William Kingdon. *Elements of Dynamic, Book IV and Appendix*. Ed. Robert Tucker. London: Macmillan, 1887.

Cole, F.N. "On Rotations in Space of Four-Dimensions." *American Journal of Mathematics* 12 (1889): 191-210.

Cole, F.N. "Fourth Summer Meeting of the American Mathematical Society." *Bulletin of the New York Mathematical Society* 4 (October 1897): 1-11.

Cole, K.C. "A Theory of Everything." *The New York Times Magazine* (18 October 1987): 23-26.

Collins, Graham P. "Quantum Black Holes Are Tied to D-Branes and Strings." *Physics Today* (March 1997): 19-22.

Cook, Roland. "Do We Need a New Physics?" Online. WWW. Accessed 04-01-98. Available at <www.zynet.co.uk/imprint/ online/Time-cook1.html>

Coolidge, Julian Lowell. *A History of Geometrical Methods*. New York: Dover, 1963; Reprint of Oxford: at the Clarendon Press, 1940.

Corben, H.C. "A Generalization of the Special Relativity Theory." *Nature* 157 (20 April 1946): 157.

Corben, H.C. "A Generalization of Special Relativity Theory." *Nature* 158 (April 1946): 516.

Corben, H.C. "A Classical Theory of Electromagnetism and Gravitation." *Nature* 156 (29 September 1945): 388-389.

Corben, H.C. "A Classical Theory of Electromagnetism and Gravitation. I. Special Relativity." Physical Review 79 (1 March and 16 March 1946): 225-234.

Corben, H.C. "Special Relativistic Field Theories in Five Dimensions." *Physical Review* 120 (1 December and 15 December 1946): 947-953.

Corben, H.C. "A Unified Field Theory with Varying Charge and Rest Mass." *Nuovo Cimento* 9 (March 1952): 235-252.

Costa de Beauregard, Olivier. "Quantum Paradoxes and Aristotle's Twofold Information Concept." *Mind at Large*. Tart, Puthoff and Targ: 177-187; Also, *Quantum Physics and Parapsychology*. Oteri: 91-108.

Costa de Beauregard, Olivier. "The Expanding Paradigm of the Einstein Theory." *The Iceland Papers*. Puharich: 161-191.

Cox, Homersham. "Homogeneous Coordinates in Imaginary Geometry and their Applications to Systems of Forces." *Quarterly Journal of Mathematics* 18 (1881): 178-215.

Cox, Homersham. "On the Application of Quaternions and Grassmann's Ausdehnungslehre to different kinds of Uniform Space." Read 20 February 1882. *Transactions of the Cambridge Philosophical Society* 13 (1882): 69-143.

Coxeter, H.S.M. *Twelve Geometric Essays*. London: Feffer and Simons, 1966.

Coxeter, H.S.M. *Non-Euclidean Geometry*. 5th edition. Toronto: University of Toronto Press, 1968.

Coxeter, H.S.M. "Cases of Hyperdimensional Awareness." *Consciousness and Reality*. Musès: 89-94.

Craig, Thomas. "Note on the Projection of the General Locus of Space of Four Dimensions into Space of Three Dimension." *American Journal of Mathematics* 2 (1879): 252-259.

Craig, Thomas. "Displacements depending on One, Two and Three Parameters in a Space of Four Dimensions." *American Journal of Mathematics* 20 (1898): 135-156.

Cramer, John G. "The Transactional Interpretation of Quantum Mechanics." *Reviews of Modern Physics* 58 (July 1986): 647-688. Online. WWW. Accessed 7-3-96. Available at <mist.npl.washington.edu/npl/int-rep/tiqm/Tl-toc.html, to .../Tlref.html>

Crathorne, Arthur R. "The Fourth Dimension, the Playground of Mathematics." *The Fourth Dimension Simply Explained*. Manning: 154-162.

Crawley, Edwin S. "Geometry: Ancient and Modern." *Popular Science Monthly* 53 (January 1901): 257-266.

Crew, Henry. *The Rise of Modern Physics*. Baltimore: Williams and Wilkins, 1935.

Crowe, Michael J. *A History of Vector Analysis: The Evolution of the Idea of a Vectorial System*. New York: Dover, 1967; Reprint of University of Notre Dame Press, 1967.

Cuny, Hilaire. *Albert Einstein: The Man and His Theories*. New York: Fawcett World Library, 1965.

Cutter, Edward H. "Fourth Dimension Absurdities." *The Fourth Dimension Simply Explained*. Manning: 60-69.

Dahlberg, Andrew. "A Topological Analysis of the Geometry Rendered by the Kaluza-Klein Theory of Hyperspace." Online. WWW. Accessed 09-17-97. Available at <www.mity.org/classes /hyperspace2/projects/klein/>

Dampier, Sir William Cecil. *A History of Science and Its Relations with Philosophy and Religion*. 4th edition. Postscript by I. Bernard Cohen. Cambridge: At the University Press, 1961.

Daniels, Michael. "Introduction to Parapsychology." Online. WWW. 1996. Accessed 06-17-97. Available at <www.mdani.demon.ac.uk/para/ paraintro.htm>

Daniels, Norman. "Lobachevsky: Some Anticipations of Later Views on the Relation between Geometry and Physics." *Isis* 66 (1975): 75-85.

Dantzig, D. Van and J.A. Schouten, "Four-Dimensional Interpretation of the Unified Field Theory." K. *Akad. Amsterdam, Proc.* 34 (1931): 1398-1407.

Dantzig, D. Van. "Theorie der projektiven Zusammenhangs n-dimensionaler Räume." *Mathematische Annalen* 106 (1932): 400.

Dantzig, D. Van. "Zur allgemein projecktiven differentialgeometrie." *K. Acad. Amsterdam, Proceedings* 35 (1932): 524, 535; 37 (1934): 150.

Darnton, Robert. *Mesmerism and the End of the Enlightenment in France*. New York: Schocken Books, 1976.

Darrieus, G. "Remarkable Form of the Maxwell-Lorentz Equations in a Five-Dimensional Universe." *Le Journal de Physique et Le Radium* 8 (November 1927): 444-446.

Davies, Paul C.W. *Superforce: The Search for a Grand Unified Theory of Nature*. New York: Simon and Schuster, 1984.

Davies, Paul C.W. and J.R. Brown. Eds. *The Ghost in the Atom*. New York: Cambridge University Press, 1988.

Davies, Paul. "Inertia Theory Magic Roundabout: Paul Davies on the Meaning of Mach's Principle." Online. WWW. Accessed 0528-97. Available at <www.physics.adelaide.edu.au...pcwd/ Guardian/1994/940922Mach.html >

Davies, Paul C.W. and Julian Brown. Eds. *Superstrings, a Theory of Everything*. Cambridge: CUP, 1988.

Davis, William R. *Classical Fields, Particles and the Theory of Relativity*. New York: Gordon & Breach Scientific Publications, 1970.

DeBroglie, Louis. "L'Univers A Cinq Dimensions et la Mécanique Ondulatoire." *Le Journal de Physique et Le Radium* 8 (Fevrier, 1927): 65-73.

DeBroglie, Louis. "Réponse a La Note de M.O. Klein." *Le Journal de Physique et Le Radium* 8 (April 1927): 244.

Dettling, J. Ray. "Toward the Theory of Everything." Online. WWW. 1996. Accessed 04-17-97. Available at <www.careerexpo.com/pub/docs/ theory.html>

Deutsch, David and Arthur Ekert. "Quantum communication moves into the unknown." Extract from *Physics World* (June 1993): Online. WWW. Accessed 04-20-97. Available at <eve.physics.ox.ac.uk/QCresearch/ communication/communication.html>

Dingle, Herbert. *The Sources of Eddington's Philosophy, the eighth Arthur Stanley Eddington Memorial Lecture*. 2 November 1954. Cambridge: Cambridge University Press, 1954.

Dobbs, Adrian. "The Feasibility of a Physical Theory of ESP." *Science and ESP*. Smythies: 225-254.

Dolman, John Jr. "The Curvature of Space." *New Science Review* 2 (July 1895): 31-41.

Domaschko, Max. "A Unified Field Theory Free of Attractive Forces." Online. WWW. Accessed 7-22-96. Available at <www.newwave.Net/ ~applogic/theory.htm>

Dombrowski, Peter. *150 Years after Gauss' <<disquisitiones generales circa superficies curvas>>*. With two original texts of Gauss. *Astérisque* 62.

Paris: Sociètè Mathématique de France, 1979; From the 1902 translation of Adam Hiltebeitel and James Morehead.

Dorman, Clark. "The Mind and Consciousness." Online. WWW. 05-13-95. Accessed 06-17-97. Available at <cns-web.bu.edu/pub/domain/conscious.html>

Dowling, Jonathan P. "Stapp's PK Theory." Online. WWW. Available at <www.hia.com/hia/pcr/pktheory.html>; Reprinted from *Physics Today* (July 1995).

Drexler, K. Eric. "The Limits to Growth." Chapter 10. *Engines of Creation*. Online. WWW. 09-23-96. Accessed 04-17-97. Available at <www.asiapac.com/ EnginesOfCreation/EOC-Chapter-10.html>

Droz-Vincent, M. "Quantification de La Théorie de Jordan-Thiry." *Les Théories Relativistes de La Gravitation*. Paris: Centre National de La Recherche Scientifique, 1962.

Drummond, Henry. *Natural Law in the Spiritual World*. Boston: Estes and Lauriat, n.d.

Dubrov, Alexander P. "Biogravitation and Psychotronics." *Future Science*. White and Krippner: 229-244.

Duncan, Ronald and Miranda Weston-Smith. Eds. *The Encyclopedia of Ignorance: Everything you ever wanted to know about the unknown*. New York: Pocket Books, 1977.

Dunne, J.W. *An Experiment with Time*. London: Macmillan, 1927.

Dunnington, Waldo G. "Note on Gauss' Triangulations." *Scripta Mathematica* 20 (1954): 108-109.

Dunnington, Waldo G. *Carl Friedrich Gauss: Titan of Science*. New York: Hafner, 1955.

Ebon, Martin. Ed. *The Signet Handbook of Parapsychology*. New York: New American Library, 1978.

Eccles, John Carey. *The Neurophysiological Basis of Mind*. Oxford: At the Clarendon Press, 1952.

Eddington, Arthur S. *Fundamental Theory*. E.T. Whittaker. Ed. Cambridge: at the University Press, 1946.

Eddington, Arthur Stanley, et.al. "Proceedings of the Meeting of the Royal Astronomical Society." *The Observatory* 40 (May 1917): 180-187.

Eddington, Arthur Stanley, et.al. "Proceedings of the Meeting of the Royal Astronomical Society." *The Observatory* 40 (December 1917): 424-427.

Eddington, Arthur Stanley. "Astronomical Consequences of the Electrical Theory of Matter." *Philosophical Magazine* 34 (1917): 163-167.

Eddington, Arthur Stanley. "Einstein's Theory of Gravitation." *The Observatory* 40 (February 1917): 93-95.

Eddington, Arthur Stanley. "Astronomical Consequences of the Electrical Theory of Matter. Note of Sir Oliver Lodge's Suggestions." *Philosophical Magazine* 34 (1917): 321-327.

Eddington, Arthur Stanley and Dr. Silberstein. "Proceedings of the Meeting of the Royal Astronomical Society." *The Observatory* 41 (June 1918): 234-239.

Eddington, Arthur Stanley. "Electrical theories of Matter and their Astronomical Consequences with special reference to the Principle of Relativity." *Philosophical Magazine* 35 (1918): 481-487.

Eddington, Arthur Stanley. "Gravitation and the Theory of Relativity." Delivered at the Royal Institution on 1 February. *Nature* 101 (7 March 1918): 15-17; and (14 March 1918): 34-36.

Eddington, Arthur Stanley. "On O'Farrell's "Relativity and Gravitation." *Nature* 101 (18 April 1918): 126.

Eddington, Arthur Stanley. *Report on the Relativity Theory of Gravitation*. London: Fleetway Press, 1918.

Eddington, Arthur Stanley. "Silberstein's Paradox and Einstein's Theory." Dated 7 August 1918. *The Observatory* 41 (September 1918): 350-352.

Eddington, Arthur Stanley. "Boundary Difficulties of Einstein's Theory." *The Observatory* 41 (October 1918): 383.

Eddington, Arthur Stanley. "The Total Eclipse of 1919 May 29 and the Influence of Gravitation on Light." *The Observatory* 42 (March 1919): 119-122.

Eddington, Arthur Stanley. "Einstein's Theory of Space and Time." *Contemporary Review* 116 (July-December 1919): 639-643.

Eddington, Arthur Stanley, et.al. "Proceedings of the Meeting of the Royal Astronomical Society." *The Observatory* 42 (December 1919): 426-431.

Eddington, Arthur Stanley. *Space, Time and Gravitation*. Cambridge: Cambridge University Press, 1920; Reprinted New York: Harper Torchbooks, 1959.

Eddington, Arthur Stanley. "A Generalisation of Weyl's Theory of the Electromagnetic Gravitational Fields." *Proceedings of the Royal Society of London* A99 (April-September 1921): 104-122.

Eddington, Arthur Stanley. *The Theory of Relativity and its Influence on Scientific Thought*. The Romanes Lecture delivered in the Sheldon Theatre, 22 May 1922. Oxford: At the Clarendon Press, 1922.

Eddington, Arthur Stanley. "The Propagation of Gravitational Waves." Read 11 October 1922. *Proceedings of the Royal Society of London* 102 (1922-1923): 268-282.

Eddington, Arthur Stanley. *The Mathematical Theory of Relativity*. Cambridge: At the University Press, 1963; Reprint of 1923.

Eddington, Arthur Stanley. *The Nature of the Physical World*. Cambridge: Cambridge University Press, 1928; Reprinted Ann Arbor: University of Michigan Press, 1958.

Eddington, Arthur Stanley. "Einstein's Field Theory." *Nature* 123 (23 February 1929): 280-281.

Edge, Hoyt L., Robert L. Morris, Joseph H. Rush, and John Palmer. *Foundations of Parapsychology: Exploring the Boundaries of Human Capability*. London: Routledge & Kegan Paul, 1986. Editor. "Investigating the Paranormal." *Nature* 252 (18 October 1974): 602-607; Online. WWW. Accessed 09-17-97. Available at <www.tcom.co.uk/hpnet/g2.htm>

Editor. "Spiritualism Again." *Popular Science Monthly* 15 (1879): 700-702.

Edwards, Paul. Ed. *The Encyclopedia of Philosophy*. Eight volumes. New York: Macmillan, 1967.

Ehlers, Jurgen. "The Nature and Structure of Space-Time." *The Physicist's Conception of Nature*. Mehra: 71-91.

Ehrenfest, Paul and G.E. Uhlenbeck. "Graphical Representation of the DeBroglie Phase Waves in the Five-Dimensional Universe of O. Klein." *Zeitschrift fur Physik* 39 (1926): 495-498.

Eiesland, John. "On Null Systems in Space of Five Dimensions and their Relations to Ordinary Space." *American Journal of Mathematics* 26 (1904): 103-148.

Einstein, Albert. "Brief Outline of the Development of the Theory of Relativity." Robert W. Lawson. Translator. 1921. Online. WWW. Accessed 0807-96. Available at <ourworld.compuserve.com/homepages/eric-baird/aebrief.htm>

Einstein, Albert and W. Mayer. "Einheitliche Theorie von Gravitation und Electrizität (I)." *Sitzungberichte der Preussischen Akademie der Wissenschaften* 25 (1931): 541-577.

Einstein, Albert and W. Mayer. "Einheitliche Theorie von Gravitation und Electrizität (II)." *Sitzungberichte der Preussischen Akademie der Wissenschaften* 11-12 (1932): 130-137.

Einstein, Albert. *The Meaning of Relativity*. 1st edition. London and Princeton: Methuen. 1922; 6th edition. Princeton University Press, 1956; And New Delhi: Oxford and IBH, 1968; Oxford Reprint of 6th edition, revised in 1956.

Einstein, Albert. "Kaluza's Theorie des Zusammeinhanger von Gravitation und Electrizität (I)." *Sitzungberichte der Preussischen Akademie der Wissenschaften* 6 (1927): 23-25.

Einstein, Albert. "Kaluza's Theorie des Zusammeinhanger von Gravitation und Electrizität (II)." *Sitzungberichte der Preussischen Akademie der Wissenschaften* 6 (1927): 26-30.

Einstein, Albert. *Ideas and Opinions*. Ed. Carl Seelig. Trans. Sonja Bergmann. New York: Dell, 1978.

Einstein, Albert. "Physics and Reality." *Ideas and Opinions*; Originally from *The Journal of the Franklin Institute* 201 (March 1936).

Einstein, Albert. "Geometry and Experience." *Ideas and Opinions*: 230-231; originally presented as a lecture before the Prussian Academy of Sciences on 27 January 1921.

Einstein, Albert. "Notes on the Origin of the General Theory of Relativity." *Ideas and Opinions*: 285-290; Adapted from *Mein Weltbild*. Amsterdam: Querido Verlag, 1934.

Einstein, Albert, Boris Podolsky and Nathan Rosen. "Can quantum mechanical description of physical reality be considered complete?" *Physical Review* 41 (15 May 1935): 777.

Einstein, Albert. *Relativity: The Special and General Theory*. New York: Crown, 1961.

Einstein, Albert. "Die Formale Grundlage der allgemeinen Relativitätstheorie." *Sitzungberichte der Preussischen Akademie der Wissenschaften* (1915): 778-786.

Einstein, Albert. "Spielen Gravitationsfeld im Aufber der materiellen Elementarteilchen eine wesentliche Rolle?" *Sitzungberichte der Preussischen Akademie der Wissenschaften*, (1919); Reprinted as "Do Gravitational Fields play an essential part in the structure of the Elementary Particles of Matter?" *The Principle of Relativity*. Albert Einstein, et. al. New York: Dover, 1952.

Einstein, Albert. "The Theory of the Affine Field." Trans. R.W. Lawson. *Nature* 112 (1923): 448-449.

Einstein, Albert. "Zur Einheitlichen Feldtheorie." *Sitzungberichte der Preussischen Akademie der Wissenschaften* (1929): 2-7.

Einstein, Albert. "Field Theories Old and New." *New York Times* (3 February 1929); Microfilm. New York: Readex Microprint, n.d.

Einstein, Albert. "Auf die Riemann-Metrik und den FernParallelismus gegründete einheitliche Feldtheorie." *Mathematische Annalen* 102 (1930): 685-697.

Einstein, Albert and Elie Cartan. *Letters on Absolute Parallelism*. Ed. Robert Debever. Princeton: Princeton University Press, 1979.

Einstein, Albert and Michele Besso. *Correspondence, 1903-1955*. Ed. Pierre Speziali. Paris: Hermann, 1972.

Einstein, Albert and A.D. Fokker. "Die Nordströmsche Gravitationstheorie von Standpunkt Absoluten Differentialkalkuls." *Annalen der Physics* 156 (1914): 321-328.

Einstein, Albert and Peter G. Bergmann. "On a Generalization of Kaluza's Theory of Electricity." *Annals of Mathematics* 39 (July 1938): 683-701.

Einstein, Albert, Peter G. Bergmann and Valentine Bargmann. "On the Five-Dimensional Representation of Gravitation and Electricity." *Theodor von Karman Anniversary Volume*. Pasadena: California Institute of Technology, 1941: 212-225.

Einstein, Albert. "Field Theory based on Riemannian Metrics and Distant Parallelism." English translation of "Auf die Riemann-Metrik und den Fernparallelismus gegrundete einheitliche Feldtheorie." Trans. Eric T. Lefevre. Online. WWW. Accessed 626-96. Available at www.lrz-muenchen.de/~u7f01bf/WWW/einstein1930.intro.html with an abstract at <www.lrzmuenchen.de/~u7f01bf/WWW/einst1930.abs.html>

Eisele, Carolyn. Ed. "The Charles Peirce-Simon Newcomb Correspondence." *Proceedings of the American Philosophical Society* 101 (October 1957): 409-433.

Eisenbud, Jule. *Parapsychology and the Unconscious*. Berkeley: North Atlantic Books, 1992.

Eisenhart, Luther Pfahler. *Riemannian Geometry*. Princeton: Princeton University Press, 1926.

Eisenhart, Luther Pfahler. *Non-Riemannian Geometry. American Mathematical Society Publication 8*. New York: American Mathematical Society, 1927.

Ellis, Major Wilmot E. "The Properties of Four-Dimensional Space." *The Fourth Dimension Simply Explained*. Manning: 242-251.

Ellis, John. "Interview." *Superstrings*. Davies and Brown: 151-169.

Elm, David. "Grand Unified Field Theory: Dual Field Theory." Online. WWW. 02-14-96. Accessed 04-28-97. Available at <www.tiac.net/davidelm/dualfd.htm>

Enderby, J.A. "It would, wouldn't it." Online. WWW. 1996. Accessed 05-05-97. Available at <ourworld.compuserve.com/homepages/Alan-Enderby/space.htm>

Enriques, Federigo. *Problems of Science*. Katherine Royce. Trans. With a note by Josiah Royce. London: Open Court, 1914; Original Italian published in 1906.

Eucken, Rudolf. *Main Currents of Modern Thought*. Meyrick Booth. Trans. New York: Scribners, 1912.

Eves, Howard. *An Introduction to the History of Mathematics*. New York: Holt, Rinehart and Winston, 1976.

Farwell, Ruth and Christopher Knee. "The End of the Absolute: A Nineteenth Century Contribution to General Relativity." *Studies in the History and Philosophy of Science* 21 (March 1990): 91-121.

Feferman, Solomon. "Penrose's Gödelian Argument: A Review of *Shadows of the Mind* by Roger Penrose." *Psyche* 2 (May 1995). Online. WWW. Accessed 06-28-96. Available at <psyche.cs.monash.edu.au/volume - 2-1/psyche-95-2-07-shadows5-feferman.html>

Feinberg, Gerald. "Foreword." *Psychic Explorations*. Mitchell: 21-24.

Feuer, Lewis S. *Einstein and the Generations of Science*. New York: Basic Books, 1974.

Feynman, Richard. "Interview." *Superstrings*. Davies and Brown: 192-210.

Fine, Arthur. *The Shaky Game: Einstein Realism and the Quantum Theory*. Chicago: University of Chicago Press, 1986.

Firth, Robert. "The Barn and the Pole." *Relativistic Paradoxes* Part A from psi.physics FAQ. Online. WWW. Accessed 7-10-96. Available at <sunsite.unc.edu/lunar/relapara.html>

Fisher, J.W. "The Wave Equation in Five-Dimensions." *Proceedings of the Royal Society* 123 (6 April 1929): 489-493.

Fitch, Lt. Col. Graham Denby. "An Elucidation of the Fourth Dimension." *The Fourth Dimension Simply Explained*. Manning: 43-51.

Fitch, Lt. Col. Graham Denby. "Non-Euclidean Geometry of the Fourth Dimension." *The Fourth Dimension Simply Explained*. Manning: 52-59.

Fiske, John. John *Fiske's Miscellaneous Writings*, Volume VI: *The Unseen World and other essays*. Boston: Houghton & Mifflin, 1902.

Flint, H.T. "Applications of Quaternions to the Theory of Relativity." *Philosophical Magazine* 39 (1920): 439-449.

Flint, H.T. and J.W. Fisher.   Contribution to Modern Ideas on the Quantum Theory." *Proceedings of the Royal Society* 115 (1927): 208-214.

Flint, H.T. and J.W. Fisher. "The Fundamental Equation of Wave Mechanics and Metrics of Space." *Royal Society, Proceedings* 117 (1 February 1928): 625-629.

Flint, H.T. "Relativity and the Quantum Theory." *Royal Society, Proceedings* 117 (February 1928): 630-637.

Flint, H.T. "The First and Second Order Equations of the Quantum Theory." *Royal Society, Proceedings* 124 (1929): 143-150.

Flint, H.T. and J.W. Fisher. "The Equations of the Quantum Theory." *Proceedings of the Royal Society* 126 (3 March 1930): 644-653.

Flint, H.T. "The Masses of the Proton and Electron." *Physical Society, Proceedings* 152 (15 April 1930): 239-242.

Flint, H.T. "A Metrical Theory in Relation to Electron and Proton." *Royal Society, Proceedings* 131 (April 1931): 170-177.

Flint, H.T. and O.W. Richardson. *Wave Mechanics.* 2nd edition. London: Methuen, 1931.

Flint, H.T. "The Uncertainty Principle in Modern Physics." *Nature* 129 (21 May 1932): 746-747.

Flint, H.T. "A Relativistic Basis of the Quantum Theory." *Proceedings of the Royal Society* 144 (29 March 1934): 413-424.

Flint, H.T. "A Relativistic Basis of the Quantum Theory (II)." *Proceedings of the Royal Society* 145 (2 July 1934): 645-656.

Flint, H.T. "A Relativistic Basis of the Quantum Theory (III)." *Proceedings of the Royal Society* 150 (1 June 1935): 421-441.

Flint, H.T. "On the Development of the Quantum Theory and a Possible Limit to its Application." *Physical Society, Proceedings* 48 (1 May 1936): 433-443.

Flint, H.T. "Ultimate Measurements of Space and Time." *Royal Society, Proceedings* 159 (2 March 1937): 45-56.

Flint, H.T. "Ratio of Masses of Fundamental Particles." *Physical Society, Proceedings* 50 (1 January 1938): 90-95.

Flint, H.T. "Difficulties and Developments in Quantum Theory." *Physical Society, Reports* 5 (1938): 407-421.

Flint, H.T. "Beginnings of the New Quantum Theory." *Physical Society, Reports* 4 (1938): 134-172.

Flint, H.T. "The Fundamental Unit of Electric Charge." *Physical Society, Proceedings* 50 (2 May 1938): 340-344.

Flint, H.T. "The Analogy between the Photon and Electron and the Derivation of the Quantum Equation." *Physical Society, Proceedings* (1 November 1938): 899-909.

Flint, H.T. "The Theory of the Electric Charge and the Quantum Theory, Part I." *Philosophical Magazine* 29 (April 1940): 330-343.

Flint, H.T. "The Theory of the Electric Charge and the Quantum Theory, Part II." *Philosophical Magazine* 29 (May 1940): 417-433.

Flint, H.T. "The Theory of the Electric Charge and the Quantum Theory, Part III." *Philosophical Magazine* 33 (May 1942): 369-383.

Flint, H.T. "The Fundamental Equation of Quantum Mechanics." *Philosophical Magazine* 34 (July 1943): 496-502.

Flint, H.T. "Quantum Equations and Nuclear Field Theories." *Philosophical Magazine* 36 (September 1945): 635-643.

Flint, H.T. "A Study of the Nature of the Field Theories of Electron and Positron and of the Meson." *Proceedings of the Royal Society* 185 (10 January 1946): 14-34.

Flint, H.T. "Energy in the Nuclear Field." *Philosophical Magazine* 38 (January 1947): 22-32.

Flint, H.T. and N. Symonds. "The Conservation of Energy, Momentum and Charge in the Nuclear Field." *Philosophical Magazine* 39 (June 1948): 413-419.

Flint, H.T. "The Quantization of Space and Time." *Physical Review* 74 (15 July 1948): 209-210.

Flint, H.T. "Coordinate Operators in Quantum Mechanics." *Nature* 163 (22 January 1949): 131-132.

Flint, H.T. and E.M. Williamson. "The Quantum Mechanics of the Electron." *Royal Society, Proceedings* 207 (6 July 1951): 380-388.

Flint, H.T. and E.M. Williamson. "Coordinate Operators and Fundamental Lengths." *Physical Review* 90 (15 April 1953): 318.

Flint, H.T. and E.M. Williamson. "The Relation of Quantum Theory to the Theory of Gravitation and Electromagnetism and an Application to the Theory of the Electron." *Zeitschrift fur Physik* 135 (1953): 260-269.

Flint, H.T. and E.M. Williamson. "Coordinate Operators and Fundamental Lengths." *Physical Review* 90 (15 April 1953): 318.

Flint, H.T. and E.M. Williamson. "A Theory of the Electron." *Nuovo Cimento* 11 (February 1954): 188-189.

Flint, H.T. and E.M. Williamson. "A Theory of the Electron." *Nuovo Cimento* 11 (May 1954): 568-569.

Flint, H.T. and E.M. Williamson. "A Note on Relativistic Cutoff." *Physical Society, Proceedings* 118 (April 1955): 354-355.

Flint, H.T. "Foundations of a Linear Field Theory." *Zeitschrift fur Physik* 147 (1955): 401-405.

Flint, H.T. and E.M. Williamson. "A Relativistic Theory of Charged Particles in an Electromagnetic and Gravitational Field." *Nuovo Cimento* 3 (March 1956): 551-565.

Flint, H.T. and E.M. Williamson. "A New Uncertainty Relation." *Nature* 178 (22 September 1956): 651.

Flint, H.T. and E.M. Williamson. "The Theory of Relativity, the Electromagnetic Theory and the Quantum Theory." *Nuovo Cimento* 8 (1 June 1958): 680-699.

Flint, H.T. *The Quantum Equation and the Theory of Fields*. London: Methuen, 1966.

Forsythe, A.R. *Geometry of Four Dimensions*. Cambridge: At the University Press, 1930.

Forwald, Haakon. *Mind, Matter and Gravitation: A Theoretical and Experimental Approach*. Parapsychology Monographs No.11. New York: Parapsychology Foundation, 1969.

Frankland, Frederick William. "On the Simplest Continuous Manifoldness of Two Dimensions and of Finite Extent." Read before the Wellington Philosophical Society on 11 November 1876. *Transactions and Proceedings of the New Zealand Institute* 9 (1876): 272-279; Also read before the London Mathematical Society on 14 December 1876. *London Mathematical Society, Proceedings* 8 (1876): 57-64; And reprinted in *Nature* 12 (April 1877): 515-517.

Frankland, Frederick William. "Simplest Continuous Manifoldness of Two Dimensions and of Finite Extent [A reply to Monro]." *Nature* 22 (24 June 1880): 170-171.

Frankland, Frederick William. "On the Doctrine of Mind-Stuff." *Transactions and Proceedings of the New Zealand Institute* 12 (1879): 205-215.

Frankland, Frederick William. [Mistakenly listed as J.W. Frankland]. "On a Mode of Explaining the Transverse Vibrations of Light." *Nature* 22 (5 August 1880): 317.

Frankland, Frederick William. Letters to Simon Newcomb, 11 January 1880, 14 April 1880. Box #22. Simon Newcomb Papers, Library of Congress.

Frankland, Frederick William. "The Doctrine of Mind-Stuff [Editor's excerpts from Frankland's essay]." *Mind* 6 (1881): 116-120.

Frankland, Frederick William. "Prof. Royce on "Mind-Stuff and Reality"." *Mind* 7 (1882): 110-114.

Frankland, Frederick William. "The Non-Euclidean Geometry Vindicated: A Reply to Mr. Skey." Read before the Wellington Philosophical Society on 13 February 1884. *Transactions and Proceedings of the New Zealand Institute* 18 (1885): 58-69; Also noted in 16 (1883): 557.

Frankland, Frederick William. "Theory of Discrete Manifolds." Read before the American Mathematical Society at its fourth Summer Meeting, in Toronto, Canada, 17 August 1897. Published as Appendix C in *Thoughts on Ultimate Problems*: 37-42. See also *Collected Essays and Citations*: 85-90.

Frankland, Frederick William. Theory of Discrete Manifolds. (n.d.) Add. Ms. 7656, #PA 1600R. Stokes Papers, University Library, Cambridge.

Frankland, Frederick William. "Theory of Discrete Manifolds, Part II." From the presentation copy to Prof. B.O. Peirce, now at Harvard University.

Frankland, Frederick William. *Collected Essays and Citations*, Volume I: Theology and Metaphysics, 1872-1906. Foxton, New Zealand: G.T. Beale, 1906; "The Rationality of the Cosmos is Objective:" 61-62; "Mathematics the Norm of all Intellection Whatever:" 73-76; "Notes on a New Theory of Time:" 77-79; "All Existence is Discrete:" 81; "F.W. Frankland's Theory of Physical Space as a Discrete Manifold:" 83; "Theory of Discrete Manifolds:" 85-90; "Theory of Discrete Manifolds, II: 91-102; "Theory of Discrete Manifolds, III:" 103-106; "Theory of Discrete Manifolds, IV:" 107-109; "Theory of Discrete Manifolds, V:" 111-116; "Theory of Discrete Manifolds, VI:" 117-120; "Theory of Discrete Manifolds, VII:" 121-123; "Theory of Discrete Manifolds, VIII:" 125-129; "Theory of Discrete Manifolds, IX:" 131-132; "Theory of Discrete Manifolds, X:" 123-127; "Theory of Discrete Manifolds, XI, Clifford on the Space-Theory of Matter:" 139-140; "Theory of Discrete Manifolds, XII:" 141-142; "On the Simplest Continuous Manifold of Two Dimensions and of Finite Extent:" 145-152.

Frankland, Frederick William. "Councillor Frederick William Frankland." *The Cyclopedia of New Zealand*. Christchurch, New Zealand: Cyclopedia Company, 1908: 695-696.

Frankland, Frederick William. Thoughts on Ultimate Problems: being a series of short studies on *Theological and Metaphysical Subjects*. 4th edition. London: David Nutt, 1911; 5th and revised edition. 1912.

Frankland, Frederick William. "All Existence is Discrete." A letter to Prof. Halsted. Wellington, New Zealand: Barnsbury Printing Works, 1906.

Frankland, Frederick William. "Why all Negative Electrons have the Same Mass." Incorporates a letter to Prof. Halsted. Foxton: Beale, 1907.

Frankland, Frederick William. "Press Notice of Third and Revised edition of *Thoughts on Ultimate Problems*, by " Leaflet in the Library of Congress. Frankland, Frederick William. "Frederick William Frankland." *Dictionary of New Zealand Biography* (1940): 279-280.

Frankland, William Barrett. *The First Book of Euclid's Elements with a Commentary*. Cambridge: At the University Press, 1905.

Frankland, William Barrett. *Theories of Parallelism: An historical critique*. Cambridge: At the University Press, 1910.

Frankland, William Barrett. "Notes on the Parallel Axiom." *Mathematical Gazette* 5 (1911-1912): 136-139.

Frankland, William Barrett. "Non-Euclidean Geometry." *Mathematical Gazette* 7 (1915): 332-333.

Franklin, Wilbur. "Metal Fracture Physics Using Scanning Electron Microscopy and the Theory of Teleneural Interactions." Online. WWW. Accessed 11-10-97. Available at <www.tcom.co.uk/hpnet /g8.htm>

Franklin, Wilbur. "The Role of Paraphysics in Physics Education." *Education in Parapsychology*. Shapin and Coly: 229-248.

Frasca, Sergio. "On Physics and Psi - I." Online. WWW. Accessed 7-1-96. Available at <www.roma1.infn.it/rog/group/frasca/b/ pepsi1.html>

Fraser, J.T. *The Voices of Time*. New York: George Braziller, 1966.

Frazier, Kendrick. *The Hundredth Monkey and other Paradigms of the Paranormal: A Skeptical Inquirer Collection*. Buffalo: Prometheus Books, 1991. Freeman, Ira. "Why is Space Three-Dimensional?" *American Journal of Physics* XXXVII. 12 (December 1969: 1222-1224; adapted from W. Büchel: "Warum hat der Raum drei Dimensionen?" *Physikalische Blätter* 19 (December 1963): 547-549.

Freundlich, Erwin. The Foundations of Einstein's Theory of Gravitation. Henry L. Brose. Trans. H.H. Turner. Intro. Albert Einstein. Pref. Cambridge: at the University Press, 1920.

Friedlander, Michael D. *At the Fringes of Science*. San Francisco: Westview, 1995.

Fullerton, George S. "On Space of Four Dimensions." *Journal of Speculative Philosophy* 18 (April 1884): 113-121.

Fullerton, George S. *The Conception of the Infinite and the Solution of the Mathematical Antinomies: A Study in Psychological Analysis*. Philadelphia: Lippincott, 1887.

Fullerton, George S. "The Doctrine of Space and Time, I." *Philosophical Review* 10 (March 1901): 113-123; "The Doctrine of Space and Time, II." (May 1901): 229-240; "The Doctrine of Space and Time, III." (July 1901): 375-385; "The Doctrine of Space and Time, IV," (September 1901): 488-504; "The Doctrine of Space and Time, V. The Real World in Space and Time." (November 1901): 583-600.

Fullerton, George S. *A System of Metaphysics*. New York: Greenwood, 1968; A reprint of Macmillan, 1904; "The Kantian Doctrine of Space:" 162-

171; "Difficulties Connected with the Kantian Doctrine of Space:" 172-183; "The Berkeleian Doctrine of Space:" 184-193; "Of Time:" 194-209; "The Real World in Space and Time:" 210-225; with Edgar A. Singer, Jr., "Note on the Physical World-Order:" 609-621.

Gardner, Martin. "Parapsychology and Quantum Mechanics." *Skeptic's Handbook*. Kurtz: 585-598; previously published in *Science and the Paranormal*.

Garrett, Eileen. *The Sense and Nonsense of Prophecy*. New York: Berkeley Publishing, 1950.

Garnett, William. "Alice through the (Convex) Looking Glass." Presidential Address before the London Branch of the Mathematical Association, 9 February 1918. *The Mathematical Gazette* 9 (1917-1919): 237-241, 249-252, 293-298.

Gauld, Alan. *The Founders of Psychical Research*. New York: Schocken, 1968.

Gauss, Karl Friedrich. *General Investigations of Curved Surfaces*. Trans. Adam Hiltebeitel. Appended. James Morehead. Hewlett. New York: Raven Press, 1965; Reprint of 1902 translation of original 1825 and 1827 papers; originally published as Carolo Frederico Gauss, "Disquisitiones generales circa superficies curvas." *Commentationes societatis regiae scientiarum Göttingensis recentiores* 6 (1928); Reprinted *in Carl Friedrich Gauss Werke. Königlichen Gessellschaft der Wissenschaften zu Göttingen* 4 (1873): 217-258.

Gauss, Karl Friedrich. "Anziege." Göttingische gelehrte Anziegen. 177 (5 November 1827): 1761-1768; Reprinted Dombrowski, *150 Years*: 82-95, 84-85. The abstract was originally delivered before the Royal Society of Göttingen on 8 October 1927. The translation used by Dombrowski is from Adam Hiltebeitel and James Morehead, 1902.

Geake, E.H. *The analogy between line geometry, point geometry in four-dimensional space and sphere geometry*. Master's Thesis at Bristol University, 1914.

Geoghegan, Edward. "The Axioms of Geometry." *Nature* 23 (10 April 1884): 551.

George, Wilma. *Biologist-Philosopher: A Study of the Life and Writings of Alfred R. Wallace*. New York: Abelard-Schuman, 1964.

Gibbs, Phil. "New and Alternative Theories of Physics." Online. WWW. Accessed 6-26-96. Available at <ourworld.compuserve.com/homepages /Phil-Gibbs/ theories.html>

Gibbs, Phil. The Cyclotron Notebooks. Online. WWW. Accessed 7-3-96: 1. "The Quantum Gravity Challenge." Online. WWW Accessed 5-16-96. 7-3-96. Available at <ourworld.compuserve.com/ homepages/Phil-Gibbs/ qugrav.htm>

Gibbs, Phil. "2. Is Space-Time Discrete." WWW. 3-3-96. Accessed 7-3-96. Available at <ourworld.compuserve.com/homepages/Phil-Gibbs/ discrete.html>

Gibbs, Phil. "3. Metaphysics of Space-Time." Online. WWW. 3-8-96. Accessed 7-3-96. Available at <ourworld.compuserve.com/homepages/ Phil Gibbs/metaphys.html>

Gibbs, Phil. "4. What about Causality?" Online. WWW 5-15-96. Accessed 73-96. Available at <ourworld.compuserve.com/ homepages/Phil-Gibbs/cause.htm>

Gibbs, Phil. "5. Universal Symmetry." Online. WWW. 6-3-96. Accessed 7-3-96. Available at <ourworld.compuserve.com/homepages/Phil-Gibbs/ symmetry.htm>

Gibbs, Phil. "6. The Superstring Mystery." Online. WWW. 6-1596. Accessed 73-96. Available at <ourworld.compuserve.com/ homepages/Phil Gibbs/strings.htm>

Gibbs, Phil. "7. Event-Symmetric Space-Time." Online. WWW. 3-8-96. Accessed 7-3-96. Available at <ourworld.compuserve.com /homepages/Phil Gibbs/event.html>

Gibbs, Phil. "8. Is String Theory in Knots?" Online. WWW. 5-15-96. Accessed 7-3-96. Available at <ourworld.compuserve.com/homepages/Phil Gibbs/knots.htm>

Gibbs, Phil. "9. The Theory of Theories." Online. WWW. 3-8-96. Accessed 7-3-96. Available at <ourworld.compuserve.com/ homepages/Phil Gibbs/theory.html>

Gibbons, G.W. "Twistor Theory." Essay review of *Twistor Geometry and Field Theory*. R.S. Ward and R.O. Wells Jr. Contemporary Physics 33, 1994: 187-188.

Gillispie, Charles Coulton. Ed. *Dictionary of Scientific Biography*. New York: Scribner's' Sons, 1974; Eight volume edition, printed in 1981.

Gliozzi, Mario. "Tullio Levi-Civita." *Dictionary of Scientific Biography* 8: 284-285.

Goe, George. "Comments on Miller's 'The Myth of Gauss' *Experiment on the Euclidean Nature of Physical Space*, Comment I." *Isis* 65 (March 1974): 83-85; And B.L. van der Waerden. "Comment II." *Isis* 65: 85.

Glashow, Sheldon L. *The Charm of Physics*. New York: Touchstone, 1991.

Glashow, Sheldon. "Interview." *Superstrings*. Davies and Brown: 180-191.

Goldfarb, Russell and Clare. *Spiritualism and Nineteenth-Century Letters*. Teaneck: Fairleigh Dickinson University Press, 1978.

Golway, James F. *Searching for Black Holes*. Videorecording. KCET TV, Los Angeles. PBS Video, 1991.

Gonseth, F. and G. Juvet. "Space Metric of Five-Dimensions of Electromagnetism and Gravitation." *Comptes Rendus* 185 (17 August 1927): 412-413.

Gonseth, F. and G. Juvet. "Schrödinger's Equation and the Equations of Electromagnetism in the Five-Dimensional Universe." *Comptes Rendus* 185 (5 September 1927): 535-538.

Gonseth, F. and G. Juvet. "The Equations of Electromagnetism." *Comptes Rendus* 185 (1 August 1927): 341-343.

Good, I.J. Ed. *The Scientist Speculates: An Anthology of Partly Baked Ideas*. New York: Basic Books, 1962.

Good, I.J. "Winding Space." *The Scientist Speculates*. Good: 330-336.

Good, I.J. "Speculations Concerning Precognition." *The Scientist Speculates*. Good: 151-157.

Grant, Edward. *Much Ado About Nothing*. Cambridge: Cambridge University Press, 1981.

Grattan-Guinness, Ivor. Ed. Psychical Research: A Guide to Its History, Principles and Practices. Wellingborough, England: Aquarian Press, 1982.

Graves, John Cowperthwaite. *The Conceptual Foundations of Contemporary Relativity Theory*. Cambridge, Massachusetts: MIT Press, 1971.

Gray, Jeremy. *Ideas of Space: Euclidean, Non-Euclidean and Relativistic*. Oxford: Clarendon Press, 1979.

Green, Michael. "Interview." *Superstrings*. Davies and Brown: 107-139.

Greenberg, Marvin Jay. *Euclidean and Non-Euclidean Geometries: Development and History*. San Francisco: W.H. Freeman, 1974.

Greene, Brian. "Superstring Theory." Online. WWW. Accessed 09-17-97. Available at <www.lassp.Cornell.edu/GraduateAdmissions/greene/greene.html>

Greenwood, Thomas. "Geometry and Reality." Read 12 June 1922. *Proceedings of the Aristotelian Society* 16 (1922): 189-204.

Gregory, Christopher. "Equations of Motion if Five-Space." *Nature* 92 (30 December 1961): 1320.

Gregory, Christopher. "Einstein, Infeld and Hoffman Problem of Motion in Five Space." *Physical Review* 125 (15 March 1962): 2136-2139.

Gregory, Christopher. "Anomalous Velocities in Quasar 3C273." *Nature* 25 (15 May 1965): 702.

Gregory, Christopher. "Extra-Dimensionality in Relativity and Spaceons." *Abstracts on the Fifth International Conference on Gravitation and the Theory of Relativity*. Tbilisi: Publishing House of Tbilisi, 1968: 245-247.

Grim, Patrick. Ed. *Philosophy of Science and the Occult*. 2nd edition. Albany, New York: State University of New York Press, 1990.

Gross, David. "Interview." *Superstrings*. Davies and Brown: 139-150.

Grünbaum, Adolph. "The Ontology of the Curvature of Empty Space in the Geometrodynamics of Clifford and Wheeler." *Space, Time and Geometry*. Synthèse Library. Ed. Patrick Suppes. Dordrecht: D. Reidel, 1973: 268-295.

Guillemin, Victor. *The Story of Quantum Mechanics*. New York: Charles Scribner's Sons, 1968.

Gurney, Edmund, F.W.H. Myers and F. Podmore. *Phantasms of the Living*. Gainesville: Scholar's Facsimiles and Reprints, 1970.

Gürsey, Feza. "Quaternionic and Octonionic Structures in Physics: Episodes in the relation between physics and mathematics." *Symmetries in Physics. Proceedings of the 1st International Meeting on the History of Scientific Ideas* held at Sant Feliu de Guixols. Catalonia, Spain, 20-26 September 1983. Eds.

Manuel G. Doncel, Armin Hermann, Louis Michel and Abraham Pais. Bellaterra, Spain: *Seminari d'Historia de les Cienceis, Universitat Autonoma de Barcelona*, 1987: 557-592.

Haan, P.A.P. den. "Gravity and Curved Space." Online. WWW. Accessed 09-17-97. Available at <eide.sci.kun.nl/~pieterh/ physics.html >

Haas, Arthur. "An Interpretation of the Fourth Dimension." *The Fourth Dimension Simply Explained*. Manning: 100-109.

Haisch, Bernhard. "Beyond E=mc^2." Online. WWW. Accessed 05-28-97. Available at <www.jse.com/haisch/sciences.html>

Haisch, Bernhard. "A Quantum Broom Sweeps Clean." Online. WWW. Accessed 05-28-97. Available at <www.jse.com/haisch/ mercury.html>

Hall, G. Stanley. *Founders of Modern Psychology*.

Hall, T. Proctor. "The Possibility of a Realization of Four-Fold Space." *Science* 19 (13 May 1892): 272-274.

Hall, T. Proctor. "The Projection of Four-Fold Figures upon a Three Flat." *American Journal of Mathematics* 15 (1893): 179-189.

Hall, T. Proctor. "Physical Theories of Gravitation." *Proceedings of the Iowa Academy of Sciences* 3 (1895): 47-52.

Hall, T. Proctor. "The Extension of Complex Algebra." *Proceedings of the Iowa Academy of Sciences* 6 (1898): 202-204.

Hall, Tord. *Carl Friedrich Gauss*. Trans. Albert Froderber. Cambridge, Massachusetts: MIT Press, 1970; Translation of *Gauss, Matematikernas Koneng*. Stockholm: Bokforlaget Prisma.

Halsted, George Bruce. "The New Ideas of Space." *Popular Science Monthly* 11 (1877): 364-365.

Halsted, George Bruce. "Four-Fold Space and Two-Fold Time." *Science* 1 (1892): 319.

Halsted, George Bruce. "How the New Mathematics Interprets the Old." Read 4 March 1893. *Transactions of the Texas Academy of Sciences* 1 (1892-1896): 89-96.

Halsted, George Bruce. "The Old and the New Geometry." *Educational Review* 6 (1893): 144-157.

Halsted, George Bruce. "Easy Corollaries in Non-Euclidean Geometry." *American Mathematical Monthly* 1 (1894): 42.

Halsted, George Bruce. "The Non-Euclidean Geometry Inevitable." *The Monist* 4 (1894): 483-493.

Halsted, George Bruce. "Some Salient Points of in the History of Non-Euclidean Geometry and Hyper-Spaces." *International Mathematical Congress* (1896): 92-95.

Halsted, George Bruce. "Subconscious Pangeometry." *The Monist* 7 (1897): 101-106.

Halsted, George Bruce. "Newcomb's Philosophy of Hyperspace." *Science* 7 (1898): 212.

Halsted, George Bruce. "Gauss and the Non-Euclidean Geometry." *Science* 12 (1900): 842-846.

Halsted, George Bruce. "Non-Euclidean Geometry." *Popular Astronomy* 8 (1900): 189-202.

Halsted, George Bruce. "The Appreciation of Non-Euclidean Geometry." *Science* 13 (1901): 462-465.

Halsted, George Bruce. "The Popularization of Non-Euclidean Geometry." *American Mathematical Monthly* 13 (1901): 31-35.

Halsted, George Bruce. "The Message of Non-Euclidean Geometry." *Science* 19 (1904): 401-413.

Halsted, George Bruce. "The Value of Non-Euclidean Geometry." *Popular Science Monthly* 67 (1905): 639-646.

Halsted, George Bruce. "Easy Non-Euclidean Geometry." *The Monist* 19 (1909): 399-402.

Halsted, George Bruce. "The Unverifiable Hypotheses of Science." *The Monist* 20 (1910): 563-573.

Hameroff, Stuart R., Alfred W. Kaszniak and Alwyn C. Scott. Eds. *Toward a Science of Consciousness: The First Tucson Discussions and Debate*. Cambridge: MIT Press, 1996.

Hameroff, Stuart and Roger Penrose. "Conscious Events as Orchestrated Space-Time Selections." Online. WWW. Accessed 05-07-98. Available at <www.u.arizona.edu/~hameroff/hardfina. html>

Hansel, C.E.M. *The Search for Psychic Power: ESP and Parapsychology Revisited*. Buffalo: Prometheus Books, 1989.

Hammond, Richard T. "New Fields in General Relativity." *Contemporary Physics* 36 (1995): 103-114.

Hardcastle, Valerie Gray. "Attention versus Consciousness: A Distinction with a Difference." Online. WWW. Accessed 05-07-98. Available at <mind.phil.vt.edu/papers/attencons.html>

Hardcastle, Valerie Gray. "Conscious Computations." Online. WWW. Accessed 05-07-98. Available at <www.phil.indiana.edu/ejap/1993.august/ hardcastle abs.html>

Hart, Hornell. *Toward a New Philosophical Basis for Parapsychological Phenomena*. *Parapsychology Monographs*, Number 7. New York: Parapsychology Foundation, 1965.

Harvey-Wilson, Simon. "Dimensions, Consciousness, UFOs, and the Paranormal." A lecture on Friday 16th June 1995 at the Australian Society for Psychical Research. Murdoch University, Western Australia. Online. WWW. Accessed 7-8-96. Available at <www.iinet.com.au/~steveb/aspr/SHWlecture-1.html>

Harre, R. and John Hawthorne. *A Selective Bibliography of Philosophy of Science*. 2nd edition. Revised J.D. Greenwood, W. Newton-Smith and R. Harre. *Study Aids 2*. Oxford: Sub-Faculty of Philosophy, 10 Merton Street, May 1977.

Hathaway, Arthur S. "Quaternions as Numbers of Four-Dimensional Space." *Bulletin of the American Mathematical Society* 4 (1897): 54-57.

Hasted, John B. "Paranormal Metal Bending." *The Iceland Papers*. Puharich: 95-110.

Hathaway, Arthur S. "Quaternion Space." *Transactions of the American Mathematical Society* 3 (January 1902): 46-59.

Hawking, Stephen. *A Brief History of Time: From the big bang to black holes*. New York: Bantam Books, 1988.

Hayhoe, Douglas. "From Ellipses to Superstrings: A Brief History of Unification." Online. WWW. Accessed 09-17-97. Available at <www.science.uwaterloo.ca/physics/p13news/number-4/ superst.html >

Heath, R.S. "On the Dynamics of a Rigid Body." *Philosophical Transactions of the Royal Society of London* 175 (1884): 281-324.

Heimann, P.M. "The Unseen Universe: Physics and the Philosophy of Nature in Victorian Britain." *British Journal for the History of Science* 6 (1972): 73-79.

Henkel, Joel. "Non-Unitary Quantum Observation as a Bridge Between Mind and Matter." Online. WWW. Accessed 7-3-96. Available at <hermesop.com/inscirev/fall95/bridge.htm>

Henkel, Joel. "Quantum Measurement as Experience." Online. WWW. Accessed 7-3-96. Available at <hermes-op.com/inscirev/fall95/quanmeas.mhtml>

Henkel, Joel. "Toward a Metaview for Interscience Discussion. Online. WWW. Accessed 7-3-96. Available at <hermes-op.com/inscirev/fall95/metaview.html>

Herbert, Nick. *Faster Than Light: Superluminal Loopholes in Physics*. New York: New American Library, 1988.

Herman, Robert. *Yang-Mills Kaluza-Klein and the Einstein Program*. Brookline, Massachusetts: Math Sci Press, 1978.

Hermann, Dieter B. "Johann Karl Friedrich Zöllner." *Dictionary of Scientific Biography* 16: 627-630.

Hermann, Dieter B. Karl Friedrich Zöllner. *Biographien hervorragender Naturwissenschaftler, Techniker und Mediziner* 57. Leipzig: B.G. Teubner, 1982.

Heron, John. "A preamble to subtle inquiry." Online. WWW. 1997. Accessed 12-04-98. Available at <zeus.sirt.pisa.it/icci/preasuin.htm>

Hesse, Mary B. *Forces and Fields*. New York: Philosophical Library, 1962.

Hessenberg, Gerhard. "Vectorielle Begründung der Differential geometrie." *Mathematische Annalen* 78 (1917): 187-216.

Heyl, Paul Renno. *The Theory of Light on the Hypothesis of a Fourth Dimension*. Philadelphia: University of Pennsylvania, Doctoral Thesis, 1897.

Heywood, Roseland. "Notes on Changing Mental Climates and Research into ESP." *Science and ESP*. Smythies: 47-60.

Hilster, David de. "Beyond Einstein: The Autodynamics Theory." Online. WWW. Accessed 7-3-96. Available at <www.areacom.it/html/ita/loris/theories.html>

Hinson, Jason W. "Relativity and FTL Travel, Part III: A Bit About General Relativity." Online. WWW. Accessed 7-29-96. Available at <sunsite.auc.dk /ftp/ pub/news.answers/Star-Trek/relativity-FTL/part3>

Hinton, Charles Howard. "What is the Fourth Dimension?" Originally published in *Cheltenham Ladies' College Magazine* and *Dublin University Magazine* (1880). Published as a pamphlet with the subtitle "Ghosts Explained" (1884); Republished in *Scientific Romances*. London: Swann and Sonnenschein, 18841886. 1: 1-32; Reprinted in *Speculations*: 1-22.

Hinton, Charles Howard. *A New Era of Thought*. London: Swann, Sonnenschein, 1888.

Hinton, Charles Howard. "Hyperbolae and the Solutions of Equations." *Bulletin of the New York Mathematical Society* 3 (June 1897): 309-321.

Hinton, Charles Howard. "Many Dimensions." *Scientific Romances*. London: Swann, Sonnenschein, 1902: 23-44.

Hinton, Charles Howard. "The Recognition of the Fourth Dimension." Read 9 November 1901. *Bulletin of the Philosophical Society of Washington* 14 (1902): 181-203; Reprinted in *Speculations*: 142-162.

Hinton, Charles Howard. *The Fourth Dimension*. London: Swann, Sonnenschein, 1904; Excerpts are reprinted in *Speculations*: 120-141.

Hinton, Charles Howard. "The Geometrical Meaning of Cayley's Formulae of Orthogonal Transformations." Read 29 November 1902. *Proceedings of the Royal Irish Academy* 24 (1902-1904): 59-65.

Hinton, Charles Howard. "The Fourth Dimension." *Harper's Magazine* 109 (July 1904): 229-233.

Hinton, Charles Howard. *Speculations on the Fourth Dimension: Selected Writings of Charles H. Hinton*. Ed. Rudolf v. B. Rucker. New York: Dover, 1980. Hirst, Thomas Archer. *Natural Knowledge in Social Contexts:*

The Journals of Thomas Archer Hirst. William H. Brock and Roy M. MacLeod. Eds. London: Mansell, 1980.

Hochheimer, Andrew H. "The Philadelphia Experiment from AZ." Online. WWW. Accessed 7-22-96. Available at <www. Wincom.net/softarts/ philexp.html>

Hodges, Andrew. "A First Few Scraps on Twistor Theory." Online. WWW. Accessed 6-17-96. Available at <www.wadham.ox.ac.uk/~ahodges/twistors .html>

Hoffman, Banesh. "Projective Relativity and the Einstein-Mayer Unified Field Theory." *Physical Review* 43 (15 April 1933): 615-619.

Hoffman, Banesh. *The Quarterly Journal of Mathematics* 7 (1936): 20.

Hoffman, Banesh. "The Vector Meson Field and Projective Relativity." *Physical Review* 72 (1947): 458.

Hoffman, Banesh. "The Gravitational, Electromagnetic and Vector Meson Fields and the Similarity Geometry." *Physical Review* 73 (1948): 30.

Hoffman, Banesh. *Albert Einstein, Creator and Rebel*. With the collaboration of Helen Dukas. New York: Viking Press, 1972.

Holton, Gerald. "On the Origins of the Special Theory of Relativity." *American Journal of Physics* 28 (1960): 627-631; partially reprinted in *Relativity Theory*. L. Pearce Williams: 100-107; also reprinted in *Thematic Origins*. Holton: 165183.

Holton, Gerald. *Thematic Origins of Scientific Thought: Kepler to Einstein*. Cambridge: Harvard University Press, 1973.

Hoppe, Edmund. "C.F. Gauss und der Euklidische Raum." *Die Naturwissenschaften* 13 (28 August 1925): 743-744; Trans. Dunnington. *Scripta Mathematica* 20 (1954): 108-109.

Hostetter, J. Clyde. "Possible Movements and Forms in a System of Four Dimensions." *The Fourth Dimension Simply Explained*. Manning: 220-231.

Howorth, Henry H. "Transcendental Premises in Science." *Nature* 107 (3 March 1921): 9-10.

Hyman, Ray. "Evaluation of Program on Anomalous Mental Phenomena." Online. WWW. 09-11-95. Accessed 05-01-97. Available at <darkwing.uoregon.edu/~uocomm/newsrel/ Hyman.html>

Hyslop, James H. "The Fourth Dimension of Space." *Philosophical Review* 5 (1896): 352-370.

Ingarden, R.S. "Equations of Motion and Field Equations in Five-Dimensional Unified Relativity Theory." *Dokl. Akad. Nauk.* SSSR 88 (1953): 773-776.

Ingleby, C.M. "The Antinomies of Kant." *Nature* 7 (6 February 1873): 262

Ingleby, C.M. "The Unreasonable." *Nature* 7 (20 February 1873): 302-303.

Ingleby, C.M. "Prof. Clifford on Curved Space." *Nature* 7 (13 February 1873): 282-283.

Ingleby, C.M. "Transcendent Space." *Nature* 1 (13 January 1870): 289; and (17 February 1870): 407.

Ingleby, C.M. Letter to C.J. Monro, 30 August 1870, #2438, Acc.1063, Monro Collection, Greater London Record Office, London, England: Ingleby to Monro, 16 October 1871, #2454: Ingleby to Monro, 29 November 1871, #2469: Ingleby to Monro, 3 February 1872, #2491: Ingleby to Monro, 24 May 1872, #2502B.

Inglis, Brian. *Natural and Supernatural: A History of the Paranormal from Earliest Times to 1914*. Revised edition. London: Hodder & Stoughton, 1977.

Irwin, H.J. *Introduction to Parapsychology*. 2nd edition. London: McFarland & Company, 1994.

Irwin, H.J. *Psi and the Mind: An Information Processing Approach*. Metuchen, New Jersey: Scarecrow Press, 1979.

Jablonski, Julia Melges. "Psychophysics: A Wholistic Approach to Energy Healing." From the book Capturing the Aura. Online. WWW. 1996. Accessed 17 June 1997. Available at <users.aol. com/Jewels321/aurachap.html> Jack, Alex. Ed. *The New Age Dictionary: A Guide to Planetary Family Consciousness*. New York: Japan Publications, 1990.

Jacoby, Harold. "Faith and the Fourth Dimension." Address delivered in St. Paul's Chapel, Columbia University, New York, 9 March 1908; *Popular Astronomy* 16 (1908): 298-300.

Jahn, Robert D. Ed. *The Role of Consciousness in the Physical World*. AAAS Selected Symposium 57. Boulder, Colorado: Praeger, 1981.

Jahn, Robert D. and Brenda J. Dunne. *Margins of Reality: The Role of Consciousness in the Physical World*. San Diego: Harcourt Brace Jovanovich, 1987.

Jaki, Stanley L. *The Relevance of Physics*. Chicago: University of Chicago Press, 1966.

James, William. *William James on Psychical Research*. Gardner Murphy and Robert Ballou. Eds. New York: Viking Press, 1960.

Jammer, Max. *Concepts of Space*. Foreword Albert Einstein. Cambridge: Harvard University Press, 1954; Reprinted 1969.

Jeans, James H. "Einstein's Theory of Gravitation." *The Observer* 40 (January 1917): 57-58.

Jeans, James H. *The Mysterious Universe*. Cambridge: at the University Press, 1931; Reprint with corrections of the 1st edition, 1930.

Jetmore, Christina, William Coleman and Becky Romanowski. "Major Events in the History of Parapsychology." Online. WWW. 4-23-96. William H. Jack. Accessed 6-24-96. Available at <www.fpc.edu/academic/behave/ psych/web93-1.htm>

Johnson, Dale M. "The Problem of the Invariance of Dimension in the Growth of Modern Topology, Part I." *Archive for the History of Exact Science* 20 (1979): 97-188; "The Problem of the Invariance of Dimension in the Growth of Modern Topology, Part II." 25 (1981): 85-267.

Johnston, Charles. "The Mind's Eye and the Fourth Dimension." *The Fourth Dimension Simply Explained*. Manning: 182-191.

Johnston, Charles. "Jones and the Fourth Dimension." *Harper's Monthly Magazine* 29 (June 1914): 117-122.

Johnston, William H. "A criticism of 'Faith and the Fourth Dimension'." *Popular Astronomy* 16 (1908): 615-617.

Jonsson, Carl Victor. "Dirac's Wave Equation in Five-Dimensional Relativity Theory." *Arkiv för Matematik, Astronomi och Fysik* 28B (1942): 1-7.

Jonsson, Carl Victor. "Studies of Five-Dimensional Relativity Theory." *Arkiv för Fysik* 3 (1951): 87-129.

Jordan, Pasqual and C. Müller. "Field Equations with a Variable 'Constant' of Gravitation." *Zeitschrift fur Physik* 2A (January 1947).

Jordan, Pasqual. "Formation of the Stars and Development of the Universe." *Nature* 164 (15 October 1949): 637-640.

Jordan, Pasqual. "Erweitung der projecktiven Relativitätstheorie." *Annalen der Physik* 6 (1947): 219-228.

Jordan, Pasqual. "Fünfdimensionale Cosmology." *Astron. Nachr.* 276 (1948): 193.

Josephson, Brian D. and Jessica Utts. "The Paranormal: The Evidence and Its Implications for Consciousness." Online. WWW. Accessed 11-10-97. Available at <www.tcm.phy.cam.uk/~bdj10/psi/tucson.psi>; a slightly shortened version published in *Times Higher Education Supplement. Special section on Consciousness* linked to the Tucson II Conference, "Toward a Science of Consciousness." (5 April 1996): v.

Josephson, Brian D. and Fotini Pallikari-Viras. "Biological Utilisation of Quantum Non-Locality." *Foundations of Physics* 21 (1991): 197-207. Online. WWW. Accessed 4-25-97. Available at <www.phy.cam.ac.uk/www/research /mm/articles/Bell.psi>

Josephson, Brian D. "Has Psychokinesis Met Science's Measure?" *Physics Today* 45 (1992): 15. Online. WWW. Accessed 6-24-96. Available at <www.phy.cam.ac.uk/www/research/mm/articles/PK-scimeasure.html>

Josephson, Brian D. and Beverly A. Rubik. "The Challenge of Consciousness Research." *Frontier Perspectives* 3 (1992): 15-19. Online. WWW. Accessed 6-24-96. Available at <www.phy.cam.ac.uk/www/research /mm/articles/athens.html>

Josephson, Brian D. "Limits to the Universality of Quantum Mechanics." Online. WWW. Accessed 6-24-96. Available at <www.phy.cam.ac.uk/www/ research/mm/articles/QMlimits.html>

Josephson, Brian D. "Physics and the Mind." Online. WWW. Accessed 6-13-96. Available at <www.phy.cam.ac.uk/www/research/mm/articles/PM. html>

Josephson, Brian D. "The Elusivity of Nature and the Mind-Matter Problem." *The Relationship between Mind and Matter*. Ed. B. Rubik. The Center for Frontier Sciences at Temple University. Online. WWW. Accessed 6-

24-96. Available at <www.phy.cam.ac.uk/www/research/mm/articles/elusivity.html>

Jourdain, Phillip E.B. "Review of Berkeley Hastings, *Mysticism in Modern Mathematics*." *Mathematical Gazette* 5 (1911-1912): 364-366.

Kafatos, Menas and Robert Nadeau. *The Conscious Universe: Part and Whole in Modern Physical Theory*. New York: Springer, 1990.

Kaku, Michio. *Hyperspace: A Scientific Odyssey through Parallel Universes, Time Warps, and the 10th Dimension*. New York: Anchor Doubleday, 1994.

Kaku, Michio. "A theory of everything?" *Mysteries of Life and the Universe*. Ed. William Shore. Harcourt Brace Jovanovich, 1992; "Black Hole, Wormholes, and the 10th Dimension." *Sunday London Times, Literary Supplement* (1994); "What Happened before the Big Bang." *London Daily Telegraph* (1995); "Hyperspace: A Scientific Odyssey Through the 10th Dimension." *Thesis Magazine*. Online. WWW. Accessed 7-29-96. Available at <www.dorsai.org/~wbai/kaku/mk-artcl.html> Kaku, Michio. "Hyperspace: A Scientific Odyssey through the 10th Dimension." *Thesis Magazine*. Online.

WWW. Accessed 626-96. Available at <www.intr.net/bertwillco/kaku1.html> Kaku, Michio. "Black Hole, Wormholes, and the 10th Dimension." *Sunday London Times, Literary Supplement* (1994). Online. WWW. Accessed 6-26-96. Available at <www.intr.net/bertwillco/kaku3.html>

Kaku, Michio. "Frequently Asked Questions." Online. WWW. Accessed 7-8-96. Available at <www.dorsai.org/~wbai/kaku/mkfaq.html>

Kaku, Michio. Abstract of "Quarks, Symmetries, and Superstrings." Online. WWW. Accessed 7-29-96. Available at <www.dorsai.org/~wbai/kaku/mkquark.html>

Kaku, Michio. Abstract of *Introduction to Superstrings*. Online. WWW. Accessed 7-29-96. Available at <www.dorsai.org/~wbai/kaku/mk-intro.html>

Kaku, Michio. Abstract of *Beyond Einstein: The Cosmic Quest for the Theory of Everything*. Online. WWW. Accessed 7-29-96. Available at <www.dorsai.org/~wbai/ kaku/mk-einst.html>

Kaku, Michio. "Explorations in Science with Dr. Michio Kaku." Online. WWW. Accessed 7-24-96. Available at www.dorsai.org/~wbai/kaku/mk-main.html.

Kaluza, Theodor. "Zur Unitätsproblem der Physik." *Sitzungberichte der Preussischen Akademie der Wissenschaften* 54 (1921): 966-972.

Kaluza, Theodor. "Zur Relativitätstheorie." *Physikalische Zeitschrift* 25 (1924): 604-606.

Kaluza, Theodor. "Über ban Energeinhalt der Atomkerne." *Physikalische Zeitschrift* 23 (1922): 474-476.

Kane, Beverly, Jean Millay, and Dean Brown. Eds. *Silver Threads: 25 Years of Parapsychological Research*. Westport, Connecticut: Praeger, 1993.

Kapp, Reginald O. "Space." *British Journal for the History of Science* 10 (1959): 1-15.

Kasner, Edward. "Note on Einstein's Theory of Gravitation and Light." *Science* 52 (29 October 1920): 413-414.

Kasner, Edward. "Einstein's Theory of Gravitation: Determination of the Field by Light Signals." *American Journal of Mathematics* 48 (January 1921): 20-28.

Kasner, Edward. "The Einstein Solar Field and Space of Six Dimensions." *Science* 53 (11 March 1921): 238-239.

Kasner, Edward. "The Impossibility of Einstein's Fields Immersed in Flat Space of Five Dimensions." *American Journal of Mathematics* 48 (April 1921): 126-129.

Kasner, Edward. "Finite Representations of the Solar Gravitational Field in Flat Space of Six Dimensions." *American Journal of Mathematics* 48 (April 1921): 130-134.

Kasner, Edward. "Geometrical Theorems on Einstein's Cosmological Equations." *American Journal of Mathematics* 48 (1921): 217-221.

Kasner, Edward. "Einstein's Cosmological Equations." *Science* 54 (30 September 1921): 304-305.

Kasner, Edward. "The Present Problems of Geometry." Harry J. Rogers. Ed. Congress of Arts and Sciences, Universal Exposition, St. Louis, 1904, *Volume I, Philosophy and Mathematics*. Boston: Houghton & Mifflin, 1905: 59-586; Reprinted in *Bulletin of the American Mathematical Society* 11 (1905): 283-314.

Kassel, Frank. *Relativity and the Critical Philosophy*. Philadelphia: 1926.

Kaufman, S. *The Dynamic Structure of Space-Time*. Online. WWW. Accessed 7-17-96. Available at <www.execpc.com/~skaufman/ book.html> to <.../book??.html>

Kaufman, William J. *Black Holes and Warped Spacetime*. Videorecording. New York: Freeman and Company, 1992.

Killing, Wilhelm Karl Joseph, "Clifford-Klein'sche Raumformen." *Clebsch, Mathematische Annalen* 39 (1891): 257-278.

Kilmister, C.W. *General Theory of Relativity*. New York: Pergamon Press, 1973.

Klein, Felix. "Ueber die sogennante Nicht-Euklidische Geometrie." *Gauss und die Anfänge der nichteuklidischen Geometrie*. Hans Reichardt. *Teubner-Archiv zur Mathematik* 4. Leipzig: B.G. Teubner, 1985: 222-238; Reprinted from *Mathematische Annalen* 4 (1871): 573-611; Zweiter Aufsatz, *Mathematische Annalen* 6 (1873): 112-145.

Klein, Felix. "Zur nichteuklidische Geometrie." *Mathematische Annalen* 37 (1890): 544-572.

Klein, Felix. "Recent Researches in Geometry." *Bulletin of the New York Mathematical Society* 2 (1893): 215-249.

Klein, Felix. *Lectures on Mathematics. The Evanston Colloquium* delivered from 28 August to 9 September 1893 before members of the Congress of Mathematics held in connection with the World's Fair in Chicago, at Northwestern University, Evanston, Illinois. Reported by Alexander Ziwet. New York: Macmillan, 1894.

Klein, Felix. *The Mathematical Theory of the Top*. Lectures delivered on the Occasion of the Sesquicentennial Celebration of Princeton University, 12-15 October 1896. New York: Scribner's Sons, 1897.

Klein, Felix. "Riemann and His Significance for the Development of Modern Mathematics."

Klein, Felix. *Vorlesungen über die Entwicklung der Mathematik im 19. Jahrhundert*. New York: Chelsea Publishing, 1967; Reproduction of Volume I, Berlin: Springer, 1926 and Volume II, 1927, in one volume.

Klein, Felix. *Vorlesungen über Nicht-Euklidische Geometry*. New York: Chelsea Publishing. Reproduction of 1927.

Klein, Martin J. and Nandor L. Balazs, "Albert Einstein." *Dictionary of Scientific Biography*. Gillespie. Volume IV: 312-333.

Klein, Oskar. "Quantentheorie und fünfdimensionale Relativitätstheorie." *Zeitschrift fur Physik* 37 (1926): 895-906.

Klein, Oskar. "The Atomicity of Electricity as a Quantum Theory Law." *Nature* 118 (9 October 1926): 516.

Klein, Oskar. "Zur fünfdimensionale Darstellung der Relativitätstheories." *Zeitschrift fur Physik* 46 (1927): 188-208.

Klein, Oskar. "Sur L'Article de M.L. DeBroglie <<L'Univers A Cinq Dimensions et la Mécanique Ondulatoire>>." *Le Journal de Physique et Le Radium* 8 (April 1927): 242-243.

Klein, Oskar. "On the Theory of Charged Fields." *New Theories in Physics*. Paris: International Institute of Intellectual Cooperation, 1939: 77-93.

Klein, Oskar. "Meson Fields and Nuclear Interaction." *Arkiv for Matematik, Astronomi och Fysik* 34A (1947): 1-19.

Klein, Oskar. "Generalizations of Einstein's Theory of Gravitation Considered from the Point of View of Quantum Field Theory." *Helvetica Physics Acta. Supplement* IV (1956): 58-71.

Klein, Stanley A. "Is Quantum Mechanics Relevant to Understanding Consciousness." Review of *Shadows of the Mind*. Roger Penrose. Psyche 2.3 (April 1995). Online. WWW. Accessed 6-28-96. Available at <psyche.cs. monash.edu.au/volume2-1/psyche95-2-03-shadows-2klein.html>

Kline, Morris. *Mathematics in Western Culture*. New York: Oxford University Press, 1961.

Kline, Morris. *Mathematical Thought from Ancient to Modern Times*. New York: Oxford University Press, 1972.

Kline, Morris. *Mathematics: The Loss of Certainty*. New York: Oxford University Press, 1982.

Kline, Morris. *Mathematics and the Search for Knowledge*. New York: Oxford University Press, 1985.

Kock, Winston E. "Unified Field Theories." *Current Science* 6 (May 1938): 546548.

Koestler, Arthur. *The Roots of Coincidence*. London: Picador, 1974.

Koslow, Arnold. Ed. *The Changeless Order*. New York: George Braziller, 1967.

Kottler, M.J. "Alfred Russell Wallace, The Origin of Man, and Spiritualism." *Isis* 65 (1964): 145-192.

Kramer, Edna. *The Nature and Growth of Modern Mathematics*. Greenwich, Connecticut: Fawcett, 1974.

Krane, Kenneth S. *Modern Physics*. New York: John Wiley and Sons, 1983.   Krippner, Stanley and Mary Lou Carlson. Eds. *Advances in Parapsychological Research. 2. Extrasensory Perception*. New York: Plenum Press, 1978.

Krippner, Stanley and Mary Lou Carlson. Eds. *Psychoenergetic Systems: The Interaction of Consciousness, Energy and Matter*. New York: Gordon and Breach Science Publishers, 1979.

Kron, G. "Invariant Form of the Maxwell-Lorentz Field Equations for Accelerated Systems." *Journal of Applied Physics* 9 (March 1938): 196-208.

Kurtz, Paul. Ed. A Skeptic's *Handbook of Parapsychology*. Buffalo: Prometheus Books, 1985.

Lanczos, Cornelius. *Space through the Ages: The Evolution of Geometrical Ideas from Pythagoras to Hilbert and Einstein*. New York: Academic Press, 1970.

Landfeld, Charles S. "Science and Spiritism in the Psychical Research of Oliver Lodge." 21-27.

Lao-Tsu, Tao Te Ching. Gia Fu-Feng and Jane English. Trans. New York: Vintage Books, 1972.

Lawrence, Mike and Sue Lawrence. "Ring Theory - A short outline." Online. WWW. Accessed 6-13-96. Available at <ourworld.compuserve. com/homepages/Mike-and-SueLawrence/a01hp011. htm>

Lawrie, Ian D. *A Unified Grand Tour of Theoretical Physics*. New York: Adam Hilger, 1990.

LeClair, Robert C. Ed. *The Letters of William James and Theodore Flournoy*. Madison: University of Wisconsin Press, 1966.

Leibniz. *Discourse on Metaphysics/Correspondence with Arnauld/Monadology*. LaSalle: Open Court, 1973. "Reply to the Thoughts of the System of Pre-Established Harmony contained in the Second Edition of Mr. Bayle's Critical Dictionary, Article Rorarius 1702." *Philosophical Papers and Letters*, Leroy Loemker. Ed. Dordrecht-Holland: D. Reidel, 1969.

LeShan, Lawrence. *Toward a General Theory of Paranormal: A Report of Work in Progress*. Monograph 9. New York: Parapsychological Foundation, 1969.

LeShan, Lawrence. *The Medium, the Mystic, and the Physicist: Toward a General Theory of the Paranormal*. New York: Ballantine, 1975.

Levere, Trevor H. *Affinity and Matter*. Oxford: Clarendon Press, 1971. Levi-Civita, Tulio. *A Simplified Presentation of Einstein's Unified Field Equations*. Trans. John Dougall. London: Blackie and Son, n.d.

Levi-Civita, Tulio. "A Proposed Modification of Einstein's Field Theory." *Nature* 123 (1929): 678-679.

Levi-Civita, Tulio and M.M.G. Ricci. "Methods de calcul différential absolu et leurs applications." Mathematische Annalen 54 (1901): 125-201: Tensor Analysis Paper: Translation, Comments and Added Material. Trans. Robert Hermann. Brookline, Massachusetts: Math-Sci Press, 1975.

Levi-Civita, Tulio. "Nozione di parallelismo in una varietà qualunque e consequente specificazione geometrica nella curvatura riemanniana." *Rendiconti del Circolo matematico di Palermo* 42 (1917): 173-215; Reprinted in *Opere Mathamatische* 4 (1917-1928): 1-39.

Levi-Civita, Tulio. *The Absolute Differential Calculus*. Ed. Enrico Persico. Trans. Miss M. Long. London: Blackie and Sons, 1927; Revision of 1923 Italian original.

Lewis, Samuel. "Relativity and Epistemology: The meaning of terms used in relativity equations." Online. WWW. 03-22-97. Accessed 06-02-97. Available at <www.sisna.com/users/lewisk/reality.html >

Lichnerowitz, André. *Théories Relativiste de La Gravitation et de L'Éectromagnétisme*. Paris: Masson, 1954: 180-214.

Liebmann, Heinrich. *Nichteuklidische Geometrie*. Berlin: Walter de Gruyter, 1923.

Lipschitz, R. "Extension of the Planet-Problem to a Space of n-Dimensions and Constant Integral Curvature." Trans. Arthur Cayley. *Quarterly Journal of Mathematics* 12 (1873): 349-370.

Lobatchewski, Nicolai. [Lobachevski]. *Geometrical Researches on the Theory of Parallels*. Original 1840 paper. Trans. G.B. Halsted. Intro. G.B. Halsted. Chicago: Open Court, 1914; originally published 1891; Reprinted as a supplement *Non-Euclidean Geometry*. Roberto Bonola.

Lodge, Oliver. "The New Theory of Gravity." *The Nineteenth Century* 86 (December 1919): 1189-1201.

Lovejoy, Arthur. *The Great Chain of Being: The Study of the History of an Idea*. New York: Harper & Row, 1960.

Lucadou, W. von and K. Kornwachs. "Can Quantum Theory Explain Paranormal Phenomena?" *Research* 1976. Morris, Roll and Morris: 187-191.

Ludwig, Jan. *Philosophy and Parapsychology*. Buffalo: Prometheus Press, 1978.

Ludwig, Kirk. "Why the Difference between Quantum and Classical Physics is Irrelevant to the Mind/Body Problem." *Psyche* 2 (September 1995). Online. WWW. 10-11-95. Accessed 0617-97. Available at <psyche.cs.monash .edu.au/v2/psyche-2-16ludwig.html>

Lyon, Nathaniel. "The Paranormal - Fact or Fiction?" Online. WWW. Accessed 2 July 1997. Available at <www.hernebay.demon.co.uk/>

Mach, Ernst. *Die Geschichte und die Wurzel des Satzes von der Erhaltung der Arbeit*. Prague: Beyer, 1872.

Mach, Ernst. *Die Mechanik in ihrer Entwicklung: HistorischKritisch Dargestellt*. Leipzig: Brockhaus, 1897; Reprint of 1883; *The Science of Mechanics*. Trans. Thomas J. McCormack. Intro. Karl Menger. LaSalle: Open Court, 1974; Reprint of the 6th American edition.

Mach, Ernst. *Popular Scientific Lectures*. Trans. Thomas J. McCormack. LaSalle: Open Court, 1943; Reprint of the 3rd edition of 1898.

Mach, Ernst. "On physiological, as distinguished from geometrical space." *The Monist* 11 (1901): 321-338; Reprinted in *Space and Geometry*: 5-37.

Mach, Ernst. "On the psychology and natural development of geometry." Trans. Thomas J. McCormack. *The Monist* 12 (1902): 481-515; Reprinted in *Space and Geometry*: 38-93.

Mach, Ernst. "Space and geometry from the point of view of physical inquiry." *The Monist* 14 (1903): 1-32; Reprinted in *Space and Geometry*: 99-143.

Mach, Ernst. *Space and Geometry in the Light of Physiological, Psychological and Physical Inquiry*. Trans. Thomas J. McCormack. LaSalle: Open Court, 1943; Reprint of 1906.

Mach, Ernst. *Analysis of Sensations*. Trans. C.M. Williams. Revised. Sydney Waterlow. New York: Dover, 1959; Reprint of the 5th edition of 1906.

Maiello, F. "Quantum Theory and Metaphysics." Online. WWW. Accessed 5-12-97. Available at <ddi.digitasl.net/~egodust/ fmpageq.html>

Maiello, F. "The Theoretical impossibility of the Existence of Matter." Online. WWW. Accessed 05-12-97. Available at <ddi.digital.net/~egodust/ fmpagezm.html>

Mandel, Heinrich. "Zur Axiomatik der fünfdimensionalen Relativitätstheorie." *Zeitschrift fur Physik* 54 (1929): 564-566.

Mandel, Heinrich. "Zur tensoriellen Form der wellenmechanischen Gleichungen des Elektrons." *Zeitschrift fur Physik* 54 (1929): 567-570.

Mandel, Heinrich. "Zur Herleitung der Feldgleichungen in der allgemeinen Relativitätstheorie." *Zeitschrift fur Physik* 52 (1927): 285-306.

Mandel, Heinrich. "Connection between Einstein's Theory of Distant Parallelism and Five-Dimensional Field Theory." *Zeitschrift fur Physik* 56 (1929): 838-844.

Manning, Henry P. *Non-Euclidean Geometry*. Boston: Ginn, 1901.

Manning, Henry P. Ed. *The Fourth Dimension Simply Explained*. London: Methuen, 1921; Reprint of New York: Munn, 1910.

Manning, Henry P. *Geometry of Four Dimensions*. New York: Macmillan, 1914.

Manning, Henry P. "Geometry of Four Dimensions." *American Mathematical Monthly* 25 (1918): 316-320.

Mansfield, Victor. "Distinguishing Synchronicity from the Paranormal: An Essay Honoring Marie-Louise von Franz." Online. WWW. 06-96. Accessed 06-02-97. Available at <lightlink.com/vic/mlv-paper.html>

Mansfield, Victor. "Time in Madhyamika Buddhism and Modern Physics." Online. WWW. 02-96. Accessed 06-02-97. Available at <lightlink.com/vic/astrol.html>

Mansfield, Victor. "An Astrophysicists' Sympathetic and Critical View of Astrology." A plenary talk presented at the Cycles and Symbols Conference, San Francisco, 14-16 February 1997. Online. WWW. 07-96. Accessed 06-02-97. Available at <lightlink.com/vic/mlv-paper.html>

Mansfield, Victor, Sally Rhine-Feather and James Hall. "The Rhine-Jung Letters: Distinguishing Parapsychology from Synchronicity" Online. WWW. 05-97. Accessed 06-02-97. Available at <lightlink.com/vic/rhine.html>

Mansfield, Victor and J. Marvin Spiegelman. "On the Physics and Psychology of the Transference as an Interactive Field." Online. WWW. 03-12-96. Accessed 06-17-97. Available at <www.lightlink. com/vic/field.html>

Marchant, James. A.R. *Wallace: Letters and Reminiscences*. New York: Arno, 1975.

Margenau, Henry. "ESP in the Framework of Modern Science." *Science and ESP*. Smythies: 209-224.

Margenau, Henry. "ESP and Modern Physics." *Encounters with Parapsychology*. McConnell: 119-128.

Marshall, Ninian. "ESP and Memory: A Physical Theory." *British Journal for the Philosophy of Science* 10 (1960): 265-286.

Mathews, George Ballard. [G.B.M.]. "The New Physics." *Nature* 100 (25 October 1917): 155.

Mathews, George Ballard. [G.B.M.]. "Non-Euclidean Geometry." *Nature* 86 (6 April 1911): 192-193.

Mattuck, Richard D. "Random Fluctuation Theory of Psychokinesis: Thermal Noise Model." *Research 1976*. Morris, Roll and Morris: 191-195.

Mattuck, Richard D. and Evan Harris Walker. "The Action of Consciousness on Matter: A Quantum Mechanical Theory of Psychokinesis." *The Iceland Papers*. Puharich: 111-150.

Maudlin, Tim. "Between the Motion and the Act ...: A Review of *Shadows of the Mind* by Roger Penrose." *Psyche* 2 (April 1995): Online. WWW. Accessed 04-20-98. Available at <psyche.cs.monash.edu.au/v2/psyche-2-02maudlin.html>

Mauthner, Fritz. *Wörterbuch der Philosophie*. Leipzig: Felix Meiner, 1924. Maxwell, James Clerk. Letter to C.J. Monro, 15 March 1871; partially reprinted in Lewis Campbell and William Garnett, *The Life of James Clerk Maxwell, with selections from his correspondence and occasional writings*. London: Macmillan, 1884: 290.

Maxwell, James Clerk. Notecard to Peter G. Tait, 11 November 1874, Add. 7655/Ig/72. Maxwell Papers, University Library, Cambridge, England. Maxwell, James Clerk. Letter to Peter G. Tait, 9 November 1876, Add.7655/#38. Maxwell Papers, University Library, Cambridge, England.

May, Edwin C., Jessica M. Utts and S. James Spottiswoode. "Decision Augmentation Theory: Towards a Model of Anomalous Mental Phenomena." *Journal of Parapsychology* (September 1995).

May, Edwin C., Jessica M. Utts and S. James Spottiswoode. Abstract of "Decision Augmentation Theory: Towards a Model of Anomalous Mental Phenomena." Online. WWW. Accessed 04-1098. Available at <www.mindspring.com/~rhine/jpab9509.html>

May, Edwin C. "The American Institutes for Research Review of the Department of Defense's STAR GATE Program: A Commentary." *The Journal of Parapsychology* 60 (March 1996): 323; Also, online. WWW. Accessed 04-10-98. Available at <uqbar.jsasoc.com/~csl/aircom.html>

May, Edwin C. "Interview with Dr. Edwin C. May of the Cognitive Sciences Laboratory." *RPKP* (12 May 1996). M.R. Watkins. Online. WWW. 6-24-96. Accessed 6-30-96. Available at <alathea.ukc.ac.uk/Dept/CPRS/RPKP/may>

McAulay, Alexander. "Quaternions." *Nature* 47 (15 December 1892): 151. McAulay, Alexander. *Utility of Quaternions in Physics*. London: Macmillan, 1893.

McAulay, Alexander. "Octonions [An Abstract]." Presented 12 December 1895. *Proceedings of the Royal Society of London* 59 (1896): 169-181.

McAulay, Alexander. *Octonions: A Development of Clifford's Bi-Quaternions*. Cambridge: at the University Press, 1898.

McAulay, Alexander. *Algebra after Hamilton, or Multenions*. Edinburgh: 1908.

McAulay, Alexander. "Quaternions, Supplementary Considerations." *Encyclopaedia Britannica*. 11th edition. 22: 722-723.

McAulay, Alexander. "Quaternions Applied to Physics in Non-Euclidean Space." *Papers and Proceedings of the Royal Society of Tasmania* (1914): 119.

McAulay, Alexander. "Inertial Frame given by a Hyperbolic Space-Time." *Philosophical Magazine* 41 (1921): 141-143.

McConnell, R.A. Editor. *Encounters with Parapsychology*. Pittsburgh: R.A. McConnell, 1982.

McCullough, Daryl. "Can Humans Escape Gödel: A Review of Shadows of the Mind by Roger Penrose." *Psyche* 2 (June 1995). Online. WWW. Accessed 0420-98. Available at <psyche.cs.monash.edu.au/v2/psyche-2-04mccullough.html>

McDermott, Drew. "[STAR] Penrose is Wrong." Review of *Shadows of the Mind*. Roger Penrose. *Psyche* 2.17 (October 1995). Online. WWW. Accessed 7-8-96. Available at <psyche.cs.monash.edu.au/volume2-1/psyche 95-2-17shadows-9mcdermott.html>

Medhurst, R.G. and K.M. Goldney. *Crookes and the Spirit World*. New York: Taplinger, 1972.

Mehra, Jagdish. Ed. *The Physicist's Conception of Nature*. Dordrecht-Holland: D. Reidel, 1973.

Mehra, Jagdish. *Einstein, Hilbert and the Theory of Gravitation*. Boston and Dordrecht-Holland: D. Reidel, 1974.

Meksyn, D. "Unified Field Theory, Part I: Electromagnetic Field." *Philosophical Magazine* 17 (1934): 99-112

Meksyn, D. "A Unified Field Theory. Part II: Gravitation." *Philosophical Magazine* 17 (February 1934): 476-482.

Meksyn, D. "Neutrons." *Nature* 131: 366.

Merz, John Theodore. *A History of European Scientific Thought in the Nineteenth Century*. New York: Dover, 1965; An unabridged and unaltered republication of Edinburgh and London: Blackwood, and Sons, 1904-1912;

originally published as *A History of European Thought in the Nineteenth Century.*

Merzbacher, Eugen. *Quantum Mechanics*. New York: John Wiley and Sons, 1960.

Michal, A.D. and J.L. Botsford. "Extension of the New Einstein Geometry." *National Academy of Sciences, Proceedings* 18 (August 1932): 554-562.

Miller, Arthur I. "The Myth of Gauss' Experiment on the Euclidean Nature of Space." *Isis* 63 (1972): 345-348.

Miller, Arthur I. "Reply to Comments on Miller's 'The Myth of Gauss' Experiments on the Euclidean Nature of Space'." *Isis* 65 (March 1974): 86-87.

Minkowski, Hermann. "Raum und Zeit." Address delivered before the 80th Assembly of the German Natural Scientists and Physicians, Cologne, 21 September 1908. *Physikalische Zeitschrift* 10 (1909): 104-111; *Jahresbericht der Deutschen Mathematiker Vereinigung* 18 (1909): 75-88.

Mishlove, Jeffrey. *The Roots of Consciousness: Psychic Liberation through History, Science and Experience*. New York: Random House; And San Francisco: Bookworks, 1975.

Mishlove, Jeffrey. *The Roots of Consciousness: The Classic Encyclopedia of Consciousness Studies Revised and Expanded*. Tulsa: Council Oak Books, 1993. Mishlove, Jeffrey. "Intuition: A link between Psi and Spirituality." Online. WWW. 02-01-97. Accessed 05-09-97. Available at <www.intuition. org/revision.htm>

Misner, Charles, Kip S. Thorne, and John A. Wheeler. *Gravitation*. San Francisco: W.H. Freeman, 1973.

Mitchell, Edgar D. *Psychic Exploration: A Challenge for Science*. John White. Ed. New York: Putnam's Sons, 1974.

Monro, C.J. "On the Simplest Continuous Manifoldness of Two Dimensions and of Finite Extent." *Nature* 17 (26 April 1877): 547; A second note appeared under the same title in *Nature* 22 (3 July 1880): 218.

Monro, C.J. "On Flexure of Spaces." Read 13 January 1878, with a comment by A. Cayley. *Proceedings of the London Mathematical Society* 9 (November 1977-November 1878): 171-176.

Monro, C.J. "Inside Out." *Nature* 18 (30 May 1878): 115.

Monro, C.J. Letter to James Clerk Maxwell, 10 September 1871, Acc.#1063, #2109. The Monro Collection. Greater London Record Office, London, England.

Moravec, Hans. "Roger Penrose's Gravitronic Brains." Review of *Shadows of the Mind* by Roger Penrose. *Psyche* 2 (May 1995). Online. WWW. Accessed 78-96. Available at <psyche.cs.monash.edu.au/volume2-1/psyche95-2-06shadows-4-moravec.html>

Morinaga, K. "Geometrical Interpretations of Wave Geometry." *Hiroshima Journal of Science* 7 (March 1937): 173-177.

Morinaga, K. "General Parallel Displacement which makes ds = 0 Invariant." *Hiroshima Journal of Science* 7 (March 1937): 169-172.

Morris, J.D., W.G. Roll and R.L. Morris. Eds. *Research in Parapsychology, 1976*. London: Scarecrow Press, 1977.

Mosteller, Frederick. "Comment*." Statistical Science* 6 (1991); Online. WWW. Accessed 02-12-98. Available at <www.stat.ucdavis.edu/users/utts/91rmpc7.html>

Mukergee, Madhusree. "Explaining Everything." *Scientific American* 274 (January 1996): 88-94.

Mulhauser, Gregory R. "On the End of a Quantum Mechanical Romance." *Psyche* 2 (November 1995). Online. WWW. Accessed 6-28-96. Available at <psyche.cs.monash.edu.au/volume2-1/psyche95-2-19-decoherence-1mulhauser.html>

Mundle, C.W.K. "The Explanation of ESP." *Science and ESP*. Smythies: 197-208. Musès, Charles, and Arthur M. Young. *Consciousness and Reality: The Human Pivot Point*. New York: Outerbridge and Lazard, 1972.

Nash, Carroll B. *Science of Psi: ESP and PK*. Springfield, Illinois: Charles T. Thomas, 1978.

Nash, Carroll B. *Parapsychology: The Science of Psiology*. Springfield, Illinois: Charles Thomas, 1986.

Needleman, Joseph. *A Sense of the Cosmos: The encounter of modern science and ancient truth*. Garden City, New York: Doubleday, 1975.

Nelson, Buddy. "Rethinking Newton's Laws: Lockheed Scientist and Colleagues Propose New Theory of Matter." Online. WWW. 03-02-94.

Accessed 05-28-97. Available at <www.lmsc.lockheed.com/newsbureau/pressreleases/1994/9405.html>

Nelson, Godfrey K. *Spiritualism and Society*. New York: Schocken, 1969.

Neville, E.H. *The Fourth Dimension*. Cambridge: At the University Press, 1921. Newcomb, Simon. "Elementary theorems relating to the geometry of a space of three dimensions and of uniform positive curvature in the fourth dimension." *Crelle's Journal für die reine und angewandte Mathematik* 83 (1877): 293-299.

Newcomb, Simon. "Note on a Class of Transformations Which Surfaces May Undergo in Space of More than Three Dimensions." *American Journal of Mathematics* 1 (1878): 1-4.

Newcomb, Simon. "Fundamental Definitions and Properties of Geometry." *Nature* 21 (January 1880): 293-295.

Newcomb, Simon. *Popular Astronomy*. London: Macmillan, 1878: *Populäre Astronomie*. Trans. Rudolf Engelmann. Leipzig: Engelmann, 1881.

Newcomb, Simon. "Speculative Science." *International Review* 12 (April 1884): 334-341.

Newcomb, Simon. "On the Fundamental Concepts of Physics." Abstract in the *Bulletin of the Washington Philosophical Society* 11 (1888/1891): 514-515; A roughly typed copy of the complete talk can be found in Box #94, Newcomb Papers, Library of Congress.

Newcomb, Simon. "Modern Mathematical Thought." *Nature* 49 (1 February 1894): 325-329.

Newcomb, Simon. *Program of the Fourth Summer Meeting of the American Mathematical Society*, August 1897. Box #143, Folder #1. Newcomb Papers, Library of Congress.

Newcomb, Simon. "The Scientific Idea." Rough notes, New Haven (17 November 1898). Newcomb Papers, Library of Congress.

Newcomb, Simon. "The Philosophy of Hyperspace." *Bulletin of the American Mathematical Society* 4 (February 1898): 187-195.

Newcomb, Simon. "Is the Airship Coming?" *McClure's Magazine* 17 (September 1901): 432-435.

Newcomb, Simon. "The Fairyland of Geometry." *Harper's Monthly Magazine* 104 (January 1902): 249-252.

Newcomb, Simon. *Sidelights on Astronomy and Kindred Fields of Popular Science*. New York: Harper, 1906.

Newman, Doug. "Consciousness Reference List." Online. WWW. 09-28-95. Accessed 06-17-97. Available at <www.maths.soton.ac.uk/staff/Dean/consref/consref.html>

Newman, James R. "William Kingdon Clifford." Scientific American 188 (February 1953): 78-84.

Newman, James R. Ed. "Commentary on William Kingdon Clifford." *The World of Mathematics*. London: Allen & Unwin, 1960: 546-547.

Newton, Isaac. *Mathematical Principles of Natural Philosophy* (*Principia*). Florian Cajori. Trans. Berkeley: University Of California Press, 1934.

Nordberg, J.T. "The solution to the question, What is time?" Online. WWW. Accessed 7-31-96. Available at <home.earthlink.net/~jtnordberg>

Nordström, Gunnar. "Über die Möglichkeit, das elektromagnetische Feld und das Gravitationsfeld zu vereinigen." *Physickalische Zeitschrift* 15 (15 May 1914): 504-506.

North, John D. *The Measure of the Universe: A History of Modern Cosmology*. Oxford: Clarendon Press, 1965.

North, John D. "William Kingdon Clifford." *Dictionary of Scientific Biography* 3: 322-323.

Novobatzky, K. "Universal Field Theory." *Zeitschrift fur Physik* 89 (2 June 1934): 373-387.

Nye, Carl. "Einstein's Special Theory of Relativity." Online. WWW. Accessed 7-31-96. Available at <www.mistral.co.uk/einstein/>

Odenwald, Sten. "Hyperspace in Science Fiction." Online. WWW. 1995. Accessed 04-17-97. Available at <www2.ari.net/home/Odenwald/anthol/scifi2.htm>

Odenwald, Sten. "A scientific journey into Hyperspace!" Online. WWW. 1995. Accessed 04-17-97. Available at <www2.ari.net/home/Odenwald/anthol/hypspace.htm>

Oliver, J.E. "Review of W.W.R. Ball's Mathematical Recreations." *Bulletin of the New York Mathematical Society* 2 (December 1892): 42-46.

O'Neill, Terry and Stacey L. Tipp. Eds. *Paranormal Phenomena: Opposing Viewpoints*. San Diego: Greenhaven Press, 1991.

Oppenheim, Janet. *Other Worlds: Spiritualism and Psychical Research in England, 1850-1914*. New York: Cambridge University Press, 1988.

O'Regan, Bernard. "The Emergence of Paraphysics: Theoretical Foundations." *Psychic Exploration*. Mitchell: 450-467.

Ornstein, Robert E. *The Psychology of Consciousness*. 2nd edition. New York:

Harcourt, Brace, Jovanovich, 1977; the 1st first edition was published in 1972. Oss, Rosene Van. "D'Alembert and the Fourth Dimension." *Historia Mathematica* 10 (1983): 455-457.

Oteri, Laura. Ed. *Quantum Physics and Parapsychology*. Proceedings of an International Conference held in Geneva, Switzerland, 26-27 August 1974. New York: Parapsychology Foundation, 1974.

Ouspensky, P.D. *Tertium Organum, The Third Cannon of Thought: A Key to the Enigmas of the World*. New York: Vintage Books, 1970; Reprint of original Manas Press, 1920.

Pacotte, J. "Potential Vector in a Field of Five Dimensions." *Comptes Rendus* 186 (6 February 1928): 362-364.

Pais, Abraham. *'Subtle is the Lord...' The Science and the Life of Albert Einstein*. Oxford: Oxford University Press, 1982.

Pais, Abraham. "The Energy-Momentum Tensor in Projective Relativity." *Physica's Grav.* 8 (December 1941): 1137-1160.

Palfreman, Jon. "William Crookes: Spiritualism and Science." *Ethics in Science and Medicine* 3 (1976): 211-227.

Palmer, John. "Extrasensory Perception: Research Findings." *Extrasensory Perception*. Krippner: 59-244.

Parker, Barry. *Einstein's Dream: The Search for a Unified Theory of the Universe*. New York: Plenum Press, 1986.

Paton, J. "Unifying field theory." Essay review of M. Le Bellac's *Quantum and Statistical Field Theory*. *Contemporary Physics* 33 (1992): 191-195.

Patrizi, Francesco. "On Physical Space." Benjamin Brickman. Ed. *The Journal of the History of Ideas*, 4 (1943): 224-245.

Pauli, Wolfgang. "Über die Formuliering der Naturgesetze mit fünf homogen Koordinäten, I. Klassische Theorie." *Annalen der Physik* 18 (1933): 305-366.

Pauli, Wolfgang and J. Solomon. "Dirac's Equations and the Unitary Theory of Einstein and Mayer." *Journal de Physique et Le Radium* 3 (October 1932): 452-463.

Pauli, Wolfgang and J. Solomon. "Unitary Field Theory and Dirac's Equations." *Journal de Physique et Le Radium* 3 (December 1932): 582-589.

Pauli, Wolfgang and S. Kusaka. *Physical Review* 63 (1943): 400.

Pauli, Wolfgang. *Theory of Relativity*. Trans. G. Field. New York: Dover, 1981; Originally "Relativitätstheorie." *Encyklopädie de mathematischen Wissenschaften* 19. Leipzig: B.G. Teubner, 1921.

Pearson, Karl. Letters to Robert J. Parker. 1879 and 4 June 1885. File #922, Pearson Papers. University College, London.

Pearson, Karl. "On the Distortion of a Solid Elastic Sphere." *Quarterly Journal of Pure and Applied Mathematics* 10 (1879): 375-381.

Pearson, Karl. "On the Motion of Spherical and Ellipsoidal Bodies in Fluid Media." *Quarterly Journal of Pure and Applied Mathematics* 20 (1883): 60-80; Part II: 184-211.

Pearson, Karl. "Note on Twists in an Infinite Elastic Solid." *Messenger of Mathematics* 13 (1883): 79-95.

Pearson, Karl. "On a certain Atomic Hypothesis." Read 2 February 1885. *Transactions of the Cambridge Philosophical Society* 14 (1883-1889): 71-120. Pearson, Karl. "Clifford's Common Sense of the Exact Sciences." *Nature* 32 (2 July 1885): 96.

Pearson, Karl. "On plane waves of the third order in an isotropic elastic medium, with special reference to certain optical phenomena." Read 25 May 1885. *Proceedings of the Cambridge Philosophical Society* 5 (1883-1886): 296-309.

Pearson, Karl. "The Elements of Dynamic- Part 1. Kinematic, Book IV, and Appendix, by W.K. Clifford." *The Athenaeum* 3116 (16 July 1887): 86-87.

Pearson, Karl. "Matter and Soul." Originally delivered before the Sunday Lecture Society at St. George's Hall, 6 December 1885. *The Ethic of*

Freethought and other Addresses and Essays, 2nd edition. London: Adam and Charles Black, 1901; First edition published in 1888: 21-44.

Pearson, Karl. "The Prostitution of Science." *The Ethic of Freethought*: 44-65. Pearson, Karl. "On a certain Atomic Hypothesis." *Proceedings of the London Mathematical Society* 20 (8 November 1888): 38-63.

Pearson, Karl. "On the Generalised Equations of Elasticity and their Application to the Wave Theory of Light." *Proceedings of the London Mathematical Society* 20 (9 May 1889): 297-350.

Pearson, Karl. "On the Scope and Method of Modern Science." Pearson Papers. University College, London.

Pearson, Karl. "Ether Squirts: Being an attempt to specialize the form of ether Motion which forms an Atom in a Theory propounded in former papers." *American Journal of Mathematics* 13 (1891): 309-362.

Pearson, Karl. Letter to Lord Kelvin "On the Nature of Physical Space." c.1891. File #104. Pearson Papers, University College, London.

Pearson, Karl. *The Grammar of Science.* London: Walter Scott, 1892; 3rd edition. London: Adam and Charles Black, 1911; New York: Meridian Books, 1961 (Reprint of 3rd edition of 1911).

Pearson, Karl. Letter to Robert J. Parker, dated 6-4-85, file #922. Pearson Papers. University College, London.

Peirce, C.S. "The Charles Peirce-Simon Newcomb Correspondence." Carolyn Eisele. Ed. *Proceedings of the American Philosophical Society* 101 (October 1957): 409-433.

Peirce, Charles S. "A review of W.K. Clifford's The Common Sense of the Exact Sciences." *The Nation* 41 (3 September 1885): 203.

Peirce, Charles S. Letters to Simon Newcomb. See Carolyn Eisele.

Peirce, Charles S. "The Architecture of Theories." *The Monist* 1 (January 1891): 161-176.

Peirce, Charles S. *Collected Papers of Charles Sanders Peirce*. Eds. Charles Hartshore and Paul Weiss (Volumes 1-6), and Arthur W. Burks (Volumes 7-8). Cambridge: Harvard University Press, 1935/1958.

Peirce, Charles S. *The New Elements of Mathematics*. Ed. Carolyn Eisele. New Jersey: Humanities, 1976.

Peirce, Charles S. "Letters to Francis Russell." 1 January 1909 (L387): 975-978; 15 April 1909: 979-982; 24 April 1909: 982-984; 5 June 1909: 984-987. *New Elements* 3.

Penrose, Roger. *The Emperor's New Mind: Concerning Computers, Minds, and the Laws of Physics*. New York: Penguin, 1991.

Penrose, Roger. "Beyond the Doubting of a Shadow." Reply to Commentaries on Shadows of the Mind. *Psyche* 2.23 (January 1996). Online. WWW. Accessed 7-10-96. Available at <psyche.cs.monash.edu.au/volume2-1/psyche96-2-23-shadows-10penrose.html>

Perkins, James S. *A Geometry of Space-Consciousness*. 2nd edition. Adyar, Madras: Theosophical Publishing House, 1973.

Persinger, Michael A. "ELF Field Mediation in Spontaneous Psi Events: Direct Information Transfer or Conditioned Elicitation?" *Mind at Large*. Tart, Puthoff and Targ: 191-204.

Phin, John. "The fourth dimension and the possibility of a new sense and sense-organ." *The Seven Follies of Science*. London: Archibald Constable, 1906: 117-125.

Piaggio, H.T.H. "Review of J. Becquerel, Le Principe de Relativité et la Théorie de la Gravitation." Gauthier-Villars, 1922. *The Mathematical Gazette* 11 (1921-1922): 174-177.

Piaggio, H.T.H. "Review of J.P. Dalton, The Rudiments of Relativity." Council of Education, Witwatersrand, Johannesburg, 1921. *The Mathematical Gazette* 11 (1921-1922): 174-177.

Piaggio, H.T.H. "Review of W.H.V. Reade, *A Criticism of Einstein and His Problem.*" Oxford: Blackwell, 1922. The *Mathematical Gazette* 11 (1921-1922): 174-177.

Piaggio, H.T.H. "Review of Hermann Weyl, *Space-Time-Matter*." Trans. H.L. Brose. Methuen, 1922. *The Mathematical Review* 11 (1921-1922): 174-177.

Piaggio, H.T.H. "The Concept of Space: account of a lecture by Albert Einstein." Online. WWW. Accessed 08-07-96. Available at <ourworld. compuserve. com/homepages/eric-baird/ae-cspace. htm>

Piel, Carl. "Die Clifford'schen Parallelen und die Clifford'sche Flache." *Mathematisch-Physikalische Semesterberichte* 1.1/2 (1954): 99-108.

Pierre, John M. "Superstrings! Home Page." Online. WWW. 0726-97. Accessed 09-17-97. Available at <www.physics.ucsb.edu/~jpierre/strings/index.html >

Phillips, David T. "Physics and Parapsychology." Posted at Parapsychology Research Forum on 5 June 1996. Online. WWW. Accessed 07-01-96. Available at <www.roma1.infn.it/rog/group/frasca/b/phillips1.html>

Podmore, Frank. *Studies in Psychical Research*. New York: Putnam's Sons, 1897.

Podmore, Frank. *Mediums of the 19th Century*. Volume 1. New Hyde Park, New York: University Books, 1963; Reprint of *Modern Spiritualism*, 1902.

Podolanski, J. "Unified Field Theory in Six Dimensions." *Proceedings of the Royal Society* 201 (22 March 1950): 234-260.

Poincaré, Henri. "Non-Euclidean Geometry." Trans. W.J.L. *Nature* 45 (25 February 1892): 404-407.

Poincaré, Henri. "The Relations of Analysis and Mathematical Physics." *The Monist* 9 (1898): 247-255.

Poincaré, Henri. "On the Foundations of Geometry." The Monist 9 (1898): 143.

Poincaré, Henri. *Science and Hypothesis*. Trans. W.J. Greenstreet. Intro. J. Larmor. New York: Dover, 1952; Reprint of 1905 edition; Also Trans. G.B. Halsted. Intro. Josiah Royce. New York: Science Press, 1906; this is a translation, translator unknown, of Poincaré's *La Science et l'Hypothèse*. Paris: Ernest Flammarion, 1902.

Poincaré, Henri. *The Value of Science*. Trans. G.B. Halsted. New York: Dover, 1958; Reprint of New York: Science Press, 1906.

Poincaré, Henri. *Science and Method*. Trans. Francis Maitland. New York: Dover, n.d.; Reprint of 1908.

Poincaré, Henri. "Mathematical Creation." *The Monist* 20 (1910): 321-335.

Pollock, Dave. "A top-down experiential unification basis for science, spirit and self: Abstract." Online. WWW. 09-14-96. Accessed 05-30-97. Available at <www.snowcrest.net/davep/ index.html>

Pollock, Frederick. "William Kingdon Clifford." *The Fortnightly Review* 25 (1879): 667-687; Revised and published as the introduction to Clifford. *Lectures and Essays*. Volume 1: 1-43.

Popper, Karl R. *Quantum Theory and the Schism in Physics*. Totowa, New Jersey: Rowman and Littlefield, 1982.

Power, Edwin Albert. "Exeter's Mathematician - W.K. Clifford, F.R.S. 1845-79." *Advancement of Science* 26 (March 1970): 318-328.

Power, Edwin Albert. "The Application of Quantum Electrodynamics to Molecular Forces." Copy obtained from Power without references: 369-379.

Power, Edwin Albert. "The insights of John Wheeler." Essay review of John Wheeler, *At Home in the Universe*. *Contemporary Physics* 35, 1994: 217-219. Pratt, J.G. "Some Notes for the Future Einstein of Parapsychology." *Journal of the ASPR* 68 (April 1964): 133-155; Reprinted in *Encounters with Parapsychology*. McConnell: 129-144.

Pratt, J.G. "Book Review of Edgar Mitchell, Psychic Explorations." *Journal of Parapsychology* 69 (March 1975): 67-73.

Pratt, J.G. "Soviet Research in Parapsychology." *Handbook*. Wolman: 883-903.

Preble, W.P. "Four-Fold Space." *Science* 19 (27 May 1892): 304-305.

Pribram, Karl H. Editor. *Learning as Self-Organization*. Mahwah, New Jersey: Lawrence Erlbaum Associates, 1996.

Price, H.H. "Psychical Research and Human Personality." *Science and ESP*. Smythies: 33-46.

Puharich, Andrija. *Beyond Telepathy*. Garden City, New York: Anchor Press/Doubleday, 1973.

Puharich, Andrija. *The Iceland Papers: Select Papers on Experimental and Theoretical Research on the Physics of Consciousness*. Amherst, Wisconsin: Essentia Research Associates, 1979.

Puthoff, Harold. "A new Rosetta Stone in physics?" Online. WWW. Accessed 04-04-97. Available at <www.livelinks.com/sumeria/free/zpe1.html>

Puthoff, Harold. "CIA-Initiated Remote Viewing at Stanford Research Institute." Online. WWW. Accessed 07-24-97. Available at <www.mindspring.com/~biomind/Pages/CIA-InitiatedRV.html>

Puthoff, Harold and Russell Targ. "Psychic Research and Modern Physics." *Psychic Exploration*. Mitchell: 524-542.

Radin, Dean. "Frequently Asked Questions about Parapsychology." Online. WWW. 12-17-95. Accessed 6-13-96. Available at <eeyore.lvhrc. nevada.edu/~cogno/para1.html>; Part 2. <.../para2.html>; Part 3. <.../para3. html>

Radin, Dean. "Psi hits and myths." 1993 Presidential Address to the Parapsychological Association. Online. WWW. Accessed 1108-96. Available at <eeyore.lv-hrc.nevada.edu/~cogno/prestalk. html>

Radin, Dean. "A Future History." Online. WWW. Accessed 11-0896. Available at <eeyore.lv-hrc.nevada.edu/~cogno/future.html>

Raman, Varadaraja V. "Theodor Kaluza." *Dictionary of Scientific Biography*. Gillespie. Volume VII: 211-212.

Randall, Allan F. "Quantum Superposition, Necessity and the Identity of Indiscernibles." Online. WWW. Accessed 7-3-96. Available at <www.io.org/ ~randall/indiscernibles/>

Randi, James. "The Role of Conjurer's in Psi Research." Skeptic's Handbook. Kurtz: 339-350.

Rao, K. Ramakrishna. *Experimental Psychology. A Review and Interpretation*. Springfield, Illinois: Charles C. Thomas, 1966.

Rao, Ramakrishna K. "Theories of Psi." *Extrasensory Perception*. Krippner. 1977: 245-295.

Rauscher, Elizabeth A. "Some Physical Models Potentially Applicable to Remote Perception." *The Iceland Papers*. Puharich: 49-94.

Raychandhuri, A.K. *Theoretical Cosmology*. Oxford: Clarendon Press, 1979. Reichardt, Hans. *Gauss und die nicht-euklidische Geometrie*. Leipzig: Teubner, 1976.

Reichardt, Hans. *Gauss und die Anfänge der nicht-euklidische Geometrie*. Leipzig: Teubner, 1974.

Reichardt, Hans. *Gauss und die Anfänge der nicht-euklidischen Geometrie, mit originalarbeiten von J. Bolyai, N.I. Lobatschewski und Felix Klein*. TeubnerArchiv zur Mathematik 4. Leipzig, B.G. Teubner, 1985.

Reichenbach, Hans. "The Philosophical Significance of the Theory of Relativity." *Einstein: Philosopher and Scientist*: 287-311.

Reichenbach, Hans. *The Rise of Scientific Philosophy*. Berkeley: University of California Press, 1951; And London: Cambridge University Press, 1951. Reichenbacher, E. "Electromagnetism and the Fifth World Dimension." *Physikalische Zeitschrift* 29 (1 December 1928): 908-911.

Reinsel, Ruth. "Parapsychology: An Empirical Science." *Philosophy*. Grim: 187-204.

Ricci, Gregorio and Tulio Levi-Civita. "Mèthodes de calcul differentiel absolu et leurs applications." *Mathematische Annalen*, (Leipzig) 54 (1901): 125-401.

Richards, Joan L. "The Reception of a Mathematical Theory: Non-Euclidean Geometry in England, 1868-1883." *Natural Order: Historical Studies of Scientific Culture*. Eds. Barry Barnes and Steven Shapin. Beverly Hills: Sage Publications, 1979: 143-165.

Richards, Joan L. "Projective Geometry and Mathematical Progress in MidVictorian Britain." *Studies in the History and Philosophy of Science* 17 (1986): 297-325.

Richards, Joan L. *Mathematical Visions: The Pursuit of Geometry in Victorian England*. Boston: Harcourt Brace Jovanovich, 1988.

Rider, Thomas J. "The Observer Dependent Principle in Cosmology." Online. WWW. 4-4-96. Accessed 6-26-96. Available at <lincoln.midcoast. com/~trider/>

Riemann, Bernhard. *Gesammelte Mathematische Werke*. Ed. Heinrich Weber. 2nd edition (1892). With Nachtrage herausgegeben von M. Noeth und W Wirtinger (1902); Reprinted New York: Dover, 1960.

Riemann, Bernhard. "On the Hypotheses which lie at the Bases of Geometry." Trans. W.K. Clifford. *Nature* 7 (1 May 1873): 14-17; Continued Nature 8 (8 May 1873): 36-37; Reprinted in Clifford, *Mathematical Papers*: 55-71.

Riemann, Bernhard. "Ueber die Hypothesen welche der Geometrie zu Grunde liegen." *Mathematische Werke*: 272-287.

Riemann, Bernhard. "Fragment philosophischen Inhalts." *Mathematische Werke*: 507-538.

Riencourt, Amaury de. *The Eye of Shiva: Eastern Mysticism and Science*. New York: William Morrow, 1980.

Roberts, Samuel. "Remarks on Mathematical Terminology, and the Philosophic Bearing of Recent Mathematical Speculations concerning the Realities of Space." *Proceedings of the London Mathematical Society* 14 (9 November 1882): 5-15.

Robertson, H.P. "Geometry as a Branch of Physics." *Problems of Space and Time*. Ed. J.J.C. Smart. New York: MacMillan, 1964: 231-247; Also in *Einstein: Philosopher and Scientist*: 313-330.

Rodwell, G.F. "On Space of Four Dimensions." *Nature* 8 (1 May 1873): 8-9.

Rogers, Reginald Arthur Percy. "Space-Time and PsychoPhysics." *Hermathena* 20 (1926-1930): 1-16.

Rogo, D. Scott. *Parapsychology: A Century of Inquiry*. New York: Dell, 1975.

Roll, W.G. "The Psi Field." *Proceedings of the Parapsychological Association*, 1957-1964 1 (1966): 32-65.

Roll, William G. *Theory and Experiment in Psychical Research*. New York: Arno Press, 1975.

Roney-Dougal, Serena. *Where Science and Magic Meet*. 2nd edition. Rockport, Massachusetts: Element Books, 1993.

Rosen, Nathan and Itzhak Goldman. "A Universe Embedded in a Five-Dimensional Flat Space." *General Relativity and Gravitation* 2 (1971): 367384.

Rosen, Nathan and Elhanan Leibowitz. "Five-Dimensional Relativity Theory." *General Relativity and Gravitation* 4 (1973): 449-474.

Rosen, Nathan and Elhanan Leibowitz. "Periodic Fields in Five-Dimensional Relativity." *General Relativity and Gravitation* 5 (1974): 409-427.

Rosenfeld, Albert. Ed. *Mind and Supermind*. New York: Holt, Rinehart and Winston, 1977.

Rosenfeld, B.A. "Nikolai Ivanovich Lobachevsky." *Dictionary of Scientific Biography* 7&8: 428-435.

Rosenfeld, B.A. *A History of Non-Euclidean Geometry*. Abe Shenitzer and Hardy Grant. Trans. London: Springer-Verlag, 1988.

Rosenfeld, L. "L'Univers a cinq dimensions et la Mecanique Ondulatoire." *Academie Royale Belgique* 13 (1927): 304-325, 326-328, 447-458, 573-580, 661-682.

Ross, G.G. "A 'theory of everything?'" *Contemporary Physics* 34 (1993): 79-88.

Ross, Hugh. "Mathematical Breakthroughs Establish God's Extra-Dimensional Might." Online. WWW. Accessed 06-02-97. Available at <www.surf.com/~westley/4q95faf/4q95dmsn.html>

Ross, Kelley L. "The Ontology and Cosmology of Non-Euclidean Geometry." Online. WWW. Accessed 09-17-97. Available at <www.friesian.com/curved-1.htm>

Rucker, Rudy. *The Fourth Dimension: A Guided Tour of the Higher Universes.* Boston: Houghton-Mifflin, 1984.

Rush, Joseph H. "Parapsychology: A historical perspective." Chapter 2. *Foundations of Parapsychology.* Edge, Morris, Rush, and Palmer: 9-44.

Rush, Joseph H. "Physical and quasi-physical theories of Psi." *Foundations of Parapsychology.* Edge, Morris, Rush, Palmer: 276-292.

Rush, Joseph H. *New Directions in Parapsychological Research.* Monograph No.4. New York: Parapsychological Foundation, 1964.

Russell, Bertrand A.W. "The Logic of Geometry." *Mind* 5 (1896): 1-23.

Russell, Bertrand A.W. *An Essay on the Foundations of Geometry.* Intro. Morris Kline. New York: Dover, 1956; unaltered reproduction of the original, Cambridge University Press, 1897.

Russell, Bertrand A.W. "Motion in a Plenum." Not published (1897). #230.030030. Bertrand Russell Archives, McMasters University, Hamilton, Ontario.

Russell, Bertrand A.W. "Les Axiomes propres a Euclide sont-ils empiriques?" *Revue de Metaphysiques* 6 (1898): 759-776.

Russell, Bertrand A.W. "On Causality as used in Dynamics." Not published (March 1898). #230.030050. Bertrand Russell Archives, McMasters University, Hamilton, Ontario.

Russell, Bertrand A.W. "Sur les Axioms de la Géométrie." *Revue de Metaphysiques* 7 (1899): 684-707.

Russell, Bertrand A.W. "On the Constituents of Space and Their Mutual Relations." Unpublished (n.d.). #230.030140. Bertrand Russell Archives, McMasters University, Hamilton, Ontario.

Russell, Bertrand A.W. "Non-Euclidean Geometry." *The Atheneum* 4018 (29 October 1904): 592-593.

Russell, Bertrand A.W. *Collected Works of Bertrand Russell*. Volume I: *The Cambridge Essays, 1888-1899*. Eds. Kenneth Blackwell, et.al. London: Allen Unwin, 1983.

Russell, Bertrand A.W. "Is Position in Time and Space Absolute or Relative?" *Mind* 10 (July 1901): 293-317.

Russell, Bertrand A.W. *The Principles of Mathematics*. 2nd edition. New York: W.W. Norton, 1938; Reprint of 1902.

Russell, Bertrand A.W. "Non-Euclidean Geometry." *Encyclopaedia Britannica*. 10th edition. 17 (1902): 664-674.

Russell, Bertrand A.W. and A.N. Whitehead. "Geometry VI: Non-Euclidean Geometry." *Encyclopaedia Britannica*. 11th edition. 11 (1910): 724-730.

Rust, C.E. "A Comment on Pseudo-Geometry." *The Monist* 18 (1908): 631-632. Ryder, L.H. "General Relativity comes of age." Essay review of G. Ellis, A. Lanza. and J. Miller. *The Renaissance of General Relativity and Cosmology. Contemporary Physics* 34 (1994): 333-336.

S. "Four-Dimensional Space." See Sylvester.

S. (Ludwig Silberstein) "The Relativity Theory of Gravitation." *The Observatory* 42 (April 1919): 171-176.

Sachs, Mendel. *Quantum Mechanics from General Relativity: An Approximation for a Theory of Inertia*. Dordrecht: D. Reidel, 1986.

Sachs, Mendel. *Einstein versus Bohr: The Continuing Controversies in Physics*. La Salle, Illinois: Open Court, 1988.

Sachs, Mendel. *Dialogues on Modern Physics*. River Edge, New Jersey: World Scientific Publishing, 1998.

Sachs, Mendel. *General Relativity and Matter: A Spinor Field Theory from Fermis to Light-Years*. Boston: D. Reidel, 1982.

Salam, Abdus. "Interview." *Superstrings*. Davies and Brown: 170-179.

Salmon, Wesley C. "The Curvature of Physical Space." *Foundations of Space-Time Theories*. Vol. VIII. *Minnesota Studies in the Philosophy of Science.* Eds. John Earman, Clark Glymour and John Stackel. Minneapolis: University of Minnesota Press, 1975: 281-302.

Salvetti, C. "Clifford's Algebra and the Meson." *Nuovo Cimento* 5 (August 1947): 250-282.

Samsonovich, Alexei. "Binding Problem for Consciousness." Online. WWW. Accessed 05-1-97.

Sanford, Edmund C. "The Possibility of a Realization of Four-Fold Space." *Science* 19 (10 June 1892): 332.

Sarfatti, Jack. "The Physical Roots of Consciousness." *Roots of Consciousness.* 1st edition. Mishlove.

Sarfatti, Jack. "Introduction by Jack Sarfatti on the logic of the original EPR paper." Version 0.3. Online. WWW. Accessed 6-2496. Available at <www.hia.com/hia/pcr/eprbell.html>

Sarfatti, Jack. "The Quantum and Beyond." New version 1.2. Online. WWW. 5-96. Accessed 6-13-96. Available at <www.hia.com/hia /pcr/qmbeynd.html>

Sarfatti, Jack. "Quantum Back Action; Does Consciousness Require a Violation of Orthodox Quantum Mechanics?" Version 0.4. Online. WWW. 3-22-96. Accessed 6-24-96. Available at <www.hia.com/hia/pcr/qmotion .html>

Sarfatti, Jack. "Sarfatti's Online Guide to The Feynman Lectures on Gravitation." Version 0.3. Online. WWW. Accessed 7-8-96. Available at <www.hia.com/hia/pcr/ feyngrav.html>

Sarfatti, Jack. "Sarfatti's Illuminati: Footnotes to In the Thick of Things." Online. WWW. Accessed 8-12-96. Available at <www.hia.com/pcr/ si03f.html>

Sarfatti, Jack. "How Thoughts Shape Matter: A Commentary." Online. WWW. Accessed 05-1-97. Available at <www.hia.com/hia/pcr/bohmtht. html>

Sarfatti, Jack. "Unconscious Quantum." Online. WWW. Accessed 04-25-97. Available at <www.hia.com/hia/pcr/unconq.html>

Sarfatti, Jack. "EPR." Online. WWW. Accessed 05-22-97. Available at <www.hia.com/hia/ pcr/bohmtht.html>

Sarfatti, Jack. "Scientific Commentary." *Space-Time and Beyond*. Toben and Wolf: 125-159.

Sarfatti, Jack. "Evidence for Quantum Brain Fluctuations." Online. WWW. Accessed 06-01-97. Available at <www.clas.ufl.edu/anthro/noetics/ quantumbrain.html>

Sarfatti, Jack. "Qualia as the Atoms of Consciousness." Online. WWW. 09-2996. Accessed 04-25-97. Available at <www.hia.com/hia/pcr/kaufman1. html>

Sarfatti, Jack. "Commentary on "On Jungian Transpersonal Psychology, Psychokinesis, Precognition and Remote-Viewing in Post-Modern Physics." Online. WWW. Accessed 09-17-97. Available at <www.hia.com/hia/pcr/ mansfield.html >

Sarfatti, Jack. "Microtubules." Online. WWW. 03-22-96. Accessed 04-25-97. Available at <www.hia.com/hia/pcr/mts.html>

Sarfatti, Jack. "Vector Quantum Mechanics of Mind." Review of Jeff Tollaksen preprint. Online. WWW. 1995. Accessed 05-01-97. Available at <www.his.com/hia/pcr/2vecmind.html>

Scaruffi, Piero. "Towards unification of Cognitive and Physical Sciences: Cognition as a property of Matter." Online. WWW. Accessed 06-01-97. Available at <www-ksl.stanford.edu/people/scaruffi/statemnt.html>

Schaff, William L. "Mathematical Curiosities and Hoaxes." *Scripta Mathematica* 6 (1939): 49-55.

Schatzer, Laro. "The Speed of Light - A Limit on Principle? A physicist's view on an old controversy." Online. WWW. 7-3-96. Accessed 7-31-96. Available at <monet.physik.unibas.ch/ ~schatzer/space-time.html>

Schidlof, A. "A Construction giving the Mass of a Charged Material Point in the Universe of Five Dimensions." *Comptes Rendus* 185 (5 December 1927): 1262-1263.

Schidlof, A. "Interpretation of Masses of the Electron and Proton in a Universe of Five Dimensions." Comptes Rendus 185 (31 October 1927): 889-891. Schlipp, Paul A. Ed. *Einstein: Philosopher and Scientist*. LaSalle: Open Court, 1949.

Schmeidler, Gertrude. *Parapsychology: Its Relation to Physics, Biology, Physiology, and Psychology*. Metuchen, New Jersey: Scarecrow Press, 1976.

Schmidt, Helmut. "Toward a Mathematical Theory of Psi." *Journal of the ASPR* 69 (1975): 301-319.

Schmidt, Helmut. Abstract of "Collapse of the State Vector and Psychokinetic Effect." *Foundations of Physics* 12 (June 1982). Online. WWW. Accessed 7-1-96. Available at <alethea.ukc.ac.uk/Dept/ CPRS/RPKP/collapse>

Schmidt, Helmut. "The Strange Properties of Psychokinesis." *Journal of Scientific Exploration* 1 (1987). Online. WWW. 7-1-96. Available at <alethea.ukc.ac.uk/CPRS/RPKP/strange>

Schofield, Alfred Taylor. *Another World; or the Fourth Dimension*. 5th edition. London: George Allen & Unwin, 1920; 1st edition published in 1880.

Schouten, J.A. "Die direkte Analysis zur neueren Relativitätstheorie ver handelingen." *Kon. Akad. Amsterdam*. 12 (1918).

Schouten, J.A. "On the number and the enclosing Euclidean space with the least possible number of dimensions." *Kon. Akad. van Weten. te Amst*. 31 (1919): 607-613.

Schouten, J.A. and D. Van Dantzig. "Four-Dimensional Interpretation of the Unified Field Theory." *K. Acad. Amsterdam, Proc*. 34 (1931): 1398-1407.

Schouten, J.A. and D. Van Dantzig. "Unitary Field Theory." *K. Acad. Amsterdam Proc*. 35 (1932): 642-655.

Schouten, J.A. and D. Van Dantzig. "Unitary Field Theory and Dirac's Equation." *K. Acad. Amsterdam, Proc*. 35 (1932): 843-852.

Schouten, J.A. and D. Van Dantzig. "Generalle Feldtheorie." *Zeitschrift fur Physik* 38 (18 October 1932): 639-667.

Schouten, J.A. "Unitary Field Theory." *Zeitschrift fur Physik* 81 (3 March 1933): 129-138.

Schouten, J.A. "Space-Time in the Unitary Field Theory." *Zeitschrift fur Physik* 81 (22 March 1933): 405-417.

Schouten, J.A. and D. Van Dantzig. "On projective Connections and Their Application to the General field Theory." *Annals of Mathematics* 34 (1933): 271.

Schouten, J.A. and J. Haantjes. "Auto-Geodetic Lines and World-Lines." *Zeitschrift fur Physik* 89 (2 June 1934): 357-369.

Schouten, J.A. "La Théorie projective de la Relativité." *Ann. Inst. H. Poincaré* 5 (1935): 51.

Schouten, J.A. and D. Van Dantzig. *Annals of Mathematics* 34 (1939): 37.

Schrödinger, Erwin. *Space-Time Structure*. Cambridge, England: At the University Press, 1960.

Schrödinger, Erwin. *What is Life & other Scientific Essays*. Garden City, New York: Anchor Doubleday, 1956.

Schrödinger, Erwin. "The Present Situation in Quantum Mechanics: A translation of Schrödinger's 'Cat Paradox Paper'." Trans. John D. Trimmer. *Proceedings of the American Philosophical Society* 124: 323-38; Also, *Quantum Theory and Measurement*. Eds. J.A. Wheeler and W.H. Zurek. Princeton: Princeton University Press, 1983. Online. WWW. Accessed 7-8-96. Available at <www.hia. com/hia/pcr/qcat.html>

Schubert, Hermann. "The Fourth Dimension: Mathematical and Spiritualistic." *The Monist* 3 (1892-1893): 402-449; Reprinted in *Mathematical Essays and Recreations*. Trans. Thomas J. McCormack. Chicago: Open Court, 1904; Reprint of 1899: 64-111.

Schubert, Hermann. "On the Nature of Mathematical Knowledge." *The Monist* 6 (1895): 294-305.

Schwarz, John. "Interview." *Superstrings*. Davies and Brown: 70-89. Schweber, Silvan S. *QED and the Men Who Made It: Dyson, Feynman, Schwinger, and Tomonaga*. Princeton: Princeton University Press, 1994.

Schwinger, Julian. Ed. *Selected Papers on Quantum Electrodynamics*. New York: Dover, 1958.

Seal, Horace. "Space and Time." *Westminster Review* 152 (December 1899): 675-679.

Sen, D.K. *Fields and/or Particles*. New York: Academic Press, 1968.

Sen, N.R. "Das Keplerproblem der fünfdimensionale Wellenmechanik und der Einfluss der Gravitation aus die Balmerformel." *Zeitschrift fur Physik* 66 (1930): 686-692.

Settanni, Harry. *The Philosophical Foundations of Paranormal Phenomena*. New York: University Press of America, 1992.

Schaffranke, Rolf. "Post-relativistic Physics." Online. WWW. Accessed 04-04-97. Available at <www.livelinks.com/sumeria/ free/schaffra.html>

Schmeidler, Gertrude. "Respice, adspice, and prospice." *Proceedings of the Parapsychology Association* 8 (1972): 117-145.

Schmeidler, Gertrude. Editor. *Parapsychology: Its Relation to Physics, Biology, Physiology, and Psychology*. Metuchen, New Jersey: Scarecrow Press, 1976.

Schroeder, Lynn and Sheila Ostrander. *Psychic Discoveries Behind the Iron Curtain*. New York: Bantam, 1971.

Schwarz, John H. "The Second Superstring Revolution." Online. WWW. Accessed 05-22-97. Available at <theory.caltech.edu/ people/jhs/strings/ index.htm>

Segre, Corrado. "On Some Tendencies in Geometric Investigations." *Bulletin of the American Mathematical Society* 11 (1904): 442-468.

Shapere, Dudley. "Causal Efficacy of Space." *Philosophy of Science* 31 (1964): 111-120.

Shapin, Betty and Lisette Coly. *Concepts and Theories of Parapsychology*. Proceedings of an International Conference held in New York, 6 December1980. New York: Parapsychology Foundation, 1981.

Shapin, Betty and Lisette Coly. Eds. *Education in Parapsychology*. Proceedings of the International Conference held in San Francisco, 11-16 August 1975. New York: Parapsychology Foundation, 1976.

Shapin, Betty and Lisette Coly. *Brain/Mind in Parapsychology*. Proceedings of the International Conference held in Montreal, Canada, 24-25 August 1978. New York: Parapsychology Foundation, 1979.

Sharpe, Kevin J. "Relating the Physics and Religion of David Bohm." *Zygon* 25 (March 1990): 105-122.

Shepard, Leslie A. *Encyclopedia of Occultism and Parapsychology*. 3rd edition. New York: Gale Research, 1993.

Silberstein, Ludwik. "General Relativity without the Equivalence Hypothesis." Written 25 March 1918. *Philosophical Magazine* 36 (1918): 94-

128. Silverman, A.C. "The Fairyland of the Fourth Dimension." *The Fourth Dimension Simply Explained*. Manning: 232-241.

Sims, Marsha. "Princeton University Mind-Matter Experiments Reported." Press Release for the Society for Scientific Exploration. Online. WWW. Accessed 05-01-97. Available at <www.jse.com/PR-Princeton-92.html>

Sims, Marsha. "Former Directors Dispute Conclusions of Government ESP Program Review by CIA." Press Release for the Society for Scientific Exploration. Online. WWW. Accessed 05-0197. Available at <www.jse.com/PR-CIA-96.html>

Sinclair, Upton. "Excerpts from Mental Radio." *Encounters with Parapsychology*. McConnell: 33-40.

Sirag, Saul-Paul. "Consciousness: A Hyperspace View." *Roots of Consciousness*. 2nd ed. Mishlove: 327-365.

Sitter, Willem de. "Space, Time, and Gravitation." *The Observatory* 39 (October 1916): 412-419.

Sitter, Willem de. "The Motion of the Perihelion on the Classical Theory of Relativity." *The Observatory 40* (August 1917): 302-303.

Siu, R.G.H. *The Tao of Science: An Essay on Western Knowledge and Eastern Wisdom*. Cambridge: M.I.T. Press, 1957.

Skey, William. "Notes upon Mr. Frankland's paper 'On the Simplest Continuous Manifoldness of Two Dimensions and of Finite Extent'." *Transactions and Proceedings of the New Zealand Institute* 13 (1880): 100-109.

Sklar, Lawrence. *Space, Time and Spacetime*. Berkeley: University of California Press, 1974.

Smart, J.J.C. Ed. *Problems of Space and Time*. New York: Macmillan, 1964.

Smart, J.J.C. "Space." *Encyclopedia of Philosophy* 7: 506-511.

Smed, Jouni A. "Out of Body Experience FAQ." Online. WWW. Accessed 8-12-96. Available at <www.spiritweb.org/Spirit/obefaq.html>

Smokler, Howard E. "W.K. Clifford's Conception of Geometry." *Philosophical Quarterly* 16 (1966): 249-257.

Smokler, Howard E. "William Kingdon Clifford." *Dictionary of Philosophy*. Edwards.

Smythies, J.R. Ed. *Science and ESP*. New York: Humanities Press, 1967; London: Routledge. Kegan Paul, 1971.

Smythies, J.R. "Is ESP Possible?" *Science and ESP*. Smythies: 1-14.

Snyder, Hartland S. "Quantized Space-Time." *Physical Review* 71 (1 January 1947): 38-41.

Solomon, J. "Einstein and Mayer's Theory and Dirac's Equation." *Comptes Rendus* 195 (22 August 1932): 461-462.

Sommerville, Duncan M.Y. *Bibliography of Non-Euclidean Geometry*. New York: Chelsea; A supplemental reprint of *Bibliography of Non-Euclidean Geometry, including the Theory of Parallels, the Foundations of Geometry, and Space of n Dimensions*. London: 1911.

Sommerville, Duncan M.Y. *The Elements of Non-Euclidean Geometry*. New York: Dover, 1958; Reprint of London: G. Bell and Sons, 1914.

Sonnenmoser, Kurt. "The Twin Paradox." Relativistic Paradoxes Part B from psi-physics FAQ. Online. WWW. Accessed 7-10-96. Available at <sunsite.unc.edu/lunar/relaparb.html>

Sorokin, Pitirim A. "The Integral Theory of Truth and Reality." Excerpted from The Crisis of Our Age. Online. WWW. 02-01-97. Accessed 05-09-97. Available at <www.intuition.org/sorokin.htm>

Souriau, Jean-Marie. "Relativité Multidimensionelle non Stationaire." *Les Théories Relativistes de La Gravitation*. Paris: Centre National de la Recherche Scientifique (1962): 293-297.

Spretnak, C. *States of Grace*. San Francisco: Harper, 1993.

Squires, E.J. "Consciousness in the Quantum World." Essay review of H.P. Stapp, *Mind, Matter and Quantum Mechanics*. *Contemporary Physics* 34 (1993): 329-331.

Squires, E.J. *Conscious Mind in the Physical World*. Adam Hilger, 1990.

Stahl, Daniel G. "A Grand Unified Field Theory." Online. WWW. 08-27-97. Accessed 10-01-97. Available at <members.aol.com/SOARINDY /GUT.html>

Stallo, Johann Bernhard. *The Concepts and Theories of Modern Physics*. Ed. Percy W. Bridgman. Cambridge, Massachusetts: The Belknap Press of Harvard University Press, 1960; Reprint of the 2nd edition of 1884; 1st edition published in 1881.

Stanford, Rex G. "Conceptual Frameworks of Contemporary Psi Research." *Handbook of Parapsychology*. Wolman: 823-858.

Stapp, Henry P. "Why Classical Mechanics Cannot Naturally Accommodate Consciousness but Quantum Mechanics Can." Psyche 2 (May 1995). Online. WWW. Available at <psyche.cs. monash.edu.au/volume2-1/psyche95-2-05qm-stapp-1stapp.html>

Stapp, Henry P. An abstract of *Chance, Choice, and Consciousness: The Role of Mind in the Quantum Brain*. Comments by Jack Sarfatti. Version 0.4. Online. WWW. 11-28-95. Accessed 6-26-96. Available at <www.hia.com/hia/ pcr/newstapp.html>

Stapp, Henry P. Stapp's *Physics Today* letter and LBL-33789 preprint of 12/6/93. *Physics Today* (July 1995). Online. WWW. Accessed 7-21-96. Available at <www.hia.com/hia/pcr/pktheory. html>

Stapp, Henry P. Excerpts from "Theoretical model of a purported empirical violation of the predictions of quantum theory." *Physical Review* A 50 (July 1994). Online. WWW. Accessed 7-8-96. Available at <alethea.ukc.ac.uk/Dept/ CPRS/RPKP/stapp>

Stapp, Henry P. "Stapp's PK Theory." Online. WWW. Available at <www.hiua.com/hia/pcr/pktheory.html>; Reprinted from Stapp's letter to *Physics Today* (July 1995) and LBL-33789 preprint of 6 December 1993.

Stapp, Henry. "Why Classical Mechanics Cannot Naturally Accommodate Consciousness but Quantum Mechanics Can: Part 1." Jack Sarfatti. Editor for Internet. Online. WWW. 02-08-95. Accessed 05-01-97. Available at <www.hia.com/hia/pcr/ stapp1.html>

Stapp, Henry P. *Mind, Matter, and Quantum Mechanics*. New York: Springer Verlag, 1993.

Stenger, Victor J. "Quantum Quackery." Response from Menas Kafatos. Commentaries by Jack Sarfatti. Online. WWW. 5-17-96. Accessed 6-26-96. Available at <www.hia.com/hia/pcr/stenger. html and stenger2.html>

Stenger, Victor J. "ESP and Cold Fusion: Parallels in Pseudoscience." Online. WWW. Accessed 7-3-96. Available at <www.phys.hawaii.edu/vjs/www/cold.txt>

Stenger, Victor J. *Physics and Psychics: The Search for a World Beyond the Senses*. Buffalo: Prometheus Books, 1990.

Stephen, Leslie. "William Kingdon Clifford." Dictionary of National Biography, from early times to 1900. Sir Leslie Stephen and Sir Sydney Lee. Eds. London: Geoffrey Cumberlege, Publisher to the University, Oxford University Press, 1917. Volume 4: 538-541.

Stevens, Paul. "Elements of a typical psychokinesis experiment." Online. WWW. 02-28-96. Accessed 04-24-97. Available at <moebius.psy.ed.ac.uk/micropk.html>

Stoker, J.J. *Differential Geometry. Pure and Applied Mathematics*, Volume XX. New York: Wiley & Sons, 1969.

Stokes, George J. "Space." *Encyclopedia of Religion and Ethics*. Ed. James Hastings. Assist. John A. Selbie and Louis H. Gray. New York: Scribners, 1951; Originally published Edinburgh: T & T Clark, 1920. 11: 758-764.

Story, William Edward. "Hyperspace and Non-Euclidean Geometry." *Mathematical Review* 1 (1897): 169-184.

Stringham, W.I. "Rotation in Four-Dimensional Space." *Johns Hopkins University Circular* 1 (March 1880): 49.

Stringham, W.I. "On the Rotation of a Rigid System in Space of Four Dimensions." *Proceedings of the American Association* 33 (1884): 55-56. Struik, D.J. and N. Weiner. "Relativity Theory of Quanta." *Journal of Math and Physics, MIT* 7 (September 1927): 1-23.

Struik, D.J. and N. Weiner. "The Fifth Dimension in Relativistic Quantum Theory." *National Academy of Sciences, Proceedings* 14 (March 1928): 262-268.

Sulton, Christine. *Building the Universe*. New York: Basil Blackwell and New Scientist, 1985.

Swenson, Lloyd S. Jr. *The Genesis of Relativity*. New York: Burt Franklin, 1979.

Sylvester, James Joseph [signed S.]. "Four-Dimensional Space." *Nature* 31 (March 1885): 481.

Sylvester, James Joseph. "Inaugural Address to the Mathematics Section." *Reports of the British Association, (Exeter)* 39 (1869): 19; Reprinted as "A Plea for the Mathematician." *Nature* 1 (30 December 1869): 237-239; Reprinted *Laws of Verse*. London: Longmans, 1870; and *Collected Papers*: 650-719. Sylvester, James Joseph. "Mr. Clifford's Discovery." *The Athenaeum* 2208 (19 February 1870): 263.

Sylvester, James Joseph. Correspondence with Arthur Cayley. 23 February 1866, 23 March 1866, 7 July 1866, 14 June 1867, 22 September 1883, 22 September 1883 (#2), 13 December 1883, 20 January 1884, 29 January 1884, 2 November 1884, 16 April 1885, 17 April 1885, 22 April 1885, 20 June 1885, [c. April] 1885, 16 April 1885. Sylvester Papers, St. Johns College, Cambridge University.

Taimni, I.K. *Science and Occultism*. Adyar, Madras, India: Theosophical Publishing House, 1974.

Tait, Peter G. *Lectures on Some Recent Advances in Physical Science*. 2nd edition. London: Macmillan, 1876.

Tait, Peter G. "Zöllner's Scientific Papers." *Nature* 17 (28 March 1878): 420-422.

Tait, Peter G. "Clifford's Dynamic." *Nature* 18 (23 May 1878): 89-91. Tait, Peter G. "On Knots." *Transactions of the Royal Society of Edinburgh* 28 (1876-1877): 145-190; Revised. *Scientific Papers* 1: 273-317.

Tait, Peter G. "On Knots. Part II." Read 2 June 1884. *Transactions of the Royal Society of Edinburgh* 32 (1887): 327-339; Scientific Papers 1: 318-334. Tait, Peter G. "On Knots. Part III." Read 20 July 1885. *Transactions of the Royal Society of Edinburgh* 32 (1887): 493-506; *Scientific Papers* 1: 335-347.

Tait, Peter G. "Clifford's Exact Sciences." *Nature* 32 (11 June 1885): 124-125. Tait, Peter G. "On the Importance of Quaternions in Physics." *Philosophical Magazine* 29 (1890); 84-97; *Scientific Papers* 2: 297-308.

Tait, Peter G. "On the Intrinsic Nature of the Quaternion Method." Read 2 July 1894. *Proceedings of the Royal Society of Edinburgh* 20 (1894): 276-284; *Scientific Papers* 2: 392-398.

Tait, P.G. and Balfour Stewart. *The Unseen Universe or Physical Speculations on a Future State*. New (4th) edition. London: Macmillan, 1901; First edition. London: Macmillan, 1875.

Targ, Russell and Harold E. Puthoff. *Mind-Reach: Scientists Look at Psychic Ability*. Delacorte Press, 1977.

Targ, Russell, Harold E. Puthoff and Edwin C. May. "Direct Perception of Remote Geographical Locations." *Mind at Large*. Tart, Puthoff and Targ: 77-106.

Tart, Charles T., Harold E. Puthoff and Russell Targ. Eds. *Mind at Large: Institute of Electrical and Electronic Engineers Symposia on the Nature of Extrasensory Perception.* New York: Praeger, 1979.

Tart, Charles T. "Improving Real-Time ESP by Suppressing the Future: TransTemporal Inhibition." *Mind at Large*. Tart, Puthoff, Targ: 137-174.

Takeno, H. "Projective Wave Geometry." *Hiroshima Journal of Science* 6 (March 1936): 147-172.

Talbot, Michael. *Mysticism and the New Physics*. New York: Bantam, 1981.

Talbot, Michael. *The Holographic Universe*. New York: Harper Perennial, 1991.

Tart, Charles. Ed. *Altered States of Consciousness*. Garden City, New York: Doubleday Anchor, 1972.

Taylor, John G. *Black Holes*. New York: Avon Books, 1973.

Taylor, John. *Superminds: A Scientist Looks at the Paranormal*. New York: Warner Books, 1975.

Tavani, F. "Physical Space and Hyperspaces." Read 19 December 1921. *Proceedings of the Aristotelian Society* 22 (1921-1922): 55-68.

Tee, Garry J. "Two New Zealand Mathematicians." *History of Mathematics. Proceedings of the First Australian Conference*. Ed. John S. Crossley. Clayton, Victoria, Australia: Monash University, Department of Mathematics, 1981: 180-199.

Tee, Garry J. "Mathematics in the Pacific Basin." *British Journal for the History of Science* 21 (1988): 401-417.

Thalbourne, Michael A. *A Glossary of Terms Used in Parapsychology*. London: Heinemann, 1982.

Thiele, Joachim. "Karl Pearson, Ernst Mach, John B. Stallo: Briefe aus den Jahren 1897 bis 1904." *Isis* 69 (1978): 535-542.

Thirring, W. "Remarks on Five-Dimensional Relativity." *The Physicist's Conception of Nature.* Mehra: 199-201.

Thiry, Yves. "Les Équations de La Théorie Unitaire de Kaluza." Comptes Rendus, Academie des Science 226 (1948): 216-218.

Thiry, Yves. "Sur La Régularité des Champs Gravitationnel et Électromagnétique dans Les Théories Unitaire*." Comptes Rendus, Academie de Science* 226 (7 June 1948): 1881-1882.

Thiry, Yves. Étude *Mathématique des Équations d'une Théorie Unitaire à Quinze Variables de Champs*. Thesis. Paris: Gauthier Villiers, 1951.

Thiry, Yves. "Sur les Théories Pentadimensionelles." *Les Théories Relativistes de la Gravitation*. Paris: Centre National de la Recherche Scientifique, 1962: 284-291.

Thomas, Clark M. "Relativity Beyond Relativity." Online. WWW. Accessed 5-5-97. Available at <home.earthlink.net/~cmthomas/Einstein. html>

Thomson, William (Lord Kelvin). "The Wave Theory of Light." Lecture delivered before the Academy of Music, Philadelphia, under the auspices of the Franklin Institute, 29 September 1884. *Journal of the Franklin Institute* 118 (November 1884): 321-341; Reprinted in Sir William Thomson. *Popular Lectures*. London: Macmillan, 1888.

Thouless, Robert H. *From Anecdote to Experiment in Psychical Research.* London: Routledge & Kegan Paul, 1972.

Toben, Bob and Fred Wolf. *Space-Time and Beyond*. New York: E.P. Dutton, 1975.

Tolman, Richard C. Relativity, *Thermodynamics and Cosmology*. Oxford: At the Clarendon Press, 1934.

Tomei, Joseph S. "Towards a Theory of Hyper-Spatial Mechanics." Online. WWW. 1994. Accessed 7-8-96. Available at <www.sunsite.unc.edu/lunar/ hmo.html>

Tufillaro, Nicholas B. "Nonlinear Dynamics - Flows in 3-D and Beyond." Online. WWW. Accessed 7-22-96. Available at <cnlswww.lanl.gov/nbt/ nldspeech.html>

Tonnelat, Marie Antoinette. *The Principles of Electrodynamic Theory and of Relativity.* Trans. A.J. Knodel. Dordrecht-Holland: D. Reidel, 1966(a).

Tonnelat, Marie Antoinette. *Einstein's Unified Field Theory*. Trans. Richard Akerib. New York: Gordon and Breach, 1966(b).

Tonnelat, Marie Antoinette. *Les Théories Unitaire de L'Électromagnétisme et de La Gravitation*. Paris: Gauthier Villiers, 1965.

Torretti, Roberto. "On the Subject of Objective Space." *Third International Kant Congress* (1972): 568-575.

Torretti, Roberto. *Philosophy of Geometry from Riemann to Poincaré*. Dordrecht: Reidel, 1979.

Torretti, Roberto. *Relativity and Geometry*. New York: Pergamon, 1983. Toth, Imré. "Non-Euclidean Geometry before Euclid." *Scientific American* 221 (1969): 87-98.

Toth, Imré. "Spekulation über die Moglichkeit eines nicht euklidischen Raumes vor Einstein." *Einstein Symposium Berlin. Lecture Notes in Physics, 100*. Eds. N. Nelkowski, A. Hermann, H. Poser, R. Shrader, and R. Seiler. Stuttgart: Springer-Verlag, 1980: 46-83.

Toth, Imré. "La revolution non-euclidienne." *La Recherche in Histoire des Sciences*. Éditions du Seuil La Recherche. Paris: Socièté d'Éditions Scientifiques, October 1983: 241-265.

Toth, Imré. "Gott und Geometrie: Eine viktorianische Kontroverse." *Evolutionstheorie und ihre Evolution*. Vortragseihe der Universität Regensburg zum 100. Todestag von Charles Darwin, U.R. *Shriftenreihe der Universität Regensburg* 7 (Dezember 1982): 141-204.

Trautman, Andrezj. "Theory of Gravitation." *The Physicist's Conception of Nature*. Mehra: 179-198.

Trudeau, Richard J. *The Non-Euclidean Revolution*. Intro. H.S.M. Coxeter. Stuttgart: Birkhäuser, 1987.

Tucker, Robert. "Professor Clifford's Mathematical Papers." *Nature* 20 (26 June 1879): 195.

Tunzelmann, G.W. de. *A Treatise on Electrical Theory and the Problem of the Universe*. London: Charles Griffin and Company, 1910.

Tunzelmann, G.W. de. "Physical Relativity Hypotheses Old and New." *Science Progress* 13 (1918-1919): 474-482.

Tunzelmann, G.W. de. "The General Theory of Relativity and Einstein's Theory of Gravitation." *Science Progress* 13 (1918-1919): 652-657.

Turner, F.M. *Between Science and Religion*. New York: Yale University Press, 1974.

Tveter, Donald R. "The Quantum Basis of Intelligence?" Online. WWW. 4-296. Accessed 5-5-96.

Tyndall, John. "Science and the Spirits." Fragments of Science. Volume I. New York: Appleton, 1899.

Utts, Jessica and Brian D. Josephson. "The Paranormal: The Evidence and its Implications for Consciousness." A shortened version of this article was published in the *Times Higher Education Supplement's* special section on Consciousness linked to the Tucson II conference "Toward a Science of Consciousness." (April 5, 1996): v. Online. WWW. Accessed 6-2496. Available at <www.tcm.phy.cam.ac.uk/~bdj10/psi/tucson.psi>

Utts, Jessica. "An Assessment of the Evidence for Psychic Functioning." Online. WWW. 1995. Accessed 05-01-97. Available at <wwwstat.ucdavis.edu /users/utts/air.htm>

Utts, Jessica. "Replication and Meta-Analysis in Parapsychology." *Statistical Science* 6 (1991): 363-403; Online. WWW. Accessed 0212-98. Available at <www.stat.ucdavis.edu/users/utts/91rmp.html>

Utts, Jessica. "Rejoinder." Statistical Science 6 (1991); Online. WWW. Accessed 02-12-98. Available at <www.stat.ucdavis.edu/users/utts/91rmp-r.html>

Utts, Jessica. "Response to Ray Hyman's Report of September 11, 1995 "Evaluation of Program on Anomalous Mental Phenomena." Online. WWW. 09-15-95. Accessed 05-01-97. Available at <www-stat.ucdavis.edu/users/ utts/response.html>

Varičak, Vladimir. "Über die nichteuklidische Interpretation der Relativtheorie." *Jahresberichte der Mathematische Vereinigung* 21 (1914): 103-127.

Vasiliev, L.L. "Parapsychology in the Soviet Union." *Encounters with Parapsychology*. McConnell: 44-55.

Veblen, Oswald and Banesh Hoffman. "Projective Relativity." *Physical Review* 36 (1 September 1930): 810-822.

Veblen, Oswald. *Projektive Relativitätstheorie*. Berlin: J. Springer, 1933.
Veblen, Oswald. "Spinors in Projective Relativity." National *Academy of Science, Proceedings* 19 (November 1933): 979-989.

Vliet, Alan T. Van. "Einstein's Vision." Online. WWW. Accessed 05-12-97. Available at <home.pinehurst.net/vanvliet/einstein.htm>

Vranceanu, G. "La Théorie Unitaire des Champs et Les Hypersurfaces Non-Holonomous." *Comptes Rendus* 200 (1935): 2056-2058.

Vranceanu, G. "Les Espaces Non-Holonomes et Leurs Applications Mécaniques." *Mém. Science et Mathematic* 76 (1936).

Vranceanu, G. "Non-Holonomic Unified Field Theory." *Le Journal de Physique et de Le Radium* 7 (December 1936): 514-526.

Waerden, B.L. van der. "Comments on Miller's 'The Myth of Gauss' Experiment on the Euclidean Nature of Space [Comment II]." *Isis* 65 (March 1974): 85-87.

Wake, C. Staniland. "The Notion of a Four-Fold Space." *Science* 19 (10 June 1892): 331-332.

Walker, Evan Harris. "Foundations of Paraphysical and Parapsychological Phenomena." *Quantum and Parapsychology*. Oteri: 1-44.

Walker, E.H. "Consciousness and Quantum Theory." *Psychic Exploration*. Mitchell: 544-560.

Wallace, A.R. *My Life*. New York: Dodd, Mead & Co., 1905.

Walsh, Anthony A. "A note on the Origin of Modern Spiritualism." *Journal of the History of Medicine* (1973): 167-171.

Waltershausen, Wolfgang Sartorius von. *Gauss zum Gedächtnis*. Wiesbaden: Dr. Martin Sandig, 1985; Neudruck der Ausgabe von 1856 mit frdl. Genehmigung des Verlags S. Hirzel KG in Stuttgart, früher Leipzig.

Wang, K.C. and K.C. Cheng. "A Five-Dimensional Field Theory." *Physical Review* 70 (1 and 15 October 1946): 516-518.

Wasserman, G.D. "An Outline of a Field Theory of Organismic Form and Behaviour." *Extrasensory Perception. A CIBA Foundation Symposium*. Eds. G.E.W. Wolstenholme and Elaine C.P. Millar. 2nd edition. New York: Citadel Press, 1966; Originally printed 1956: 53-72.

Watson, Lyall. *Supernature*. London: Hodder Paperbacks, 1973.

Weber, Renee. "The Enfolding-Unfolding Universe: A Conversation with David Bohm." *Holographic Paradigm*. Wilbur.

Weinberg, Steven. "Newtonianism and today's physics." *Metaphysical Review* 2 (March 1, 1996). Online. WWW. Accessed 2-26-96. Available at <einstein.sr.unh.edu:1905/MetaphysicalReview/archive/Vol2/March-1996.html>

Weinberg, Steven. "Interview." *Superstrings*. Davies and Brown: 211-224.   Weitzenbock, Roland W. *Der Vierdimensionale Raum*. Stuttgart, Birkhäuser, 1956.

Wells, Herbert George. *The Time Machine*. London: Heinemann, 1895.

Wells, Herbert George. *The Wonderful Visit*. London: J.M. Dent, 1895.

Wells, Herbert George. "The remarkable case of Davidson's eyes." *The stolen bacillus and other incidents*. London: Methuen, 1895.

Wells, Herbert George. *The Invisible Man: A grotesque romance*. London: L.A. Pearson, 1897.

Wells, Herbert George. "The Plattner story," *The Plattner Story and Others*. London: Methuen, 1897.

Wells, Herbert George. "The Stolen Body." *Twelve Stories and a Dream*. London: MacMillan, 1903.

Westfall, Richard S. *Science and Religion in Seventeenth Century England*. New Haven: Yale University Press, 1958.

Weyer, Edward Moffet. "Is Euclid's Geometry merely a Theory?" *Popular Science Monthly* 68 (1911): 554-563.

Weyl, Hermann. "Gravitation und Elektrizität." *Berliner Sitzungberichte kgl. preussische Akademie* (30 May 1918): 465-480; Reprinted "Gravitation and Electricity." *Principles of Relativity*. Einstein: 201-216.

Weyl, Hermann. "Reine infinitesimale Geometrie." *Mathematische Zeitschrift* 2 (1918): 385-411.

Weyl, Hermann. "Eine neue Erweitung der Relativitätstheorie." *Annalen der Physik* 59 (1919): 101.

Weyl, Hermann. *Space, Time, Matter*. Trans. Henry L. Brose. American edition. 1922; Reprinted New York: Dover, 1952. Originally published in

German. *Raum Zeit Materie*. 1st edition 1918; 3rd edition 1919; 4th edition of 1920 was translated for American edition.

Wheatley, James O. and Hoyt L. Edge. Eds. *Philosophical Dimensions of Parapsychology*. Springfield, Illinois: Charles C. Thomas, 1976.

Wheeler, John Archibald. *Geometrodynamics*. New York: Academic, 1962. Wheeler, John Archibald. "From Relativity to Mutability." *Physicists Conception of Matter*. Mehra: 202-247.

Wheeler, John Archibald. "Curved Empty Space as the Building Material of the Physical World: An Assessment." *Logic, Methodology and the Philosophy of Science*. Eds. Ernest Nagel, Patrick Suppes and Alfred Tarski. Proceedings of the 1960 International Congress. Stanford: Stanford University Press, 1962: 361-374.

Wheeler, J. Craig. "Of Wormholes, Time Machines, and Paradoxes." *Astronomy* 24 (February 1996): 52-57.

Whitaker, Andrew. *Einstein, Bohr and the Quantum Dilemma*. Cambridge: Cambridge University Press, 1996.

White, John and Stanley Krippner. Eds. *Future Science: Life Energies and the Physics of Paranormal Phenomena*. Garden City, New York: Anchor Doubleday, 1977.

White, John. Ed. *Frontiers of Consciousness: The Meeting Ground Between Inner and Outer Reality*. New York: Julian Press, 1974.

Whitehead, A.N. "On Mathematical Concepts of the Physical World." Read 7 December 1905. *Philosophical Transactions of the Royal Society* 205A (1906): 465-525; Reprinted *An Anthology*. Ed. F.S.C. Northrop and Mason W. Gross. Cambridge: at the University Press, 1953: 11-82.

Whitehead, A.N. and Bertrand Russell. "VI. Non-Euclidean Geometry." See Russell.

Whitehead, A.N. and Bertrand Russell. "VII: Axioms of Geometry." *Encyclopaedia Britannica*. 11th edition. 11: 730-736.

Whitehead, A.N. "Space, Time, and Relativity." Proceedings of the Aristotelian Society 22 (1921-1922): 104-129.

Whitehead, A.N. "The Philosophical Aspects of the Principle of Relativity." Proceedings of the Aristotelian Society, 22 (1921-1922): 215-223.

Whitehead, A.N. and B.A.W. Russell. *Principia Mathematica*. See Russell. Whitehead, J.H.C. "The Foundations of Differential Geometry I." Science Progress 33 (January 1939): 503-509.

Whitehead, J.H.C. "The Foundations of Differential Geometry II." *Science Progress* 34 (1939): 76-82.

Whitehead, J.H.C. "Elie Joseph Cartan." *Royal Society of London: Obituary Notices of Fellows* 8 (1952-1953): 71-95.

Whiteman, J.H.M. "Parapsychology and Physics." *Handbook of Parapsychology*. Wolman: 730-756.

Whiteman, J.H.M. "Quantum Theory and Parapsychology." *Journal of the ASPR* 67 (1973): 341-360.

Whitmell, Charles T. *Space and Its Dimensions*. Cardiff: South Wales Printing, 1892.

Whitrow, G.J. "Why Space has Three Dimensions." *The British Journal for the Philosophy of Science* 6 (May 1955): 13-31.

Whittaker, Sir Edmund. *From Euclid to Eddington: A Study of the Conceptions of the External World*. New York: Dover, 1958; Unaltered reprint of Cambridge: At the University Press, 1947.

Whittaker, Sir Edmund. *A History of the Theories of Aether and Electricity. Volume II. The Modern Theories, 1900-1926*. New York: Harper Torchbooks, 1960; A reprint of the original, London: Thomas Nelson and Sons, 1953.

Why, Doctor. (Jack Sarfatti) "Quantum Mystic Physics Theater." Presented by The Physics/Consciousness Research Group. Online. WWW. Accessed 6-13-96. Available at <www. hia.com/hia/pcr/z.html#stenger>

Why, Doctor. (Jack Sarfatti) "What's DAT all about?" Online. WWW. 5-24-96. Accessed 7-3-96. Available at <www.hia.com/ hia/pcr/emay.html>

Wietleitner, H. "Zur Fruhgeschischte des Raume von mehr als drei Dimensionen." *Isis* 7 (1925): 486-489.

Wilbur, Ken. Ed. The *Holographic Paradigm and Other Paradoxes*. Boston: New Science Library, 1982.

Wilbur, Ken. Ed. *Quantum Questions: Mystical Writings of the World's Great Physicists*. Boston: New Science Library, 1984.

Wilczynski, E.J. "The Fourth Dimension." *American Mathematical Monthly* 16 (1909): 149-153.

Williams, Andy. "Consciousness, Physics and the Holographic Paradigm." Online. WWW. Accessed 04-01-98. Available at <www.livingston. net /hermital/intro.html>

Williams, Herbert W. "Some Observations on the Fourth Dimension." *Transactions and Proceedings of the New Zealand Institute* 34 (1901): 507-513.

Williams, L. Pearce. *Relativity Theory: Its Origin and Impact on Modern Thought*. New York: Wiley & Sons, 1968.

Williams, L. Pearce. *The Origins of Field Theory*. New York: Random House, 1966.

Williams, L. Pearce. "Michael Faraday and the Evolution of the Concept of the Electric and Magnetic Fields." *Nature* 187 (27 August 1960): 730-733.

Williams, Selma S. and Pamela Williams Adelman. *Riding the Nightmare: Women and Witchcraft from the Old World to Colonial Salem*. New York: Harper Perennial, 1975.

Williamson, E.M. "Energy in the Nuclear Field." *Nuovo Cimento* 10 (February 1953): 113-126.

Willink, Arthur. *World of the Unseen*. New York: Macmillan, 1893.

Wilson, David B. "The Thought of Late Victorian Physicists: Oliver Lodge's Ethereal Body." *Victorian Studies* 15 (1971): 29-48.

Wilson, Edwin Bidwell. "Projective and Metric Geometry." Read before the American Mathematical Society on 29 December 1903. *Annals of Mathematics* 5 (April 1904): 145-150.

Wilson, Edwin Bidwell. "The So-Called Foundations of Geometry." *Grünert's Archiv der Mathematik und Physik* 3 (1904): 104-122.

Wilson, Edwin Bidwell. "On the Principle of Relativity." *Philosophical Magazine* 6 (1908): 419-422.

Wilson, Edwin Bidwell. "On the Theory of Double Products and Strains in Hyperspace." *Transactions of the Connecticut Academy of Arts and Sciences* 14 (September 1908): 1-57.

Wilson, Edwin Bidwell and Gilbert N. Lewis, "The Space-Time Manifold of Relativity: The Non-Euclidean Geometry of Mechanics and Electromagnetics." *Proceedings of the American Academy of Arts and Sciences*, (Boston) 43 (November 1912): 389-507.

Wilson, William. "The Quantum Theory and Electromagnetic Phenomena." *Proceedings of the Royal Society* 107 (1922): 478483.

Wilson, William. "Relativity and Wave Mechanics." Proceedings of the Royal Society 118 (April 1928): 441-448.

Wilson, William and J. Cattermole. "The Elementary Particle." *Philosophical Magazine* 30 (January 1939): 84-93.

Wilson, William. "Dimensions of Physical Quantities." Philosophical Magazine 33 (January 1942): 26-33.

Wilson, William. *Relativity and Quantum Dynamics. Volume III. Theoretical Physics*. London: Methuen and Company, 1940.

Wilson, William and H.T. Flint. "The Fundamental Unit of electric Charge." *Physical Society, Proceedings* 50 (2 May 1938): 340-344.

Withers, John William. *Euclid's Parallel Postulate: Its Nature, Validity, and Place in Geometrical Systems*. Chicago: Open Court, 1905.

Witten, Edward. "Interview." *Superstrings*. Davies and Brown: 90-106.

Wodkiewicz, Krzysztof. "Non-local and local ghost fields in quantum correlations." *Contemporary Physics* 36 (1996): 139-147.

Wolf, Fred Alan. *Taking the Quantum Leap: The New Physics for Nonscientists.* New York: Harper & Row, 1981.

Wolf, Fred Alan. "From Parallel Universes." Online. WWW. Accessed 05-01-97. Available at <www.hia.com/hia/pcr/parauniv.html>

Wolman, Benjamin B. Ed. *Handbook of Parapsychology*. New York: Van Nostrand Reinhold, 1977.

Wolstenholme, G.E.W. and Elaine C.P. Millar Eds. *CIBA Foundation Symposium on ESP*. Boston: Little, Brown, 1956.

Woods, Frederick Shenstone. "Forms of Non-Euclidean Space." Delivered 2 to 5 September 1903 at the Summer Meeting of the American Mathematical Society at the Massachusetts Institute of Technology. *The*

Boston Colloquium, Lectures on Mathematics. Eds. Edward Burr Van Vleck, Henry Seely and F.S. Woods. New York: Macmillan, 1905: 32-74.

Woods, Frederick Shenstone. "Space of Constant Curvature." *Annals of Mathematics* 3 (1902): 71-112.

Wormley, S. "Special and General Relativity Resources." Selections from

George F.R. Ellis and Ruth M. Williams. Flat and Curved Space-Times. New York: Oxford University Press, 1988. Online. WWW. Accessed 6-13-96. Available at <www.cnde.iastate.edu/staff/swormley/bkr/bkr.91.08>

Wolstenholme, G.E.W. and Elaine C.P. Millar. Eds. *CIBA Foundation Symposium on ESP*. Boston: Little, Brown, 1956.

Wortz, Edward C.A.S., Bauer, R.F. Backwelder, J.W. Eerkens and A.J. Saur, "An Investigation of Soviet Psychical Research." *Mind at Large*. Tart, Puthoff and Targ: 235-260.

Wundt, Wilhelm. "Spiritualism as a Scientific Question: An Open Letter to Prof. Hermann Ulrici of Halle." *Popular Science Monthly* 15 (1879): 578-593. Yano, Kentano. "Théorie de la Relativité." *Comptes Rendus* 204 (1 February 1937): 332-334.

Yano, Kentano. "Einstein and Bergmann's New Unitary Field Theory." *Imperial Academy of Tokyo, Proceedings* 14 (November 1938): 325-332.

Yano, Kentano. "Unitary Non-Holonomous Field Theory." Parts I and II." *Physics and Mathematics Society of Japan, Proceedings* 19 (October 1937): 867-896; (November 1937): 945-976.

Zahar, Elie G. "The Mathematical Origins of General Relativity and of Unified Field Theory." *The Einstein Symposium Berlin. Lecture Notes in Physics*, 100 H. Eds. Nelkowski, et.al. Stuttgart: Springer-Verlag, 1979: 370-396.

Zahar, Elie G. "Poincaré's Independent Discovery of the Relativity Principle." *Fundamenta Scientiae* 4 (1983): 147-175.

Zaycoff, R. "Connection between Field and Matter." *Zeitschrift fur Physik* 82 (29 April 1923): 267-278.

Zaycoff, R. "Integral Theory of Field and Matter." *Zeitschrift fur Physik* 85 (14 October 1933): 788-794.

Zedler, Beatrice H. "Royce and James on Psychical Research." 235-252.

Zinchenko, W.P., A.N. Leontiev, B.M. Lomov and A.R. Luria. "Parapsychology in the Soviet Union." *Signet Handbook*. Ebon: 425-451.

Zöllner, Johann Karl Friedrich. *Über die Natur der Cometen*. Leipzig: Engelmann, 1872.

Zöllner, Johann Karl Friedrich. "On Space of Four Dimensions." Translator and Reviewer unnamed. *Quarterly Journal of Science* 15 (1878): 227-237.

Zöllner, Johann Karl Friedrich. *Wissenschaftliche Abhandlungen*. Leipzig: L. Staackmann, 1878: Volume 3, *Die Transcendentale Physik und die sogenannte Philosophie*. Berlin: Verlag von Karl Siegismund, 1878; *Transcendental Physics*. Trans. C.C. Massey. Boston: Colby and Rich, 1881.

Zöllner, Johann Karl Friedrich. *Naturwissenschatliche und christliche Ofenbarung*. Gera: C.B. Griesbach's Verlag, 1880.

Zukav, Gary. *The Dancing Wu Li Master: An Overview of the New Physics*. New York: William Morrow, 1979.

"William Kingdon Clifford." *The Encyclopedia of Philosophy*. Edwards. Volume 2: 123-125.

Unnamed Reviewer. "Professor Clifford's Elements of Dynamic." *The Saturday Review* 45 (22 June 1878): 792-794.

Author Unknown. "The Universe as a Hologram." Online. WWW. Accessed 04-04-97. Available at <www.livelinks.com/sumeria/phys/hologram.htm>

Author Unlisted. "Physics Facts and Speculation." Online. WWW. Accessed 07-08-96. Available at <www.mtnmath.com/physics.htm#contemp.view>

www.ingramcontent.com/pod-product-compliance
Lightning Source LLC
Chambersburg PA
CBHW060515160726
47991CB00001B/37